NEUROSCIENCE
INTELLIGENCE
UNIT

The Role of Oxidative Stress in Neuronal Death

Irène Ceballos-Picot
Hôpital Necker
Paris, France

Springer-Verlag
Berlin Heidelberg GmbH

LANDES
BIOSCIENCE
AUSTIN, TEXAS
U.S.A.

Neuroscience Intelligence Unit

The Role of Oxidative Stress in Neuronal Death

LANDES BIOSCIENCE
Austin, Texas, U.S.A.

Originally published by Springer-Verlag, Heidelberg, Germany in 1997
Softcover reprint of the hardcover 1st edition 1997

Springer

International ISBN 978-3-662-22518-9

Library of Congress Cataloging-in-Publication Data

Ceballos-Picot, Irène, 1957-
Oxidative stress in neuronal death / Irène Ceballos-Picot.
p. cm. — (Neuroscience intelligence unit)
Includes bibliographical references and index.
ISBN 978-3-662-22518-9 ISBN 978-3-662-22516-5 (eBook)
DOI 10.1007/978-3-662-22516-5
1. Nervous system--Pathophysiology. 2. Neurons. 3. Apoptosis. 4. Active oxygen in the body. 5. Nervous system--Degeneration. 6. Oxidation, Physiological. 7. Stress (Physiology) I. Title. II. Series.
[DNLM: 1. Nervous System Diseases--physiopathology. 2. Oxidative stress--physiology. 3. Neurons--physiology. 4. Cell Death--physiology. WL 140 C387o 1997]
RC347.C43 1997
616.8'047—dc21
DNLM/DLC 97-19204
for Library of Congress CIP

Publisher's Note

Landes Bioscience produces books in six Intelligence Unit series: *Medical, Molecular Biology, Neuroscience, Tissue Engineering, Biotechnology* and *Environmental.* The authors of our books are acknowledged leaders in their fields. Topics are unique; almost without exception, no similar books exist on these topics.

Our goal is to publish books in important and rapidly changing areas of bioscience for sophisticated researchers and clinicians. To achieve this goal, we have accelerated our publishing program to conform to the fast pace at which information grows in bioscience. Most of our books are published within 90 to 120 days of receipt of the manuscript. We would like to thank our readers for their continuing interest and welcome any comments or suggestions they may have for future books.

Shyamali Ghosh
Publications Director
Landes Bioscience

Dedication

To Jean-Patrice and Antoine
To my parents
To my teachers and co-workers

CONTENTS

PREFACE

Oxidative stress refers to the cytotoxic consequences of reactive oxygen species (ROS) which are generated as by-products of normal or aberrant metabolic processes that utilize molecular oxygen, or changes in protection systems. Thus, generation of ROS is a part of normal life and their interaction with host defense systems appears to exert a significant influence on the normal and abnormal functioning of the central nervous system. Oxidative stress could also be a secondary consequence of defects in energy metabolism involving mitochondria. A subbtle shift toward the prooxidant status that occurs in brain aging could contribute to the pathogenesis and neuropathology of neurodegenerative disorders exemplified by Alzheimer's disease, Parkinson's disease and amyotrophic lateral sclerosis. The attractive feature of the oxidative stress hypothesis is that it can account for cumulative damage associated with the delayed onset and the progressive nature of these conditions. Oxidative damage may constitute a common pathogenic link between these seemingly diverse neurodegenerative disorders.

This book examines current knowledge and recent advances of fundamental processes involved in neuronal death, particularly oxidative stress as a causal, or at least an ancillary, factor in neuronal death. This book also presents an up-to-date account of the current state of knowledge about of oxidative stress in neuronal apoptosis and its role in the neuropathogenesis of age-related neurodegenerative diseases. The first chapters are devoted to the biology of oxidative stress, biochemical consequences of damaging free radical reactions in the brain, and their participation in neurodegenerative processes. Another part of the book will describe the role of ROS in activation of transcription factors and its oxidative influence on gene expression, with links to alterations in cell signalling processes and to neuronal death. Some chapters will address recent progress in understanding the relationships between its oxidative stress, excitotoxicity, mitochondria in neuronal apoptosis and the potential role for apoptosis in neurodegeneration. The central role of oxidative stress in the major age-related neurodegenerative disorders is critically reviewed and potential prevention and therapy by antioxidants is examined.

ABBREVIATIONS

AA: arachidonic acid
Aβ: amyloid peptide
AD: Alzheimer's disease
AGE: advanced glycated end products
ALS: amyotrophic lateral sclerosis
AMPA: α-amino-3-hydroxy-5-methyl-4-isoxasole
AP-1: Activator protein 1
ApoE: apolipoprotein E
APP: amyloid peptide precursor
ARE: antioxidant response element
ATP: adenosine triphosphate
ATPases: adenosine triphosphatases
BDNF: brain-derived neurotrophic factor
bFGF; basic fibroblast factor
BHA: butylated hydroxyanisole
CI: complex I
CII: complex II
CIV: cytochrome oxidase
CA: cornu ammonis
CAT: catalase
Cdk: cyclin-dependent kinase
ChAT: choline acetyl transferase
CK: creatine kinase
CNS: central nervous system
Cox: cytochrome oxidase
CSF: cerebrospinal fluid
CS: citrate synthase
CuZnSOD: copper-zinc superoxide dismutase
DA: dopamine
DCF: dichlorofluorescin
DMPO: 5,5-dimethyl-1 pyrroline-N-oxide
DOPAC: dihydroxyphenylacetic acid
DS: Down's syndrome
e^-: electron
EAA: Excitatory amino-acids
EC-SOD: extracellular superoxide dismutase
EPR: electron paramagnetic resonance
EpRE: electrophile response element
ERK2: extracellular signal-regulated protein kinase
ETC: electron transport chain

FAD: familial Alzheimer disease
FALS: familial amyotrophic lateral sclerosis
GC: guanylate cyclase
Glu: glutamate
GS: glutamine synthetase
GSH: reduced glutathione
GSHPx: glutathione peroxidase
GSSG: oxidized glutathione
GSSG-Rd: glutathione reductase
G6PD: glucose 6-phosphate dehydrogenase
H: hydrogen
HDL: high density lipoprotein
He: Helium
HNE: hydroxynonenal
HO: heme oxygenase
HO_2°: hydroperoxyl radical
H_2O_2: hydrogen peroxide
HVA: homovanillic acid
ICE: interleukin converting enzyme
IEGs: immediate early genes
IL: interleukin
KA: kainic acid
KPI: Kunitz protease inhibitor domain
LPS: lipopolysaccharides
LRP: Low density lipoprotein receptor-related protein
MAO: monoamine oxidase
MAPK: mitogen-activated protein kinase
MCA: middle cerebral artery
mGluR: glutamate metabotropic receptor
MnSOD: manganese superoxide dismutase
MPTP: 1-methyl-4-phenyl-1,2,3,6-tetrahydropyridine
Mt: Mitochondria
MTT: 3-(4,5-dimethylthiazol-2-yl)-2,5-dimethylthiazolium bromide
NAC: N-acetylcysteine
NF: neurofilaments
NF-κB: nuclear factor-κB
NFT: neurofibrillary tangles
NGF: nerve growth factor
7-NI: 7-nitroindazole
NM: neuromelanin
NMDA: N-methyl-D-aspartate
NO°: nitric oxide

NO_2°: nitrogen dioxide
NOS: nitric oxide synthase
NT-3: neurotrophin-3
$^{\circ}$OH: hydroxyl radical
$ONOO^-$: peroxinitrite anion
$O_2^{\circ-}$: superoxide anion
O_3: ozone
6PGD: 6-phosphogluconate dehydrogenase
PARS: poly-ADP-ribose synthetase
PBN: N-tert-butyl-2-sulfophenylnitrone
PD: Parkinson's disease
PHF: paired helical filaments
PLA_2: phospholipase A_2
PTK: Protein tyrosine kinase
PTP: protein tyrosine phosphatase
PUFA: polyunsaturated fatty acid
RAGE: receptor for advanced glycated end products
RCR: respiratory control ratio
Ref-1: redox factor-1
ROS: Reactive oxygen species
Se: selenium
SN: substantia nigra
SOD: superoxide dismutase
SP: senile plaques
SR: scavenger receptor
SRE: serum response element
SRF: serum response factor
TBARS: thiobarbituric-reactive materials
TCF: ternary complex factor
TH: tyrosine hydroxylase
TNF: tumor necrosis factor
Trk: tyrosine kinase
VDCC: voltage-dependent calcium channels
VLDL: very low density lipoprotein

CHAPTER 1

Introduction

Age-related neurodegenerative diseases remain one of greatest challenges of contemporary neurology. The debilitating consequences of age-related brain deterioration are widespread and extremely costly in terms of quality of life and longevity. Some of the most conspicuous and well-known examples of degenerative conditions of the central nervous system (CNS) whose frequency of occurrence increases with aging include Parkinson's disease (PD), amyotrophic lateral sclerosis (ALS) and Alzheimer's disease (AD) (Table 1.1).

Although extensively studied in terms of their causative factors, there is no definitive consensus on the etiology of any of these conditions; the etiology of neurodegenerative diseases remains particularly enigmatic. However, evidence for defects in energy metabolism, excitotoxicity, and for oxidative damage is increasingly compelling. The brain is exposed thoughout life to oxidative stress, and a number of diseases of the nervous system have been hypothesized to involve reactive oxygen species (ROS), free radical processes and oxidative damage either as a cause or a consequence of the disease process.

These diseases commonly are associated with movement and/or severe cognitive impairments; in some cases, significant overlap in the morphopathology of these conditions occurs. All of these disorders involve site-specific loss of certain neuronal populations. Although the type and site of neurons lost vary in each disorder, there is a growing recognition that there is some morphological pathology overlap between them. For instance, Down's syndrome (DS), PD, ALS, and AD all show varying forms of neurofibrillary pathology. PD patients develop Lewy bodies, AD patients develop neurofibrillary tangles, and ALS patients develop variants of Lewy bodies as well as structures similar to neurofibrillary tangles, termed spheroids and hyaline bodies.[1,2] In addition, they also have other features of pathology in common, such as granulovacuolar degeneration.[3] PD, ALS, and AD are among the most common illnesses among a larger family of neurodegenerative disorders that also show neurofibrillary pathology, such as supranuclear palsy, Pick's disease, diffuse Lewy body disease, and Gerstmann-Straussler syndrome.[2,3]

Because of the similarities in the pathologies, it is reasonable to speculate that although these neurodegenerative disorders may have separate etiologies, they may share the same terminal biochemical defects that result in neuronal damage and death. Defining the cause of one of these disorders may, therefore, be of value in our understanding of the biochemical mechanisms involved in the others. In each of these three illnesses, damage from free radicals appears to play an important role, causing cumulative oxidation that interferes with cellular function and ultimately kills the neurons. Many investigators have speculated that neurons are particularly prone to neurodegeneration because of their requirement for many

The Role of Oxidative Stress in Neuronal Death, by Irène Ceballos-Picot.

Table 1.1. Pathological or physiological conditions of brain and neural tissue in which oxidative stress is thought to play a role

Alzheimer's disease	Oxygen inhalation therapy
AIDS (HIV infection)	Parkinson's disease
Amyotrophic lateral sclerosis	Progeria
Apoptosis	
Down's syndrome	Schizophrenia
Epileptic seizures	Spinal cord damage
Head trauma	Ischemia
Inflammatory diseases	Tardive dyskinesia
Neurotoxins (MPTP)	Vitamin E deficiency

mitochondria to support their large energy needs. Processes that interfere with mitochondrial function, for instance, by inhibiting the complexes involved in oxidative phosphorylation, increase free radical formation. A defect in energy metabolism may lead to neuronal depolarization, activation of N-methyl-D-aspartate (NMDA) excitatory amino acids receptors, and increases in intracellular calcium concentration $[Ca^{2+}]i$ enhance free radical generation. Mitochondrial (Mt) DNA is particularly susceptible to oxidative stress, and there is evidence of age-dependent damage and deterioration of respiratory enzyme activities with normal aging. This may contribute to the delayed onset and age dependence of neurodegenerative diseases. There is evidence for increased oxidative damage to macromolecules in ALS, PD, AD and DS. In addition, because mitochondria have large supplies of calcium, loss of mitochondrial function also causes release of calcium into the cytoplasm, which can stimulate proteases as well as the DNases involved in apoptosis. Moreover, the brain and neural tissue contain sources of oxidative stress unique to these tissues: excitatory amino acids (EAA) and neurotransmitters whose metabolism produces ROS. It is likely that there is a complex interplay between these mechanisms.

That aging in general and aging of the central nervous system in particular may, in part, relate to the damage inflicted by oxygen free radicals and their intermediates has considerable experimental support. One of the potential major causes of age-related destruction of neuronal tissue ROS, known as pro-oxidants, that are an attribute of normal aerobic life. The steady-state formation of pro-oxidants is balanced by a similar rate of their consumption by antioxidants that are enzymatic and/or nonenzymatic. "Oxidative stress" results from imbalance in this pro-oxidant-antioxidant equilibrium in favor of the pro-oxidants. Oxidative stress is the usual phrase used to identify the association of toxic free radicals with damage to cells and tissues. During aging, however, the ROS generation may increase as a consequence of normal aging processes, exposure to toxins and ultraviolet light, and stress; alternatively, the defense systems that have evolved to combat oxidative stress (e.g., antioxidative enzymes, free radical scavengers, metal chelating agents) may diminish. The result of these changing processes is oxidative stress associated with an accelerated rate of accumulated damage and associated pathophysiology in advanced age.

Current evidence in clinical research does not show unequivocal distinction between causal or associative relationships of oxidative stress to the disease process. With neurodegenerative diseases, we must then add onto this background of age-related loss of Mt and antioxidant function, potential damage from the action of the harmful processes characteristic of each illness. The combination of these processes results in an increased burden on the neuron. Damaged mitochondria release calcium and free radicals. In some cases the damage is exacerbated by toxins like 1-methyl-4-phenyl-1,2,3,6-tetrahydropyridine (MPTP), which results in increased release of calcium and increased production of free radicals. Amyloid peptide (Aβ) and mitochondrial damage may produce an additive effect that increases calcium levels or free radical production to a point where they are harmful. Thus, while the differing presentations of the various neurodegenerative illnesses probably result from differences in etiologies, there also appear to be common pathways through which the neurodegeneration proceeds.

The possibility of deferring age-related degenerative changes in the brain has become a major interest of both experimentalists and clinicians. A number of synthetic molecules that have antioxidative activity are being tested for their ability to alter the degenerative processes of the aging brain. Remarkably, the brain is not enriched with antioxidative defense systems as one might expect, considering its importance to survival. It may, however, be within our capability to defer loss of brain function and thereby postpone death with agents that forestall the degeneration and death of the neurons in the CNS. Specifically, what, if any, natural or synthetic antioxidants will best meet this hope is still the goal of future research. Potential preventive or therapeutic approaches include glutamate release inhibitors, excitatory amino acid antagonists, strategies to improve mitochondrial function, free radical scavengers, and trophic factors. All of these approaches appear promising in experimental studies and are now being applied to human studies.

The present book addresses recent progress in understanding the role of oxidative stress in neuronal death and apoptosis in both normal aging and in the major neurodegenerative diseases.

CHAPTER 2

Biology of Oxidative Stress

INTRODUCTION TO OXYGEN CHEMISTRY AND REACTIVE OXYGEN SPECIES

The element oxygen (O) exists in air as a molecule (O_2) known as dioxygen or molecular oxygen. It was first isolated and characterized between 1772-1774 by the individual skills of the great European scientists Lavoisier and Scheele. Dioxygen, hereafter referred to as oxygen, appeared in significant amounts on the surface of the Earth some 2.5 x 10^9 years ago, and geological evidence suggests that this was due to the photosynthetic activity of micro-organisms (blue-green algae). The slow and steady rise in atmospheric oxygen concentration was accompanied by the formation of the ozone layer in the stratosphere. Both oxygen and the ozone layer acted as critical filters against the intense solar ultraviolet light reaching the surface of the Earth. The universe exists predominantly as hydrogen (H) and helium (He) with the Earth as a unique center of oxidation in an otherwise reducing universe.

The percentage of oxygen in dry air is now around 21%, making it, after nitrogen (78%), the second most abundant element in the atmosphere. When the Earth's atmosphere changed from a highly reducing state to the oxygen-rich state that we know today, anaerobic life forms ceased to exist or retreated to places where oxygen was excluded. The slow change from anaerobic life necessitated the evolution of specialized antioxidants to protect against the toxic properties of oxygen. Aerobic life uses oxygen to oxidize (burn) carbon and hydrogen-rich substrates (foods) to obtain the chemical energy and heat essential for life. Unfortunately, when we oxidize molecules with oxygen, the oxygen molecule itself becomes reduced and forms intermediates (Eqs.1-4).

$O_2 + e^- + H^+ \rightarrow HO_2^{\circ}$ hydroperoxyl radical (1)

$HO_2^{\circ} \rightarrow H^+ + O_2^{\circ -}$ superoxide radical

$O_2^{\circ -} + 2H^+ + e^- \rightarrow H_2O_2$ hydrogen peroxide (2)

$H_2O_2 + e^- \rightarrow OH^- + {}^{\circ}OH$ hydroxyl radical (3)

${}^{\circ}OH + e^- + H^+ \rightarrow H_2O$ (4)

When oxygen is reduced by the stepwise addition of electrons (Eqs.1-4) two free radicals (HO_2°, ${}^{\circ}OH$) are formed, together with hydrogen peroxide (H_2O_2). At a physiological pH value of 7.4 the hydroperoxyl radical (HO_2°) with pK_a of 4.8 (the pH value at which equal concentrations of both acid (HO_2°) and base ($O_2^{\circ -}$) are present) dissociates to give the superoxide anion radical ($O_2^{\circ -}$).

The Role of Oxidative Stress in Neuronal Death, by Irène Ceballos-Picot.

$HO_2^{\circ} \rightarrow H^+ + O_2^{\circ-}$ (5)

A free radical may be defined as any chemical species capable of independent existence that contains one or more unpaired electrons.[4]

REACTIVE OXYGEN SPECIES: BIOCHEMISTRY

O_2 is a powerful oxidizing agent because of its ability to readily accept electrons (e^-). Within atoms and molecules, e^- occupy regions known as orbitals with each orbital capable of containing two e^-s (a diradical), one of which spins in a clockwise direction and the second in an anticlockwise direction. When an orbital contains only a single e^-, the e^- is said to be unpaired, and an atomic or molecular species with an unpaired e^- is referred to as a free radical. O_2 qualifies as a radical because it possesses two unpaired e^-, each in a different orbital and both spinning in the same direction (Fig. 2.1.).

SUPEROXIDE ANION

Superoxide ($O_2^{\circ-}$) anion is a radical formed when one electron enters one of the II*2p orbitals of oxygen. The chemistry of superoxide differs greatly depending on its solution environment. In aqueous solution $O_2^{\circ-}$ is a weak oxidizing agent able to oxidize molecules such as ascorbic acid and thiols. However, $O_2^{\circ-}$ is a much stronger reducing agent, and is able to reduce several iron complexes such as cytochrome *c* and ferric-EDTA. Superoxide rapidly disappears in aqueous solution due to its dismutation reaction in which H_2O_2 and O_2 are formed

$O_2^{\circ-} + O_2^{\circ-} + 2H^+ \rightarrow H_2O_2 + O_2$ (6)

Superoxide dismutase (SOD) discovered by McCord and Fridovich in 1968[5] greatly accelerates the above reaction. Several different forms of the enzyme are known to exist in plant, microbial and mammalian cells, differing in structure and transition metal ions at their native centers, but catalyzing exactly the same chemical reaction shown in Eq. 6. The protonated form of $O_2^{\circ-}$, the hydroperoxyl radical, is both a more powerful oxidant and reductant than $O_2^{\circ-}$, but little HO°_2 will be present at pH 7.4.

HYDROGEN PEROXIDE

Any system producing superoxide will, as a result of the dismutation reaction, also produce H_2O_2. Many enzymes such as urate oxidase, glucose oxidase and D-amino acid oxidase produce H_2O_2 directly by the transfer of two electrons to oxygen. H_2O_2 is a weak oxidant and weak reducing agent that is relatively stable in the absence of transition metal ions. The molecule has an uncharged covalent structure. It readily mixes with water, and is treated as a water molecule by the body, rapidly diffusing across cell membranes. The redox properties of H_2O_2, and its ability to form highly reactive free radicals ($^{\circ}OH$) in the presence of transition metal ions, has necessitated the evolution of body defenses against it. Unwanted H_2O_2 is removed from cells by the action of enzymes such as catalase (CAT), glutathione peroxidase (GSHPx), and certain other peroxidases.

HYDROXYL RADICALS

The hydroxyl radical ($^{\circ}OH$) is a major product arising from the high energy ionization of water (radiolysis)

$H_2O \rightarrow {}^{\circ}OH + H^{\circ} + E_{aq} + H_2O_2$ (7)

Most of our definitive knowledge about the hydroxyl radical comes from studies by

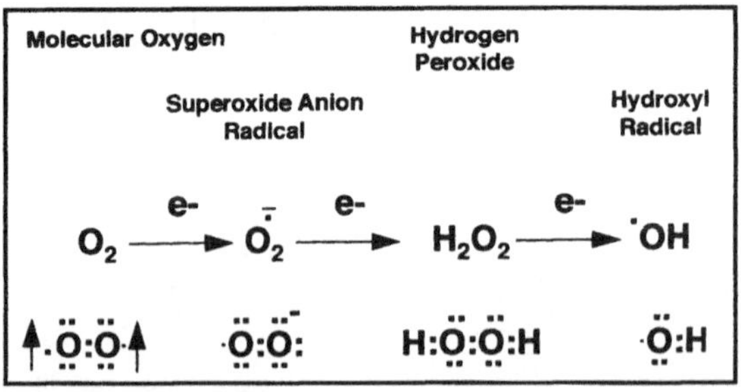

Fig. 2.1. The three-electron (e^-) reduction of molecular ground state oxygen to the hydroxyl radical. The arrows (both pointing in the same direction) associated with the oxygen molecule indicate the two unpaired e^-, located in different orbitals, which have the same spin direction.

radiation chemists. The $^{\circ}OH$ radical is an extremely aggressive oxidant that can attack most biological molecules at an almost diffusion-controlled rate.

Singlet Oxygen

Singlet oxygen ($^1\Delta g\, O_2$) is not a free radical as it does not contain an unpaired electron. Nevertheless, it is a highly reactive form of oxygen in which the spin restriction of oxygen (two unpaired electrons with parallel spins) is removed, thereby increasing its oxidizing ability. In addition to $^1\Delta g\, O_2$, another form of singlet oxygen exists (Σg^+). However, this state is extremely energetic and rapidly decays to the $^1\Delta g$ form in biological systems. Formation of singlet oxygen is extremely important when considering photo-chemical reactions.

Ozone

The pale blue gas ozone (O_3) provides an important protective shield (global antioxidant) in the stratosphere against solar radiation. At ground level, however, ozone is a toxic and unwanted pollutant. Ozone is produced in polluted urban air. It is extremely damaging to the lung, readily oxidizing proteins, DNA and lipids.

Oxides of Nitrogen

Nitric oxide (NO°) and nitrogen dioxide (NO°_2) contain odd numbers of electrons, and are therefore free radicals, whereas nitrous oxide (N_2O) does not. Nitrogen dioxide is a dense brown poisonous gas, and a powerful oxidizing agent. It was first recognized as a distinct gas in 1772 by Joseph Priestley, who prepared an iron complex of it. Recently, biological interest in nitric oxide has centered around the observation that the vascular endothelium and other cells in the body produce small amounts of the gas from the amino acid Λ-arginine.[6] At present nitric oxide is indistinguishable from the vasodilator, endothelium derived relaxing factor (EDRF) which can react with another endogenous free radical, superoxide, to produce a reactive intermediate peroxynitrite ($ONOO^-$).[7] Peroxynitrite is a powerful oxidant, able to damage many biological molecules, and can decompose to release hydroxyl radicals independently of metal catalysis (Eq. 8).

$$ONOO^- + H^+ \rightarrow {}^{\circ}OH + NO^{\circ}_2 \quad (8)$$

Reaction of Iron with Oxygen

During the 1890s the chemist H.J.H. Fenton described a reaction between iron salts and hydrogen peroxide that caused oxidative damage to organic molecules,[8] such as tartaric acid. This reaction is known as the 'Fenton reaction' and is widely represented as follows:

$$Fe^{2+} + H_2O_2 \rightarrow Fe^{3+} + OH^- + {}^{\circ}OH \quad (9)$$

$$Fe^{3+} + H_2O_2 \rightarrow Fe^{2+} + HO_2^{\circ} + H^+ \quad (10)$$

overall reaction:

$$\text{Iron salt} + 2H_2O_2 \rightarrow 2H_2O + O_2 \quad (11)$$

Equations 9 and 10 are gross over-simplifications of the chemistry involved, particularly when such reactions are considered in biological systems.

At physiological pH values (7.4) ferrous ions (FeII), in the presence of oxygen and phosphate ions, exist only transiently before auto-oxidizing to the ferric state (FeIII). In the process of changing from the ferrous to the ferric state, an electron is transferred from iron to oxygen to make superoxide as follows:

$$Fe^{2+} + O_2 \rightarrow Fe^{2+}O_2 \leftrightarrow Fe^{3+}O_2^{\circ-} \rightarrow Fe^{3+} + O_2^{\circ-} \quad (12)$$

THE OXYGEN PARADOX

Oxygen presents a unique problem for aerobic organisms. It plays a vital role as the terminal electron acceptor during respiration, yet, due to its biradical nature, oxygen and its metabolites are potentially cytotoxic, clastogenic, and deleterious.

There are a large number of physiological and pathological sources of oxygen radicals and related oxidants. Exogenous agents such as photo-chemical smog, ozone, pesticides, xenobiotics, and ionizing radiation are all known to generate radicals/oxidants. Significantly, it has also been well documented that a variety of endogenous processes are significant generators of radicals/oxidants.[9,10]

During reduction of molecular oxygen to water in cellular respiration, partially reduced oxygen species are produced (Fig. 2.2).[11,12]

These ROS are also formed in the cyclooxygenase pathway and by cellular enzymes such as cytochrome P450 oxidase, xanthine oxidase and NADPH oxidase. In cellular and acellular systems, the presence of iron or other divalent cations can convert $O_2^{\circ-}$ and H_2O_2 to $^{\circ}OH$, one of the most reactive ROS. Another potential source of $O_2^{\circ-}$ is the peroxisomal enzyme, xanthine oxidase, which generates $O_2^{\circ-}$ and H_2O_2 during the metabolism of xanthine or hypoxanthine to uric acid. Cellular conversion of xanthine dehydrogenase to xanthine oxidase occurs in a variety of cell types after exposure to specific stimuli, including phorbol esters, tumor necrosis factor α (TNF) or ischemia-reperfusion injury. NO°, formed during oxidation of arginine, can react with $O_2^{\circ-}$ to form $ONOO^-$, a stable intermediate which decomposes into $^{\circ}OH$.

A variety of deleterious effects of ROS on neuronal function have been described, including increased permeability of the blood-brain barrier, altered tubulin formation, inhibition of mitochondrial respiration, and perturbations in synaptic transmitter and ion functions (reviewed in ref. 13). Changes in the brain with age, namely limitation of homeostatic mechanisms, increased ROS production, compromised antioxidant defenses, increased lipofuscin, accumulation of oxidatively-damaged enzymes, and loss of neurones, contribute to the enhanced vulnerability of the aged brain to injurious insults.

All cellular constituents appear to be sensitive to radical/oxidant damage. Proteins, lipids, nucleic acids and carbohydrates are all known to undergo oxidative modifications. The profile of damage products is dependent on the quantity and quality of radical/oxidant exposure and of the antioxidant defenses.

It normally would be assumed that, considering its importance to organismal physiology and survival, the CNS would be amply endowed with defense mechanisms to counter oxidative stress. Contrary to this expectation, however, the brain is highly vulnerable to attack by ROS and does not have an abundance of processes to neutralize these molecular renegades.[14,15]

Because of its high metabolic activity, the brain uses large amounts of O_2.[16] In a young individual, O_2 is used at a rate of 3.5 ml/100g neural tissue/min.[17] To put this into perspective, the brain, which accounts for only 2% of the body weight, uses 20% of the total inspired O_2 in a resting individual. In neural tissue, O_2 primarily participates in the oxidation of carbohydrates and generates an estimated steady turnover of roughly 4×10^{21} molecules of ATP/min in the total brain.[17] Despite the brain's reliance on O_2, elevated pO_2 may induce convulsions that are a consequence either of the reduction in the activity of the enzyme glutamate decarboxylase or the increased generation of oxygen

Fig. 2.2. Reactive oxygen species production by the respiratory chain.

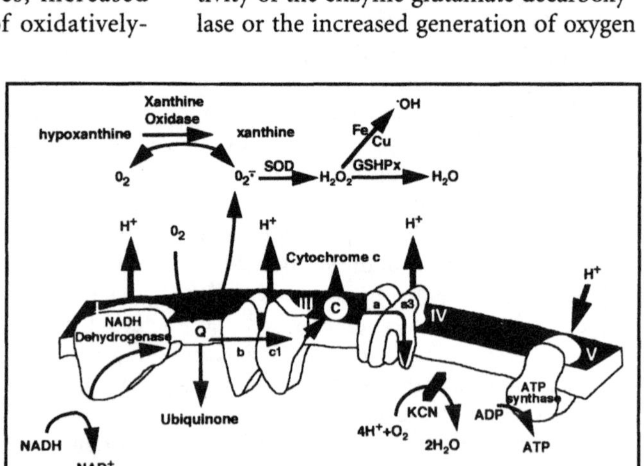

free radicals.[16] Because a small portion (< 5%) of the O_2 used by cells obviously is not involved in the production of ATP, it is used in an alternate pathway, i.e., it is reduced to reactive oxygen species that may induce tissue damage. Because the utilization of O_2 is high in the brain, it follows that generation of damaging oxygen species is also elevated. Inasmuch as the brain does not have an excess of protective mechanisms against the reactive derivatives of O_2, the damage left in the wake of these aggressive radicals is probably higher than elsewhere in the body.

The high utilization of O_2 is not the only reason the brain suffers more than its share of oxidative destruction. The brain is highly enriched with polyunsaturated fatty acids (PUFA). The unsaturated bonds in these molecules render them susceptible to oxidative damage by free radicals.[18] Cerebrospinal fluid (CSF) contains little transferrin or ceruloplasmin; these compounds serve as antioxidants because of their ability to bind iron and copper, respectively; these transition metals can be used in the generation of the most toxic oxygen-based radical, the hydroxyl radical (°OH). Furthermore, CSF contains low molecular weight iron and copper complexes that can catalyze free radical generation. Beyond this, there are other reasons why the brain exhibits a high susceptibility to oxidative damage. Neural tissue itself contains high concentrations of nonheme iron, which, as noted before, is involved in the generation of the highly toxic °OH.[19] The elevated levels of iron coupled with high concentrations of ascorbic acid in the central nervous system is another factor that puts the brain in jeopardy of damage by ROS. Although generally considered an antioxidant, ascorbic acid in the presence of free iron can be a pro-oxidant.[20] The free iron in this case is often derived from hemoglobin. Hemoglobin, normally transported in erythrocytes, sequesters iron and renders it unavailable for free radical reactions. However, hemoglobin outside of the erythrocyte, e.g., after the breakdown of the cells, potentially is highly dangerous because it can be degraded and liberate iron.[21] Thus, in the event of hemorrhage into the brain, e.g., during stroke, the neural damage due to free radical attack can be extensive.

Finally, excitatory amino acid neurotransmitters, which are abundant at some locations in the brain, generate massive numbers of free radicals after their release.[22-24] Glu receptor-mediated excitotoxicity is well known to be a causative factor in a variety of neurological disease states. Glu, which is the chief excitatory neurotransmitter in the brain, is distributed unequally in the central nervous system. Thus, the damage caused by this neurotransmitter, which is secondary to increased ROS production, is concentrated in those neural structures that contain the highest density of glutamate-containing neuronal endings.[23] With increased longevity, the damage related to Glu release accumulates, leading to the gradual morphological and physiological destruction of those neurons that are normally bombarded by this excitatory amino acid neurotransmitter.[14]

In summary, in comparison with other organs of the body, the CNS may, for a number of biochemical, physiological and anatomical reasons, be especially vulnerable to ROS-mediated injury. This involves: (1) high rate of oxidative metabolic activity; (2) high concentration of readily oxidizable substrate, in particular, membrane lipid polyunsaturated fatty acids; (3) low level of protective antioxidant enzymes, namely catalase and glutathione peroxidase; (4) endogenous generation of ROS by specific neurochemical reactions, for example, dopamine oxidation, Glu excitotoxicity; (5) specialized neuronal conduction and synaptic transmission activity dependent on efficient membrane function; (6) high ratio of membrane surface area to cytoplasmic volume; (7) extended axonal morphology prone to peripheral injury; (8) neural anatomical network vulnerable to disruption; (9) neuronal cells are non-replicating.

Unquestionably, the wear and tear on the brain inflicted by toxic free radicals over a lifetime is substantial, with the gradual deterioration of neurobiological tissue being inevitable. Surely the drop in the functional efficiency of the brain, e.g., slowed reactions, diminished memory, tremor, and gross deterioration of some neuronal functions, is in part a consequence of the necessity of most organisms to rely on oxygen for their metabolism. The oxygen free radical damage to the brain is part of the price organisms pay for using this toxic environmental molecule.

SOURCES OF REACTIVE OXYGEN SPECIES IN THE BRAIN

The brain consumes a disproportionate amount of the body's O_2, as it derives its energy almost exclusively from oxidative metabolism of the mitochondrial respiratory chain.[25] Mitochondria are found in neuronal cell bodies but are also distributed throughout the neuritic processes: dendrites, axons, and synaptic bottons, where adenosine triphosphatases (ATPases) maintain ion gradients across the neuronal membrane. Neuronal oxidative phosphorylation, which generates adenosine triphosphate (ATP) while reducing O_2 to H_2O by the sequential addition of four electrons and four H^+, varies in proportion of neuronal firing. The "leakage" of high energy electrons along the mitochondrial electron transport chain causes the formation of O_2°- and H_2O_2 (Fig. 2.2). This "accidental" ROS generation increases as the O_2 concentration rises because of the associated rise in e^- leakage.[26]

Within the brain there exists a variety of biochemical mechanisms which result in the production of ROS: (1) $O_2^{\circ-}$ is readily dismutated in the presence of the enzyme superoxide dismutase (SOD) with the resultant production of H_2O_2.[5] (2) Besides SOD, several other enzymes generate H_2O_2. These include monoamine oxidase (MAO) catalyzing oxidation of catecholamine and dopamine, tyrosine hydroxylase, glycolate oxydase and L-amino acid oxidase, which produce H_2O_2 as a normal byproduct of their activity. The accelerated oxidation of dopamine (DA) by MAO in the nerve terminals of the dopaminergic axons from the substantia nigra has been suggested to increase H_2O_2 generation in these neurons, thereby elevating oxidative stress.[27] This may lead to a production of highly toxic oxygen species beyond the capability of the antioxidative defense mechanisms of the DA neurons to cope with, thus leading to the eventual destruction of these neurons. (3) Auto-oxidation of endogenous substances, for example ascorbic acid and catecholamines, which are differentially concentrated in certain neuronal systems, yields H_2O_2.[28] (iv) Activation of macrophage-type microglial cells yields oxidants.[29] (v) Finally, the activity of some neuronal enzymes yields oxidants. The Ca^{2+}-dependent activation of phospholipase A_2 (PLA_2), releases arachidonic acid (AA), which yields $O_2^{\circ-}$ through its subsequent metabolism by lipoxygenases and cyclo-oxygenases to form eicosanoids. The formation of NO° catalyzed by nitric oxide synthase (NOS), a form of which is concentrated to certain neurons, is activated by Ca^{2+} and is regulated by neurotransmitter receptors. NO reacts rapidly with $O_2^{\circ-}$ to yield the peroxynitrite anion, which decomposes to $^{\circ}OH$. Under conditions of energy failure and elevated intracellular Ca^{2+}, xanthine deshydrogenase is converted to xanthine oxidase by Ca^{2+}-dependent proteases. The conversion of accumulated hypoxanthine and xanthine by xanthine oxidase to uric acid yields $O_2^{\circ-}$. The hydroxyl radical, the most reactive species, is not generated directly by any known enzymatic reaction. However H_2O_2 slowly decomposes to $^{\circ}OH$, a process that is markedly accelerated in the presence of Fe^{2+} by the Fenton reaction. The half-life of oxygen radicals is inversely related to their reactivity with each other or with other molecules.

As proposed by some authors, superoxide anions could be involved in the NMDA-induced neurotoxicity.[30] The β-amyloid protein that accumulates in CNS plaques during Alzheimer's disease[31] is toxic for cultured cortical neurons[32,33] and, as recently proposed, H_2O_2 could be responsible for this neurodegenerative process.[33] A neurotoxic effect of H_2O_2 originating from microglial cells has also been reported.[34]

NITRIC OXIDE: BIOLOGICAL MEDIATOR, MODULATOR AND EFFECTOR

Biosynthesis

NO° is formed from the amino acid arginine by NO synthase (NOS) with stoichiometric generation of citrulline (Fig. 2.3).

Three different forms of NOS derive from distinct genes and are related to the three fundamental physiologic roles of NO:

(1) Endothelial NOS (e-NOS, NOS-III) produces NO in blood vessels, leading to vasodilatation. e-NOS is constitutively expressed and also found in some neurons.

(2) Inducible NOS (iNOS, NOS-II), found in macrophages and other tissues, is induced by diverse stimuli that often lead to inflammation. Originally described in macrophages and hepatocytes but also found in CNS glial cells (astrocytes and microglia) where it may contribute to NO-mediated neurotoxicity, the enzyme is expressed by cells following transcriptional induction. Thus the transcriptional regulation of iNOS controls the production of NO.

(3) Neuronal NOS (nNOS, NOS-I) occurs primarily in neurons but also in skeletal muscle and some epithelia.

eNOS and nNOS catalytic activities are triggered by calcium entering cells and binding to calmodulin, which stimulates enzyme activity. Once generated, NO can interact with a number of molecular targets including haem proteins, enzymes, DNA, thiols, oxygen and superoxide. These reactions determine the profile of NO as a major biological mediator, modulator and effector molecule.

NO: A Physiological Entity in the Brain

Nitric oxide (NO) may be the most recently identified neurotransmitter, but it is already linked in an impressive fashion to more normal and abnormal functions of the nervous system than any other.[35] It plays a role in regulating agressive and sexual behavior as well as opiate and amphetamine toxicity[36] and in neurotoxicity is secondary to the excitatory amino acid Glu.[37] There is now reported a connection between NO and Parkinson's disease.[38–40] Inhibition of NO synthesis in baboons and mice prevents the loss of dopaminergic neurons in the MPTP model, which faithfully mimics clinical parkinsonism in baboons.[38,39] The critical mediator appears to be peroxynitrite, which is formed by chemical reaction of NO° with $O_2^{\circ -}$. Peroxynitrite produces oxidative damage to lipids, proteins and DNA.[7] It also results in metal-catalyzed nitration of proteins and the production of 3-nitrotyrosine.[41–43]

The NO pathway is widely distributed throughout the central nervous system,[44] where it has been implicated in the

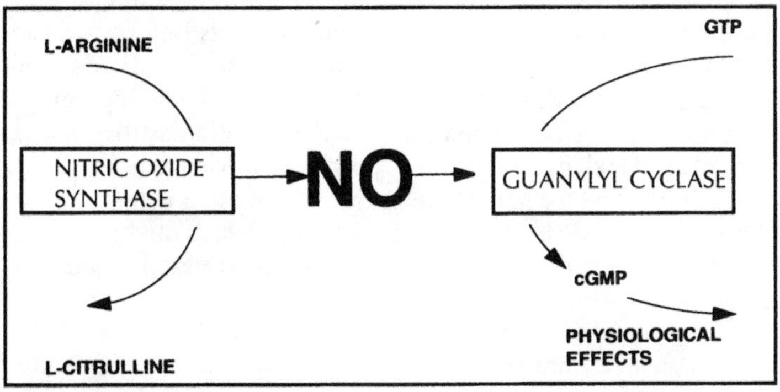

Fig. 2.3. Biosynthesis of nitric oxide. Three types of nitric oxide synthase exist: Type I (cNOS/NOSI/neuronal) constitutively expressed in neurons; type II (iNOS/NOSII/inducible) found in astrocytes and microglia; type III (eNOS/NOSIII/endothelial) constituvely expressed in endothelial cells and found in some neurons.

regulation of cerebral blood flow[45] and synaptic plasticity including long-term potentiation and[46] long-term synaptic depression in the cerebellum.[47] In the peripheral nervous system, NO has been proposed to mediate certain types of sensory transmission.[48] Thus, some of the basic functions of the nervous system, including memory, learning and pain perception, may be NO-mediated.

There is indirect evidence that NO may participate in neuroendocrine function. The presence of NOS has been detected in the hypothalamus, the posterior lobe of the pituitary gland and in discrete cells and fibers of the adrenal gland.[48] In addition, interleukin 2 has been shown to stimulate the release of corticotrophin-releasing factor via a NO-dependent mechanism.[49] These observations indicate that neuronal NO may regulate the release of hormones.

In the central nervous system NO is generated in response to the stimulation of N-methyl-D-aspartate (NMDA) receptors.[50,51] Many of the actions of NO in neurons are linked to Glu, the principal excitatory neurotransmitter in the brain. Physiologically, Glu, acting through the NMDA subtype of receptor, causes increases in cGMP levels mediated by NO. Glutamate neurons form synapses with nNOS neurons, and released NO diffuses into post-synaptic cells, binding to heme in the active site of guanylyl cyclase and activating it. When released in excess, Glu stimulation of NMDA receptors in nNOS neurons causes a massive release of NO, which kills adjacent neurons in neuronal cultures.[24] Evidence for this includes blockade of neurotoxicity by NOS inhibitors and the lack of such toxicity in cultures from mice in which the gene for nNOS has been deleted.[37] Dawson et al[24,63] also have found that SOD greatly reduces the toxicity of endogenous nitric production after NMDA receptor stimulation in their neuronal cell culture system, a finding consistent with peroxynitrite formation. Neural damage following vascular stroke elicited by occlusion of the middle cerebral artery is diminished by treatment with NOS inhibitors or in the NOS knockout mice.[52] The mechanisms whereby NO kills cells is not fully established. However, it appears that NO combines with superoxide to form the toxic peroxynitrite, which in turn may degenerate into the even more noxious hydroxyl free radical. Numerous sequelae are possible, including lipid and protein oxidation and protein nitration.[53] There is evidence that neurotoxic cell death derives primarily from DNA damage. Damaged DNA fragments activate poly-ADP-ribose synthetase (PARS), which attempts to repair DNA by attaching chains of ADP-ribose to nuclear proteins such as histones and PARS itself. Massive activation of PARS by excess amounts of NO depletes NAD, the PARS substrate, and subsequently ADP, leading to cell death by energy deprivation. PARS inhibitors block glutamate-NO-mediated neurotoxicity.[54] Inactivation of complexes of the respiratory chain by an excess of NO[55,56] could also largely contribute to cell death due to energy depletion.

Peripheral nerve injury can induce the iNOS form to high levels.[57] The macrophage remains the best characterized source of the iNOS. Its synthesis is induced by bacterial lipopolysaccharides, γ-interferon, and various interleukins with a lag time of a few hours. Transforming growth factor beta (TGF-β) downregulates expression of the iNOS. Tissue expressing the inducible isozyme can produce nitric oxide at rates a thousand-fold higher than typically produced by endothelium for signal transduction[58] and this isoform appears to be constitutively active. Microglia also produce nitric oxide via the high output nitric oxide.[59] In addition, astrocytes and glia can also produce nitric oxide.[60,61] Surprisingly, the activity of NOS is high in human skeletal muscle,[62] which might contribute to injury of the neuromuscular junction.

The Molecular Targets of NO

Nitric oxide has a high affinity to heme proteins including the soluble guanylate cyclase (GC-S), hemoglobin and myoglobin. While the interactions with hemoglobin and myoglobin lead to attenuation of the biological activity of NO, the stimulation of GC-S by this mediator gives rise to the cyclic GMP (cGMP)-mediated transduction mechanism. The cellular effects of cGMP include: the control of ion flux via cGMP-gated channels, modulation of cAMP-mediated responses and increase in cGMP- and cAMP-dependent protein kinase phosphorylation.[63,64] The ability of NO to interact with some enzymes, including complex I and complex II of the mitochondrial respiratory electron transport system and the citric acid cycle enzyme, aconitase, has been linked to its cytostatic/cytotoxic properties.[65]

Macrophage-derived NO has been shown to inhibit DNA synthesis. This may be due, in part, to NO-induced inhibition of ribonucleotide reductase, the rate-limiting enzyme of DNA biosynthesis.[66] Interestingly, exogenous NO and some NO-generating compounds, when applied at high concentrations, may cause mutagenicity by deamination of DNA.[67]

The S-nitrosylation of sulfhydryls by nitrogen oxides may provide the means for the storage of bioaccessible NO[68] and modulation of the activity of some receptors, ion channels and enzymes.

Nitric oxide reacts very rapidly with superoxide and initially this reaction was believed to attenuate potentially harmful effects of the latter molecule.[69] The product of this reaction, however, is peroxynitrite, which is highly reactive and capable of causing direct and indirect oxidation of many biological molecules.[70]

NO: Pathological Significance

The widespread involvement of NO in the physiological regulation of cellular function suggests that disturbances in its generation, release or action can have important repercussions for the neuropathogenesis of a number of disease states.The generation of NO in the CNS results from Glu-induced activation of NMDA receptors. Because glutamate is implicated in neurodegenerative disorders, such as ischemic hypoxic insults, Alzheimer's and Parkinson's diseases, and because the inhibitors of NO synthesis prevent the glutamate cytotoxicity in primary neuronal cultures and MPTP cytotoxicity in mice and baboons, it is possible that NO overproduction is involved in neuronal cell death under these conditions.

Peroxynitrite: Solution to the "Dr. Jekyll and Mr. Hyde" Dilemma of the Biological Actions of NO

One of the most important and outstanding issues in the biology of NO is the understanding of how and under what circumstances the same molecule may seemingly produce both physiological and pathological effects. Furthermore, in some animal models of neuronal damage, cellular protection can be achieved by inhibition of endogenous generation of NO as well as by administration of exogenous NO donors. The formation of $ONOO^-$ from NO and $O_2^{\circ-}$ may aid in understanding these conceptually disparate observations. Superoxide is often co-generated with NO[69] and can then attack NO at a very fast rate of $6.7 \times 10^9\ M^{-1}\ s^{-1}$. The rate of forming $ONOO^-$ depends on the product of NO and $O^{\circ}_2{}^-$ steady concentrations. If the concentrations of $O^{\circ}_2{}^-$ and NO are each increased 10-fold, then $ONOO^-$ will be formed at a 100-fold greater rate.[70]

Interestingly, the deleterious effects of $ONOO^-$ could be limited in the presence of endogenous thiols or an NO donor, S-nitroso-N-acetyl-D, L penicillamine.[53] These results suggest, that under conditions where cytotoxic $ONOO^-$ is generated, cellular cytoprotection may be achieved both by inhibiting formation of this oxidant using inhibitors of NOS or SOD and by administering exogenous NO to counteract the damaging effects of $ONOO^-$.

ANTIOXIDANT DEFENSE SYSTEMS IN THE BRAIN

Introduction

The steady-state formation of pro-oxidants is balanced by antioxidants that are enzymatic and/or nonenzymatic. To maintain homeostasis, there is a requirement for the continuous regeneration of the antioxidant capacity.

Antioxidants can act by:

(1) removing oxygen or decreasing local O_2 concentrations;

(2) removing catalytic metal ions;

(3) removing key reactive oxygen species (ROS) such as O_2 and H_2O_2;

(4) scavenging initiating radicals such as $^{\circ}OH$;

(5) breaking the chain of an initiated sequence;

(6) quenching or scavenging singlet oxygen.

Antioxidants inhibiting lipid peroxidation by the mechanisms listed in 1, 2, 4 and 6 can be called "preventative" antioxidants. Those acting by mechanism 3 are also preventative but since they are enzymes (e.g., catalase, SOD, and glutathione peroxidase) they are not consumed by the reaction. Chain-breaking antioxidants, singlet oxygen quenchers and metal chelators will be consumed whilst carrying out their protective functions.

Antioxidant protection can operate at several different levels within cells, for example by:

(1) preventing radical formation;

(2) intercepting formed radicals;

(3) repairing oxidative damage;

(4) increasing elimination of damaged molecules;

(5) nonrepair-recognition of excessively damaged molecules in order to prevent mutations occurring.

Enzymatic Antioxidants

Mammalian tissues contain three forms of superoxide dismutase (SOD), an enzyme converting $O_2^{\circ-}$ to H_2O_2. These include two CuZnSODs, one localized extracellularly (ECSOD)[71] and the other within peroxisomes and the cytoplasm. The third form of SOD contains manganese (MnSOD) and is localized almost exclusively in the matrix of mitochondria. The efficacy of SOD as an antioxidant relies on its cooperation with other enzymes, i.e., CAT and GSHPx, which metabolize the dismutation product of $O_2^{\circ-}$, H_2O_2. CAT quickly decomposes H_2O_2 to water and O_2, but there is very little CAT activity in the brain, so it is essentially inconsequential in this tissue. On the other hand, GSHPx (a selenium-dependent enzyme) probably plays a major role in disposing of H_2O_2 in neural tissue.[72] In this reaction, H_2O_2 is used to oxidize reduced glutathione, GSH (a tripeptide, glutamylcysteinyl-glycine). Under normal conditions, a balance may exist between the rate of H_2O_2 formation via dismutation of superoxide by SOD and its elimination by GSHPx and CAT to prevent oxidative stress. In the glutathione peroxidase system, the enzymatic oxydation of GSH to oxidized glutathione (GSSG) is coupled to detoxification of H_2O_2 in the cytosol and mitochondria or other peroxides, notably those derived from the oxidation of the membrane phospholipids. Levels of GSH are believed to be limiting in this process, requiring the regeneration of GSH from GSSG. This is achieved by GSSGRd which utilizes NADPH as a reducing cofactor. Both GSHPx and GSSGRd are distributed throughout the brain; although GSHPx usually is considered to be primarily localized in glial cytoplasm rather than in neurons,[83] some immunocytochemical evidence claims weak GSHPx activity in neurons as well.[84] The functions of some relevant antioxidant enzymes are outlined in Figure 2.4

Nonenzymatic Antioxidants

The SODs and other antioxidant enzymes, including CAT and GSHPx, are complemented by a number of nonenzymatic factors located both intra- and extracellularly. Sulfhydryl-containing molecules such as GSH as well as vitamins C and E, albumin, ceruloplasmin, bilirubin and uric acid are components of the nonenzymatic antioxidant system.[13,73–80] Metallothionein and heme oxygenase are proteins induced after exposure to agents that cause oxidative stress.[81,82] Heme oxygenase, an enzyme catalyzing conversion of heme to biliverdin, leads to a reduction of the cellular pool of heme containing proteins, thus removing potential pro-oxidant catalysts. Furthermore, bilirubin, the end product of the heme degradation pathway, is a molecule with antioxidant properties.

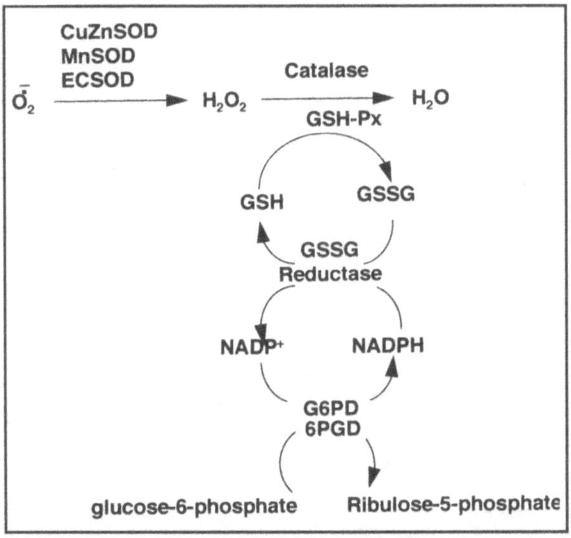

Fig. 2.4. Detoxification of reactive oxygen species by antioxidant enzymes. This interrelated system is complemented by a number of non-enzymatic antioxidants. CuZnSOD, copper zinc-containing superoxide dismutase; MnSOD, manganese-containing superoxide dismutase; ECSOD, extracellular superoxide dismutase; GSHPx, glutathione peroxidase; GSH, reduced glutathione; GSSG, oxidized glutathione; G6PD, glucose 6-phosphate dehydrogenase; 6PGD, 6-phosphogluconate dehydrogenase.

Within the hydrophobic lipid interior of membranes, different types of lipophilic radicals are formed from those seen in the intracellular aqueous milieu. Lipophilic radicals require different types of antioxidants for their removal. Vitamin E (α-tocopherol), a fat-soluble vitamin, is a poor antioxidant outside a membrane biolayer but is extremely effective when incorporated into the membrane. Although α-tocopherol can to react with singlet oxygen and ˙OH, the major antioxidant action of α-tocopherol is due to its ability to donate labile hydrogens to peroxy and alkoxy radicals, thereby preventing lipid peroxidation. Thus, α-tocopherol has been classified as a « chain breaking » antioxidant. α-Tocopherol is important for normal brain physiology as evidenced by observations that patients with a prolonged deficiency of this vitamin due to intestinal fat absorption problems suffer from neurological deficits.[85] Likewise, in cultured neurons vitamin E promotes survival and neurite outgrowth, suggesting that it may be beneficial to these cells in vivo.[73]

The function of vitamin C (ascorbic acid) in the brain is a double-edged sword with respect to free radical damage. In general, both the gray and white matter of the brain contain substantial amounts of ascorbic acid. Because of a special active transport system in the choroid plexus, ascorbic acid concentrations in the CSF can exceed those in the blood by 10-fold. Also, a second transport system is present in neural cells to further concentrate ascorbic acid.[86] Ascorbate is a free radical scavenger and functions as an antioxidant in recycling the vitamin E radical back to vitamin E, especially in the brain.

The major copper-containing protein of human plasma is ceruloplasmin. Apart from its known acute-phase reactant properties its biological functions have remained an enigma. However, the protein's ferroxidase activity makes a major contribution to extracellular antioxidant protection against lipid peroxidation and Fenton chemistry.[75]

Recently it was shown that a constituent produced in the pineal gland, i.e., N-acetyl-5-methoxytryptamine or melatonin, readily passes through the blood-brain barrier and enters neurons and glial cells where it may exert strong antioxidant activity.[87,88] Exogenously administered melatonin quickly enters the brain, and because of its lipid and water solubility, it probably distributes readily to all subcellular compartments. This lack of compartmentalization makes melatonin unique among antioxidants that are characteristically confined to one cellular compartment, e.g., vitamin E in lipid cell membranes, vitamin C in the aqueous environment of the cytosol, etc.

Preliminary studies indicate that melatonin is an efficient scavenger of both the °OH and ROO°. The findings published to date indicate that melatonin restricts lipid peroxidation by preventing the initiating events as well as interrupting the chain reaction.[14]

This combination of enzymatic and nonenzymatic sources provides an important protective system against various oxidant stresses. Normally, a balance exists between formation of ROS and antioxidants. However, oxidative stress, cell injury and disease may ensue after excessive production of oxidants or deficient functioning of antioxidant defenses or repair capacity.

HETEROGENEOUS CELLULAR LOCALIZATION OF ANTIOXIDANTS GOVERNS THE SUSCEPTIBILITY OF NEURONS TO OXIDATIVE STRESS

Introduction

Neurodegenerative disorders are characterized by the selective vulnerability of specific neuronal subsets that display different metabolic profiles. On this basis, one may speculate that certain neuronal populations may be more vulnerable to oxidative stress as a result of a greater oxidative burden or, alternatively, lower antioxidant protection.[15] Overexpression of the CuZnSOD gene, due

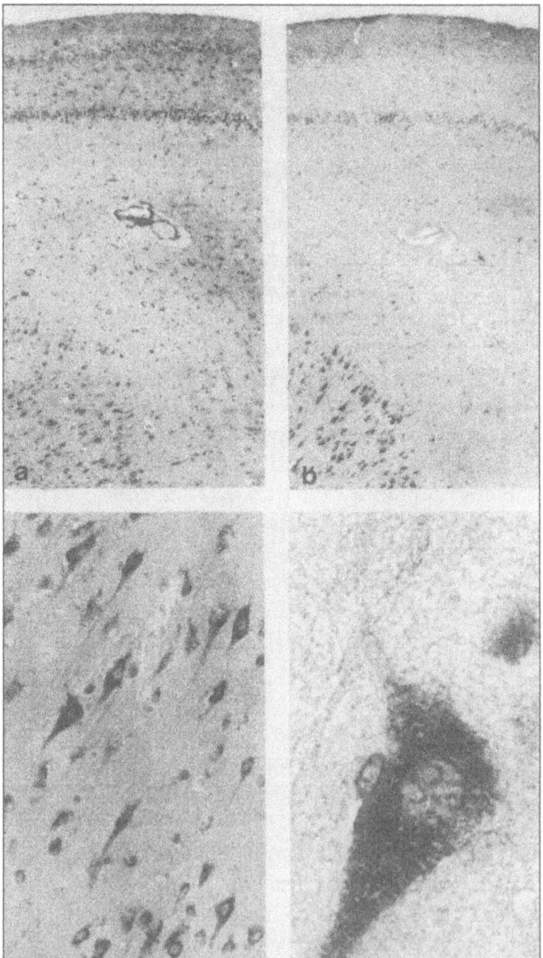

Fig. 2.5. Adjacent histological sections of hippocampus from a normal aged patient.
a. hemalum erythrosin staining. Field CA4 (x 80).
b. anti-CuZnSOD immunostaining. Note the important labeling of pyramidal cells by the immunoserum, the moderate labeling of the hippocampal granule cells and the weak labeling of other cells.
c,d. observation of the immunolabeling at a higher magnification (c, x 400; d, x 2400). The labeling is confined to the perikaryon and the proximal dendrites. From Delacourte et al.[92]

to gene dosage, has been recorded in all tissues of trisomic 21 patients and in a variety of trisomy 21 cells in culture, as well as in the brain of DS patients.[89] Moreover, elevated CuZnSOD activities have been reported in other neurodegenerative disorders. First, In brains of patients with AD increased SOD levels were found in certain regions of the brain[90] with high levels of constitutive expression in AD-vulnerable pyramidal cells of hippocampus.[91–92] Second, this enzyme is particularly highly expressed in the age-vulnerable dopaminergic melanized neurons in PD.[93,94] Third, CuZnSOD mRNA levels are increased in sporadic amyotrophic lateral sclerosis motorneurons.[95,96]

This underlines the paradoxical role of copper-zinc SOD in the nervous system and suggests that a high constitutive level of CuZnSOD expression may play a role in the selective vulnerability of certain neuronal populations.

Preferential Localization of Copper-Zinc Superoxide Dismutase in Neuronal Cells: Correlation with Selective Neuronal Vulnerability

Interestingly, in the human brain CuZnSOD gene expression is particularly high in neurons.[91,92,97] Immunohistochemistry[91,92] and cDNA-mRNA in situ

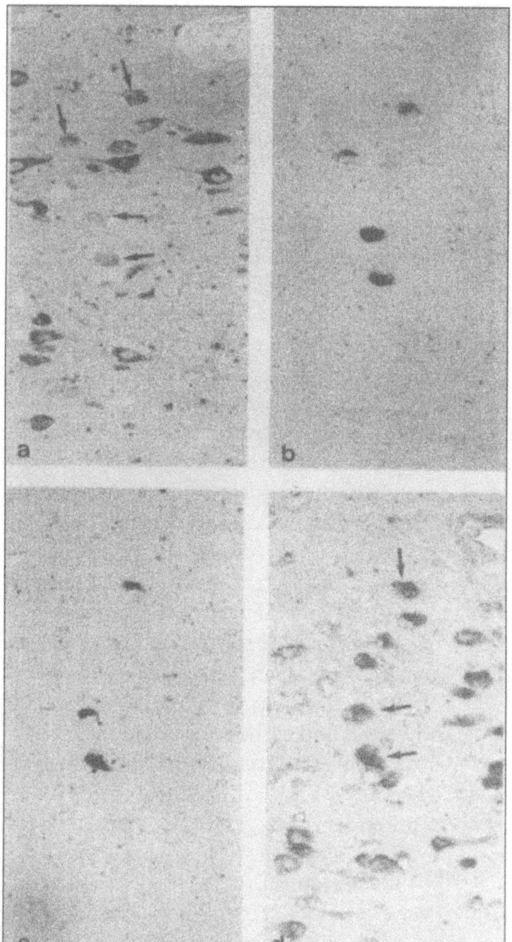

Fig. 2.6. Two photomicrographs of the same section of AD hippocampus.
a. The distribution of neurons immunolabeled with the anti-CuZnSOD antiserum.
b. The degenerating neurons immunolabeled with the anti-PHF antiserum, after elution of the first antiserum (arrows in a). c,d: two photographs of the adjacent sections.
c. Labeling with anti-PHF.
d. Elution and restaining with the anti-CuZnSOD antiserum. Note that degenerating neurons (arrows) are among the population of the CuZnSOD immunostained neurons. x 400. From Delacourte et al.[92]

hybridization[91,95–100] studies on control human hippocampus have shown that the amounts of CuZnSOD protein and mRNA are high in large pyramidal neurons and granular cells and very weak in other cells, suggesting that the production of superoxide radicals within these cells might be particularly elevated. At high magnification, the staining was apparent homogenous, excluding any obvious sublocalization of CuZnSOD to the membrane or nucleus of these cells (Fig. 2.5)

In Alzheimer's brains, cell distribution of the CuZnSOD immunostaining was similar to that observed in control brains. In one of these brains, successive immunostaining was performed with anti-CuZnSOD and anti-paired helical filaments (a marker of neurofibrillary degeneration which recognized phosphorylated tau protein). Among the populations of pyramidal cells that were intensely stained with the antiCuZnSOD antiserum, several neurons were affected by neurofibrillary degeneration (Fig. 2.6).

The same results were observed for the cellular distribution in human hippocampus of CuZnSOD mRNA by in situ hybridization. Autoradiograms of the hippocampal sections hybridized with SOD probe showed a high density of grains over the dentate gyrus and in the fields of Ammon's horn (Fig. 2.7).

The elevated abundance of CuZnSOD mRNA and protein in the pyramidal neurons suggest that they represent a subset of neurons with distinct characteristics towards oxidative stress. Thus, this study shows that large πψραμιδαλ νευρονσ which are potentially susceptible to degenerative processes in AD have the property to contain apparently higher levels of CuZnSOD than other brain cells. This might indicate that biochemical pathways leading to $O_2^{\circ-}$ generation are specially active in these cells, requiring a high CuZnSOD content to eliminate these radicals. Nevertheless, too high a level of enzyme activity as in DS might be noxious for cells and contribute to degenerative processes leading to AD pathology. In support of this hypothesis, mouse cells expressing the transfected human CuZnSOD gene show enhanced lipid peroxidation[98] and induction of glutathione peroxidase activity.[99] PC12 cells expressing the transfected human CuZnSOD gene have impaired neurotransmitter uptake resulting from modifications of the membrane properties of the chromaffin granules, likely secondary to lipid peroxidation.[100] It is interesting to underline that, in the brain areas we have studied, the cell distribution of CuZnSOD is the same as for the RNA that encodes the precursor of the β-amyloid peptide (APP)[101,102] which accumulates in the amyloid substance in AD and DS. Common mechanisms might control the expression of the APP and CuZnSOD genes, which are both located on chromosome 21, inside or in the vicinity of the obligatory region for Down's syndrome.[103]

Recent study showed that CuZnSOD is also selectively enriched in other subsets of neurons, particularly motor neurons in spinal cord and brainstem, pyramidal neurons in sectors CA1-CA4 of hippocampus, neurons in neocortex, pyriform cortex, and amygdala.[91,92,104–108] and in melanized neurons in human mesencephalon.[93,94] The presence of CuZnSOD in the axonal compartment strongly suggests that the enzyme is carried by anterograde transport, a conclusion further supported by accumulation of CuZnSOD in nerve termini.

Moreover, the increase in heme oxygenase-1 protein (HO-1 or HSP32, a protein induced during oxidative stress[109] in association with CuZnSOD and neurofibrillary pathology of AD,[110] supports the notion that oxidative intraneuronal stress mechanisms are coincident with the formation or presence of cytoskeletal pathology. HO-1 is a microsomal enzyme that oxidatively cleaves heme, a pro-oxidant, to produce biliverdin and carbon monoxide.[111] Biliverdin is converted to bilirubin, a potent antioxidant,[112] and within the brain carbon monoxide is suggested to act as a neurotransmitter.[113]

These results provide evidence that oxidative stress and free radicals play a role in AD disease etiology or the pathogenesis of neurofibrillary pathology. This reinforces

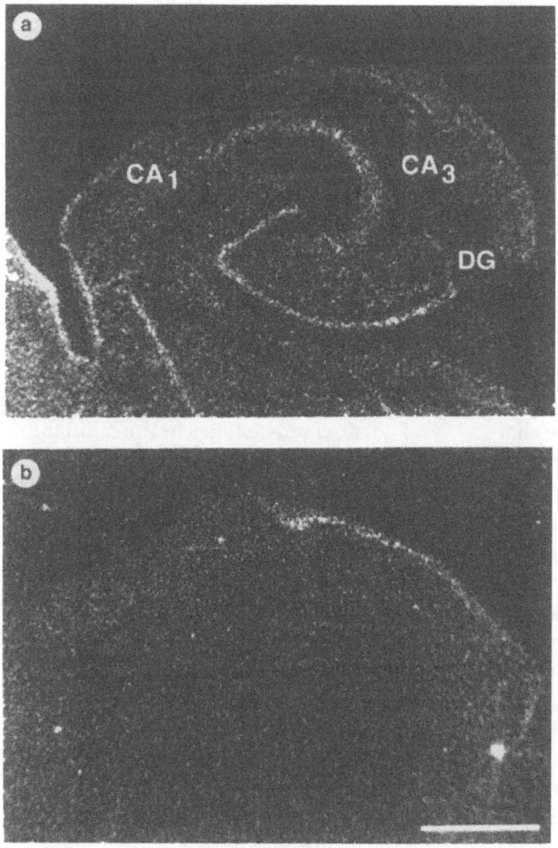

Fig. 2.7. Film autoradiograms of adjacent sagittal sections of a human hippocampus hybridized with a human ^{35}S-labeled CuZnSOD cDNA probe (a) or with a human ^{35}S-labeled TH cDNA probe (b). Coronal tissue section from mid-rostrocaudal levels of the hippocampal formation were exposed in direct contact with X-ray film for 2 days. An intense labeling is observed with the CuZnSOD cDNA probe over the pyramidal cell layers of Ammon's horn (CA), the CA1-CA4 fields, the subiculum, and in the granule cell layer of the dentate gyrus (DG). The specificity of the hybridization pattern is demonstrated by the absence of labeling with a radioactive cDNA unrelated to the hippocampus such as the tyrosine hydroxylase cDNA probe (b). Bar = 2 mm. From Ceballos et al.[97]

the hypothesis that oxidative stress may account for preferential pyramidal neurons' vulnerability in aging and in Alzheimer's disease. Indeed it was recently demonstrated that neurofibrillary tangles and senile plaques are associated with posttranslational modifications typical of nonenzymatic advanced Maillard reaction end products.[114] Such Maillard modifications are initiated and potentiated during oxidative stress and would generate crosslinked aggregates of insoluble protein characteristic of senile plaques and neurofibrillary tangles.[115] Secondly, oxidative complications would also be important in the action of apolipoprotein E (ApoE) where individuals that are heterozygous or homozygous for the ApoE isoform are at greater risk of developing AD.[116] ApoE forms sodium dodecyl sulfate resistant complexes with Aβ and the formation of ApoE-Aβ complex is completely abolished under reducing conditions.[117]

The abundance of CuZnSOD in a subset of nerve cells suggests an especially important role for this enzyme in neuroprotection or neurodegeneration. Neurons may be selectively vulnerable in neurogenerative diseases, not because CuZnSOD activity is limiting, but because it is so abundant. In this view, high, rather than limiting, levels of CuZnSOD may place motor neurons selectively at risk in FALS, pyramidal neurons at risk in AD and DS as well as melanized neurons at risk in PD.

Localization of Manganese Superoxide Dismutase in Cerebral Sites of Degeneration

To further consider this idea, we have investigated other enzymatic steps involved in cellular antioxidant defense: MnSOD which is the mitochondrial form of SOD. Human MnSOD, located mainly in the mitochondrial matrix, is a homotetramer of particular importance given that production of free radicals through the respiratory chain reaction is especially high.

In control hippocampi, the pattern of MnSOD immunoreactivity was the same as CuZnSOD. At the cellular levels MnSOD was localized predominantly to pyramidal neurons and scarcely seen in glial cells. In AD hippocampi, MnSOD immunoreactivity is enriched in senile plaques in cells with morphology corresponding to reactive glial cells (Fig. 2.8). This showed the presence of MnSOD in microglial cells in the pathological conditions of senile plaques formation involving inflammatory processes and cytokines production.

This change in the glial localization of MnSOD in the brain of AD was the main difference between controls and AD tissue and is possibly due to metabolic activation of reactive glia. Thus, this enzyme could have a role during plaque formation but we don't know if these changes in the cellular expression are causally related or represent a primary mechanism in the early stage of senile plaques formation.

As we described for AD, the distributions and intensities of these two forms of SOD immunoreactivities were different in DS as compared with controls. CuZnSOD was enriched in pyramidal neurons undergoing degeneration, whereas MnSOD was more enriched in reactive astrocytes than in neurons.[106] In senile plaques, CuZnSOD-positive globular structures were surrounded by astrocytes highly enriched in MnSOD. Amyloid cores, diffuse plaques, and microglia scarcely showed colocalization with SOD-positive structures. Interestingly, MAO_B, another enzyme producing hydrogen peroxide, is highly expressed in astrocytes in or around senile plaques in AD.[118]

The observed changes in the cellular localization of SODs in neocortex and hippocampus in cases of Alzheimer's disease and Down's syndrome support a role for oxidative injury in neuronal degeneration and senile plaque formation. The differential localization of CuZnSOD and MnSOD in cerebral sites of degeneration suggests that cellular responses to oxidative stress are antioxidant enzyme specific and cell type specific and that these two forms of SOD may have different functions in antioxidant mechanisms.

MnSOD is of particular importance given that production of ROS through respiratory chain reactions is especially high. Moreover, MnSOD activity can be induced under various stimuli such as cytokines[119]

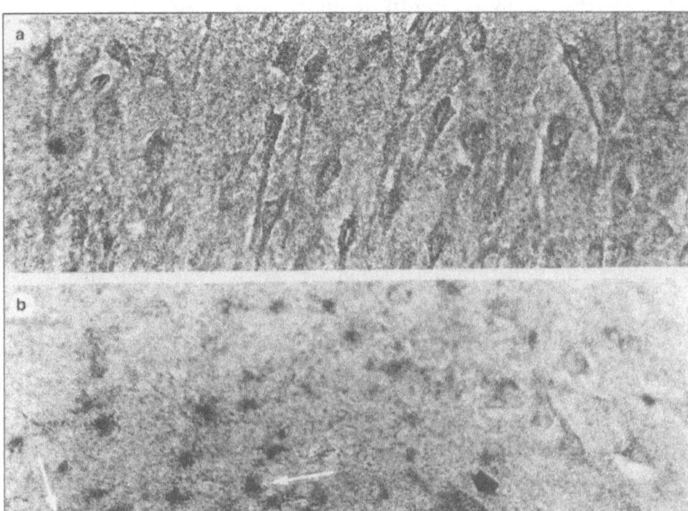

Fig. 2.8. Photo-micrograph of MnSOD immunostaining in human control hippocampus post-mortem (a) and AD hippocampus (b).
In control hippocampus, MnSOD labeling was largely limited to the cell bodies or the proximal processes of pyramidal neurons.
In AD hippocampus, intensely stained glial cells located around senile plaques were detected. Scale bar = 25 μm.

and ROS[120] and postischemic neurodegeneration.[121] This suggests that MnSOD may be activated by oxidative stress as a defense mechanism against the deleterious effects of ROS.

A neuronal distribution of MnSOD has been also described in the basal ganglia and hippocampal formation in the rat[122–124] and in the cerebral cortex and spinal cord in humans.[122] In the striatum, two populations of neurons were immunostained, large neurons with short processes corresponding to the morphology of cholinergic interneurons and other smaller cells which may correspond to striatal output neurons and other interneurons.[121,125] The MnSOD immunoreactivity was visualized in neuronal perikarya and proximal dendrites in rat[121] and in humans.[125] MnSOD content in the pyramidal neurons of CA3 was shown to be considerably higher than in other CA subsectors in the rat hippocampus.[122] Whether this selective distribution of MnSOD immunostaining is related to the differential vulnerability of the CA subsectors in pathology remains to be determined. Yet, it is interesting to notice that, in AD, the density of senile plaques (one of the pathological hallmark of the disease) is least in the CA2 subsector[126,127] where MnSOD immunoreactivity is highest. In ALS, neurons that successfully induce the expression of sufficient MnSOD can survive to the disease process, while those failing to activate adequate expression of the enzyme succumb to the toxic effects of ROS and die.[122]

The consequences of the heterogeneous distribution of MnSOD in the human brain are not known. It is however, interesting to note that oxidative stress is thought to participate in the mechanism of nerve cell death occurring in normal aging and that MnSOD is precisely localized in, among other regions, the most vulnerable cerebral regions. These results are suggestive of an important role for MnSOD in neuronal defense against oxidative stress, at least in mitochondria, given its localization in this organelle. Additionally, MnSOD may also participate in neuronal degeneration. Indeed, MnSOD catalyzes the production of H_2O_2 which may in turn be transformed into highly toxic hydroxyl radicals if it is not degraded. The free radical defense mechanisms thus imply a close relationship between MnSOD and glutathione peroxidase and catalase, which catabolize H_2O_2. However, given the fact that mitochondrial damage occurs with aging in neurodegenerative disorders[128–130] the full efficacy of the system is questionable.

Glutathione Peroxidase and Glutathione are Mainly in Glial Cells

Novel data about the cellular distribution (i.e., neurons vs glial cells) of the antioxidants underline the putative role of glial cells in the modulation of the neurotoxic effect of ROS. In the brain, the seleno-dependent GSHPx and CAT are the major protective enzymes against H_2O_2 cytotoxicity.

The distribution of glutathione peroxidase-containing cells was investigated by immunohistochemistry in the midbrain of control subjects and patients with PD.[83] GSHPx-like immunoreactivity was detected exclusively in glial cells. In the control mesencephalon, the lowest density of GSHPx-positive glial cells was found in the substantia nigra pars compacta, the dopaminergic cell group the most vulnerable to PD. Thus, dopaminergic neurons surrounded by a low density of glial cells may be less protected and thereby more susceptible to oxidative stress. Conversely, central gray substance dopaminergic neurons, surrounded by more densely concentrated glial cells, could be better protected against oxygen-derived toxicity, and this may contribute to preventing their degeneration. In PD midbrain, an increased density of GSHPx-immunostained cells was observed, surrounding the surviving dopaminergic neurons. The increase in GSHPx-containing cells was correlated with the severity in dopaminergic cell loss in the respective cell groups.

Thus, the amount of GSHPx-containing cells may be critical for a protective effect against oxidative stress, although it cannot be excluded that the level of the enzyme activity remains the crucial factor. The density of GFAP-positive cells and therefore the density of astrocytic glial cells also increased with the severity of the dopaminergic neuronal loss in parkinsonian midbrain. This suggests that the enhanced number of GSHPx-positive glial cells in the pathological brains corresponds to a gliosis following neuronal degeneration. The cause of this gliosis is not known but glial cells could produce NO[59,60] or superoxide anions when activated.[29] Glial cells are generally considered to migrate in order to occupy the empty space left by the disappearance of neurons. Alternatively, the density of glial cells may increase in the vicinity of the still remaining dopaminergic neurons to regulate a defense mechanism. As GSHPx is contained in glial cells, the increase in the number of these cells may protect dopaminergic neurons against an oxidative stress. The contribution of other factors expressed in glial cells cannot be excluded, however. For example, glial cells are known to express the nerve growth factor gene (NGF)[131], which may take part in the survival of neurons, or protect the neurons against a rise in intracellular calcium levels which leads to nerve cell death.[132]

Oxygen-derived toxics as a cause of pathological degeneration of dopaminergic neurons in PD may originate from several sources:

(1) The presence of neuromelanin. The pigment is formed from catechol autooxydation and polymerization. Its synthesis leads to oxidative products which may be toxic to neurons;

(2) oxidative degradation of DA by MAO, highly concentrated in the substantia nigra and which produces H_2O_2; and

(3) increased formation of $^{\circ}OH$, possibly as a result of the increased concentrations of iron in the substantia nigra of parkinsonian patients.[133] Thus, a low baseline GSHPx may facilitate conversion of DA to neuromelanin and contribute to vulnerability to degeneration in parkinsonism, whereas a higher baselne GSHPx may help to prevent neuromelanin synthesis, and prevent an excessive $^{\circ}OH$ formation in the surrounding dopaminergic neurons.

The immunolocalization of GSHPx was also investigated in another area of the human brain particularly vulnerable: the hippocampus. As in the midbrain, glutathione peroxidase-like immunoreactivity was exclusively located in small cells with a morphology compatible with that of astrocytic glial cells (unpublished data) (Fig. 2.9).

This finding is compatible with the preferential astroglial localization of GSH. In cellular cultures, the level of GSH was found to be lower in neurons than in astroglial cells,[77,80,134–136] or GSH could not be detected at all.[74] Furthermore, astroglial cells appear to supply cysteine for GSH neuronal synthesis.[78,79] Astrocytes have a three times higher concentration of GSH than oligodendroglial precursors in culture.[76] In vivo, an additional complexity is present. Histochemical analysis of fresh frozen tissue has shown that brain GSH in both the monkey and the rat is virtually absent from the cytosol of neuronal cell bodies, but appears to be localized to axons, nerve terminal regions, and glia.[137,138]

These data fit well with recent reports indicating that neurons are particularly sensitive to H_2O_2.[139] A recent study was undertaken to investigate the putative role of astrocytes in the modulation of the neurotoxic effect of H_2O_2.[140] The neurotoxic effect of H_2O_2 on neurons cocultured with astrocytes was strongly attenuated compared with that observed on a pure population of neurons seeded at the same density. Moreover, the protective effect of astrocytes depended on the astrocyte/neuron ratio, a significant neuroprotection being detectable for 1 astrocyte to 20 neurons. Catalase seems to be the main hydrogen peroxidase activity involved in the neuroprotective effect of astrocytes. Indeed, in the culture conditions used, this enzymatic activity was enriched in this cell type compared with neurons; its inhibition, and not that of GSHPx, reduced the disappearance rate of the oxidant. Therefore, astrocytes could delay neuronal death in pathological situations in which H_2O_2 has been, at least partially, demonstrated to be involved.

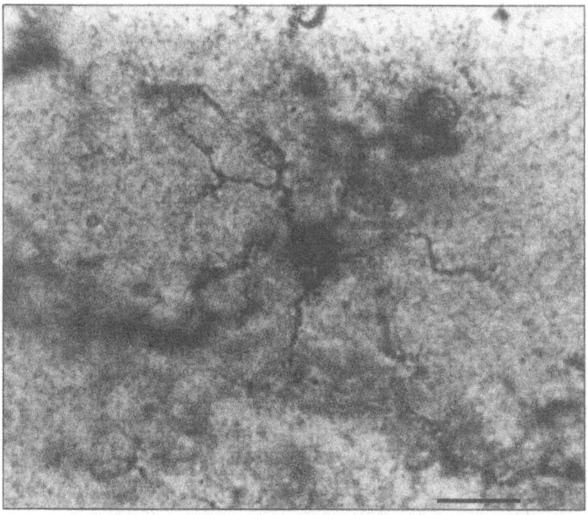

Fig. 2.9. Representative glutathione peroxidase-positive cell in human hippocampus which displays an astrocytic morphology.
Hippocampus: Scale bar = 70 µm.

Conclusion

That astroglial cells represent the major, if not exclusive, compartment of GSHPx in the human mesencephalon and hippocampus indicates that most of H_2O_2 scavenging involves glial cells. The presence of GSHPx in astrocytes is all the more interesting considering the evidence for a neuronal localization of CuZnSOD and MnSOD in human brain as described above. This heterogeneous cellular distribution of the enzymes implicated in oxy-radical detoxification in the brain suggests that the free radical defense system could employ a neuronal-glial coupling. Alternatively this cellular compartimentation of SOD and GSHPx in the brain also suggests a poor cooperation between these enzymes. This could contribute to the selective vulnerability of neurons which are not well protected against H_2O_2 toxicity.

EXCITOTOXICITY AND OXIDATIVE STRESS

Introduction

The selective vulnerability of neuronal systems is a remarkable characteristic of age-related degenerative disorders of the brain. Over the last decade, considerable circumstancial evidence has accumulated that implicates excessive stimulation of glutamate (Glu)-gated cation channels in triggering neuronal death

Glu and related acidic amino acids are thought to be the major excitatory neurotransmitters in brain and may be utilized by 40 percent of the synapses. Thus, two broad mechanisms, oxidative stress and excessive activation of Glu receptors, are converging and represent sequential as well as interacting processes that provide a final common pathway for cell vulnerability in the brain. The broad distribution in brain of the processes regulating oxidative stress and mediating glutamatergic neurotransmission may explain the wide range of disorders in which both have been implicated. Yet differential expression of components of the processes in particular neuronal systems may account for selective neurodegeneration in certain disorders.

A separate line of investigation has focused on the role of oxidative stress as the proximate cause of several of these degenerative disorders. The attractive feature of the oxidative stress hypothesis is that it can account for cumulative damage associated with the delayed onset and progessive nature of these conditions. Evidence is now emerging that activation of Glu-gated cation channels may be an important source of oxidative stress and that these two mechanisms may act in a sequential as well as a reinforcing manner, leading to selective neuronal degeneration. Understanding the relation between oxidative stress and Glu neurotransmission could lead to the development of pharmacologic interventions that disrupt this chain of pathological events without impairing excitatory neurotransmission.

Glutamate, Calcium and Reactive Oxygen Species

Although multiple factors can precipitate oxidative stress in cells, the neurotransmitter Glu is the major effector of this process in brain, primarily through activation of its ionotropic receptors. The ionotropic receptors mediating the depolarizing action of Glu have been named after their most potent agonists and can be distinguished by their pharmacological and electrophysiological properties: the NMDA (N-methyl-D-aspartate), the AMPA (α-amino-3-hydroxy-5-methyl-4-isoxasole-proprionic acid), and the KA (kainic acid) receptors. The best characterized of the three, the NMDA receptor complex, consists of a membrane-spanning channel, which is highly permeable to both $Na^+ \cdot K^+$ and Ca^{2+} in a voltage-dependent manner and possesses several regulatory sites, including glycine, Zn^{2+}, polyamine, and phencyclidine binding sites, all of which allosterically affect Glu-mediated channel opening. Both AMPA and KA receptors are linked to Na^+ permeable channels. Each exhibits significant affinity for the other's recognition site, and no antagonists selectively differentiate KA from AMPA receptors. Thus, they are often designated as « non-NMDA receptors ». Finally, a fourth class of Glu receptors, the metabotropic receptors, exert their effects by means of G protein-initiated biochemical events and not ion channels. Although the metabotropic Glu receptors do not directly mediate the neurotoxic effects of Glu, activation of subtypes of these receptors may potentiate or attenuate neurotoxicity.[141]

Glutamate-induced $[Ca^{2+}]$i increase correlates with subsequent neuronal degeneration.[142] Furthermore, cell-permeant calcium chelators reduce excitotoxic neuronal damage in vitro and ischemic neuronal injury in vivo.[143] Calcium influx via NMDA receptors is much more effective in mediating cell death than that occurring through non-NMDA receptors or voltage-dependent calcium channels.[144] This suggests compartmentalization of calcium-dependent neurotoxic processes within neurons, with a preferential localization in the submembrane space adjacent to NMDA receptors.

Increased $[Ca^{2+}]$i levels can initiate a number of deleterious processes including activation of NOS and free radical generation. Exposure of cultured hippocampal neurons to Glu resulted in a time- and concentration-dependent accumulation of peroxides.[145] NOS inhibitors abolish Glu neurotoxicity in cultured cortical and striatal neurons.[24] Pretreatment of cultures with low doses of quisqualate, which preferentially kills NOS-containing neurons, blocks subsequent Glu neurotoxicity in the cultures.[63] Antioxidants suppress glutamate-induced elevation of the $[Ca^{2+}]$i and H_2O_2, indicating that glutamate-induced ROS contribute to the elevation of the $[Ca^{2+}]$i. Previous studies in cerebellar and cortical cell cultures showed that antioxidants can protect neurons against excitotoxic insults.[22,146,147] These data are consistent with the recent study of Lafon-cazal et al,[30] who used electron paramagnetic resonance spectroscopy to demonstrate induction of $O_2^{\circ -}$ in cultured cerebellar granule neurons exposed to Glu. Neurons overexpressing superoxide dismutase are resistant to glutamate.[148]

Thus, the induction of ROS by Glu contributes to the elevation of $[Ca^{2+}]$i, which plays a major role in excitotoxic neuronal injury and death. The mechanism(s) whereby ROS promote an increase of $[Ca^{2+}]$i may involve compromise of the activites of several different $[Ca^{2+}]$i -regulating proteins known to be sensitive to oxidation, including glutamate receptors, voltage-dependent Ca^{2+} channels, plasma-membrane Ca^{2+}-ATPases, and the Na^+/Ca^{2+}exchanger.

Mechanisms of Reactive Oxygen Species Production by Glutamate

The source of the ROS induced by glutamate is not clear, but NMDA receptor activation and Ca^{2+} influx were required because the NMDA receptor antagonist and removal of calcium from the culture medium each prevented glutamate-induced peroxide accumulation.[145] The Ca^{2+}-mediated effects of Glu receptor activation leading to neuronal degeneration may involve a number of different pathways that cause oxidative stress:

(1) NMDA receptor-mediated stimulation of PLA_2 and the subsequent release of AA leads to the generation of oxygen radicals.[30] AA and oxygen radicals enhance the release of Glu and inhibit its uptake inactivation by neuronal and glial transporter processes, thus promoting a vicious cycle;[149]

(2) Ca^{2+} uptake by mitochondria, which disrupts the mitochondrial transmembrane potential with production of ROS;[150,151]

(3) activation of NOS, resulting in production of NO, which reacts with $O_2^{\circ-}$ to produce peroxynitrite;[152,153] and

(4) elevated intraneuronal Ca^{2+} activates peptidases, such as calpain I, which can catalyze the enzymatic conversion of xanthine dehydrogenase to xanthine oxidase; the catabolism of purine bases by xanthine oxidase yields $O_2^{\circ-}$. Furthermore, as lactic acid increases markedly at the same time, the acidic conditions favor the liberation of cellular stores of Fe^{2+}, which promotes the Fenton reaction to yield $^{\circ}OH$ from H_2O_2.

One or more of the above mechanisms may underlie the contribution of Ca^{2+} to the increase in levels of ROS. Using cultured cerebellar granule cells, Lafon-Cazal and coworkers have demonstrated that NMDA receptor stimulation, but not KA receptor stimulation, produces marked elevations in $O_2^{\circ-}$ and $^{\circ}OH$. Furthermore, co-incubation of cells with NMDA and the spin traps provided nearly complete protection against delayed neurodegeneration. In this system, the primary source of oxidants was likely AA metabolism as a consequence of PLA_2 activation and not peroxynitrite anions generated by NOS. One of the relevant hypotheses relates to the calcium-dependent production of NO by neurons as a cytotoxic mechanism.[152,153] Glutamate-mediated neurotransmission was first linked with NO production by neurons by Garthwaite.[50] The importance of NO as a retrograde messenger contributing to the formation of memory and development was shortly thereafter postulated by Williams et al[154] and modeled by Gally et al.[155] But excitatory amino acids are also key initiators of neuronal death, and Garthwaite et al[156] also proposed that NO production could also contribute to neuronal degeneration. Dawson et al[24,63] have proposed that NMDA receptor-mediated activation of NOS by Ca^{2+} in certain neurons causes the degeneration of surrounding neurons. They have demonstrated protection against NMDA neurotoxicity in tissue culture by treatment with NOS inhibitors or with reduced hemoglobin, which binds NO; by elimination of arginine, the substrate for NOS, from the culture medium; or by selective destruction of NOS-expressing neurons by prior treatment with quisqualic acid, to which they are differentially sensitive. However, not all laboratories have found inhibition of nitric oxide to be protective, and under some circumstances administration of nitric oxide donors is protective.[157]

These disparities may reflect the salience of the redox state of the cellular environment. As recently reported by Lipton and co-workers,[53] elevation of $O_2^{\circ-}$ favors the formation of peroxynitrite, whereas reducing conditions supports S-nitrosylation of the NMDA receptor thiol, which downregulates the receptor and confers protection against neurotoxicity.

Conclusions

Glu-mediated oxidative stress alone or in combination with other sources of oxidation has been implicated in neurodegeneration. Unrelated primary pathologic processes may compromise neuronal function, thereby enhancing the vulnerability to Glu such as amyloid peptide in Alzheimer's disease and/or compromised antioxidant defenses with age. Finally, oxidative stress from other sources such as catecholamine metabolism in PD may place neurons in jeopardy to Glu related stressors. Nevertheless, the evidence, while still largely circumstancial, is convincing that oxidative stress represents an important pathway, initiated in part by Glu, that leads to neuronal degeneration in a manner consistent with the course and pathology of several degenerative disorders of the brain.

There remains, however, a substantial gap in our knowledge between Glu receptor activation and the specific metabolic processes that promote oxidative stress at the neuronal level. To close this breach, future studies should provide a more refined molecular characterization of Glu receptor subunits that are expressed in vulnerable systems; a more precise linkage between the biophysical characteristics of these receptors and the intracellular mechanisms generating oxidants; and a more detailed understanding of oxidant defense mechanisms as they relate to the unique microenvironments of specific neuronal systems. Molecular strategies, especially transgenic methods to alter the expression of specific components involved in this complex sequence of events, will be important in these studies. Understanding the relation between Glu and oxidative stress is important, because it offers a pharmacological strategy for blocking a pathway significantly involved in neurodegeneration without interfering directly with excitatory neurotransmission.

MITOCHONDRIA AND OXIDATIVE STRESS

Although there are a number of intracellular sources of free radicals, the mitochondria (Mt) are thought to be the most important. Isolated Mt generate 0.6 to 1.0 nmol of H_2O_2 min^{-1} mg of $protein^{-1}$, which is estimated to account for 2 to 4% of oxygen uptake under physiological conditions.[11] The main sites at which superoxide is generated are ubiquinone and NADH dehydrogenase.[12] Electrons are transferred one at a time to form ubisemiquinone, which can then react with O_2 to form superoxide.[158] Evidence favoring ubiquinone as a site of free radical generation is the observation that antimycin (a complex III inhibitor) leads to increased mitochondrial H_2O_2 production, by blocking electron transfer from more proximal components of the respiratory chain. Furthermore, depletion of ubiquinone from mitochondria results in a decrease in H_2O_2 generation.[159]

Is there a relationship between excitotoxicity and free radical generation by mitochondria? The critical link may be mitochondrial calcium.[160] Calcium is transported into mitochondria by a uniporter that utilizes the large electrochemical gradient across the inner mitochondrial membrane to transport Ca^{2+} in lieu of proton uptake. Under conditions of a large Ca^{2+} load, mitochondria can buffer intracellular Ca^{2+} by fast sequestration.[161] Exposure of Mt to Ca^{2+} concentrations that are of a similar magnitude to those that occur after exposure of neurons to excitotoxins leads to mitochondrial generation of hydroxyl and carbon-centered radicals.[146] Studies have directly linked glutamate-induced increases in Mt calcium to mitochondrial free radical generation[151] and the neurotoxicity of glutamate via NMDA receptors to a decreased energy production.[162]

The genes encoding the components of the Mt respiratory chain are located both on nuclear and MtDNA. MtDNA has a mutation rate that is ten times greater than that of nuclear DNA and has less effective repair mechanisms.[128,130] Consistent with the proximity of MtDNA to a major source of cellular oxidants, there is a remarkable 15-fold increase in oxidized nucleotides in brain MtDNA with age.[130] Recent evidence that hereditary MtDNA mutations and deletions are associated with syndromes characterized by neurodegeneration may indicate that mutations acquired during aging may disrupt the efficiency of electron transport and augment oxidative stress.

CELL RESPONSES TO OXIDATIVE DAMAGE

The antioxidant compounds and enzymes they evolve remove most of the potentially damaging radicals/oxidants; however, damage to cellular proteins, lipids, nucleic acids and carbohydrates can be observed even under normal physiological conditions. Reduction of cellular components (direct repair) may be important for some biomolecules. In most cases studied to date, however, enzymatic degradation (by proteases, lipases, nucleases) appears to release damaged elements for excretion and conserve undamaged components for reutilization (indirect repair). In addition, the removal of damaged components appears to prevent or diminish the potential cytotoxicity of oxidized macromolecules.

That oxidative stress may function similarly to heat shock, which results in the overexpression of heat-shock proteins (HSPs).[163] Cells treated with mildly toxic doses of oxidants overexpress a specific set of proteins which were either not previously expressed, or which were expressed at low levels. Expression of HSPs is associated with increased resistance to subsequent oxidant challenges.[82] This suggests that HSPs play a role in protecting cells from oxidative stress. Decreased activities or constitutive levels of oxidant repair enzymes may contribute to a progressive accumulation of oxidant damage with aging.[164] Furthermore, the ability to mount an effective response to oxidative stress (induction of oxidant stress genes and proteins) may decline with age, thus predisposing older cells and organisms to oxidant damage.

Figure 2.10 provides a general scheme of cell responses to oxidative stress and possible ramifications.

ROS can react with a number of target molecules including proteins, lipids, and DNA.[81] These interactions elicit a number of signals including activation of gene regulatory factors (transcription factors) which in turn activate oxidative stress-responsive genes. Consequently, a number of proteins are produced with distinctive functions, including DNA repair enzymes, HSPs, antioxidants, proteases or protease inhibitors, cytokines and proteins such as growth factors affecting cell responses to the insult.[81,82,145,163–167] These cellular responses to ROS can lead to restoration of normal cellular function and adaptation to oxidative stress, cell death or aberrant proliferation. It is the latter two responses which can lead to a variety of disease states. The overall outcome of the cellular response to oxidative stress is complex and might depend on the status of cellular differentiation and

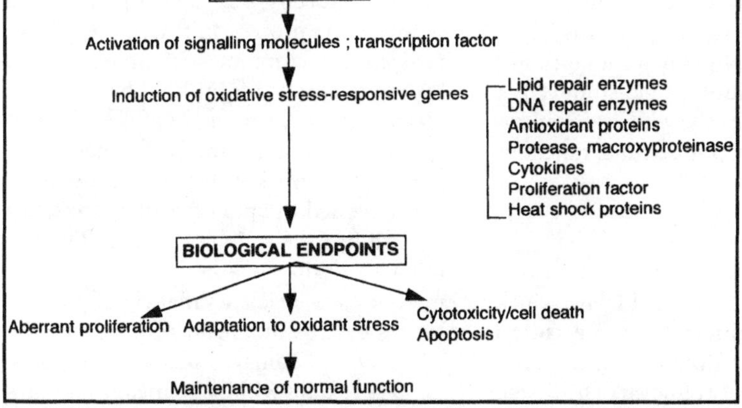

Fig. 2.10. Summary of cell responses to oxidative damage in relationship to homeostasis and the development of disease.

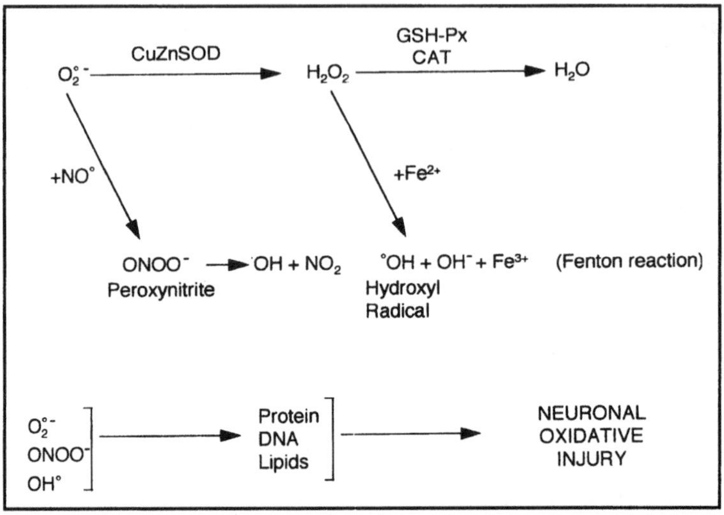

Fig. 2.11. Reactions of superoxide anion and cellular targets of derived-reactive oxygen species. (Top) The superoxide anion ($O_2^{\circ -}$) can react with CuZnSOD to be detoxified to form hydrogen peroxide (H_2O_2) that, in turn, is converted to water through the action of catalase and glutathione peroxidase. Superoxide may also combine with nitric oxide (NO) to form peroxynitrite ($ONOO^-$). Peroxynitrite may then breakdown nonenzymatically to produce hydroxyl radicals (OH°), which may be also generated from hydrogen peroxide via Fe^{2+} (Fenton reaction). (Bottom) These reactive oxygen species may cause oxidative degradation of DNA, lipids, and proteins.

phase of the cell cycle and/or antioxidant content.

THE OXIDANT STRESS: THE MOLECULAR TARGETS IN THE BRAIN

When there is a compromise in the levels of cellular antioxidants or accelerated production of reactive oxygen species, or both, a state of 'oxidative stress' results that can cause permanent cellular damage. Oxygen radicals can attack proteins, deoxynucleic acids (DNA damage lesions, impaired DNA replication, single-strand DNA breakage, mutagenesis) and lipid membranes, thereby disrupting cellular functions and integrity[10] (Fig. 2.11).

Lipoperoxidation

The brain contains large amounts of polyunsaturated fatty acids (PUFA), which are particularly vulnerable to free radical attack, because the double bonds within membranes allow easy removal of hydrogen atoms by reactive oxygen species such as °OH. The carbon radical formed in the PUFA undergoes molecular rearrangement to form more stable conjugated dienes, which can crosslink fatty acids within cellular membranes. Under aerobic conditions, lipid peroxidation continues as conjugated dienes combine with O_2 to form additional organic peroxy radicals. Peroxy radicals abstract hydrogen from adjacent fatty acid chains, thereby propagating the lipid peroxidation process. Furthermore, peroxy radicals can combine with an abstracted hydrogen atom to form lipid hydroperoxides which, in the presence of Fe^{2+}, decompose to alkoxy radicals and aldehydes. As the process continues many membrane lipids are converted to lipid peroxides. Thus, like a brush fire, the action of a single °OH can initiate a chain reaction that generates numerous toxic reactants that rigidify membranes by crosslinking, disrupt membrane integrity, and damage membrane proteins. Lipid peroxides within the membrane have a devastating effect on the functional state of the membrane because they alter its fluidity, typically decreasing it and thereby allowing ions such as Ca^{2+} to leak into the cell.[168]

Thus lipid peroxidation of cell membranes leads to a chain reaction of events that produce peroxyl radicals and lipid peroxides, both of which are highly destructive to macromolecules and disruptive to key neuronal functions.

Toxic carbonyl compounds are common products of the complex processes of lipid decomposition by radicals. Some of these carbonyls are aldehydes including malondialdehyde (MDA); other carbonyl products include the highly toxic unsaturated aldehydes such as 4-hydroxyl-2,3-trans-nonenal (Table 2.2).

Protein Oxidation and Glycoxidation

Hydroxyl radicals also attack amino acid residues in protein molecules;[169] these reactions induce extensive protein-protein crosslinking.[170] Oxidative damage to proteins by $^{\circ}OH$ is magnified by the fact that iron, often associated with proteins, generates the radical in the immediate vicinity of the protein. Once generated, the $^{\circ}OH$ readily interact because of their close proximity with the protein molecule. Damaged proteins, many of them crucial enzymes in neurons, lose their efficiency and cellular function wanes. Protein oxidation in many tissues, including the brain, has been proposed as an explanation for functional deficits associated with aging.

Hydroxyl radicals are highly toxic and can initiate lipid peroxidation as well as create protein or polynucleotide-centered free radicals by hydrogen atom abstraction. Oxyradical-induced modification of protein is associated with fragmentation, increased susceptibility to proteolysis and, especially, crosslinking reactions.[171]

Advanced glycosylation end products (AGEs) are a family of complex, evolved post-translational modifications that are initiated by condensation of reducing sugars with protein amino groups via the Maillard reaction. Oxidative stress increases the frequency of hydroxyl-radical-induced autoxidation of unsaturated membrane lipids. Reduced metal ion-mediated fragmentation of resulting lipid hydroperoxides releases reactive aldehydes which can modify proteins in an adduct-evolutionary manner that almost parallels that seen in AGE. Modification by AGE or products of lipid peroxidation will have the most effect on lysine-rich proteins, frequently with neutralization of the charge, and consequent alteration of protein-protein electrostatic interactions, and occasionally with intra- and intermolecular crosslinking

Although AGE modifications or oxidative-stress mechanisms can lead individually to pathological modifications of neuronal proteins, these two factors together result in synergistically accelerated protein damage (Fig. 2.12).

Table 2.2. Aldehyde products resulting from lipid peroxidation[a]

4-Hydroxy Alkenals	2-Alkenals	*n*-Alkenals	Others
4-OH nonenal	Acrolein	Propanal	Malonaldehyde
4-OH hexenal	Pentenal	Butanal	Butanone
4-OH 2,5 monadienal	Hexenal	Pentanal	2,4-Decadienal
	Octenal	Hexanal	
	Nonenal	Nonanal	

[a] In nerve cell membranes these products, some of which are highly toxic, can wreak havoc by altering the structure of protein-receptor molecules and destroying enzymes within the membranes.

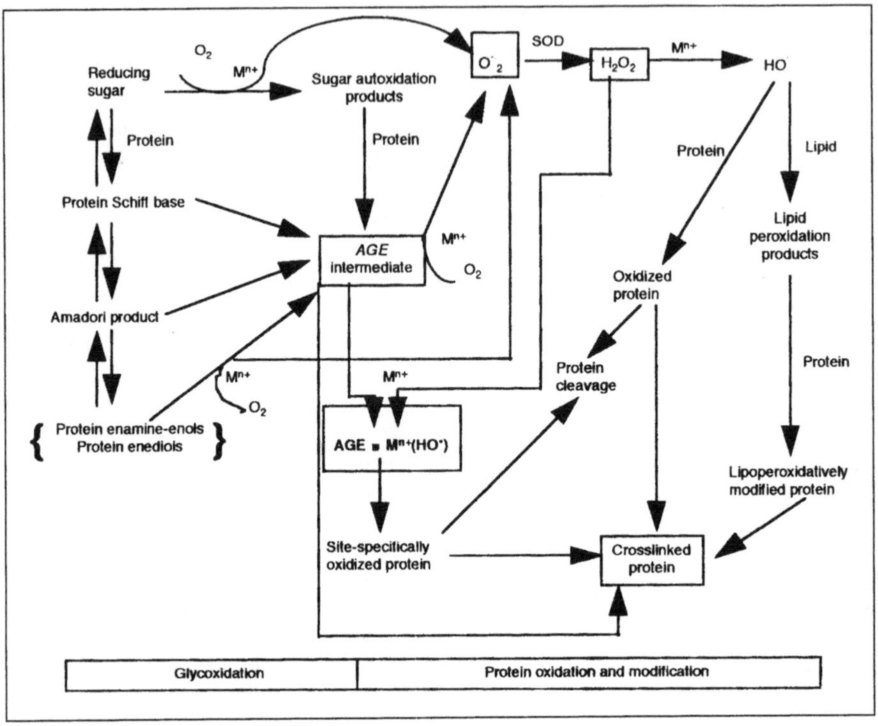

Fig. 2.12. Modification of proteins by glycation, autoxidation and reactive aldehydes that are generated from lipid peroxidation. Protein modification and protein-protein crosslinking by glycation, autoxidation and reactive aldehydes that are generated from lipid peroxidation is a synergistic process; in particular, glycation and lipoperoxidative modification involve the adduction of redox-active moieties that potentiate free-radical autoxidation. Reprinted with permission from Smith et al. Trends Neurosci 1995; 18:172. © Elsevier Science.

There is synergy of AGE and oxidative stress-induced protein damage.Indeed, formation of adducts, and evolution of AGE modification of proteins, is accelerated by oxygen in a process that is called glycoxidation. The condensation of reducing sugars with protein amino groups and subsequent Amadori arrangement leads to the equilibrium presence of redox-active enediol and enol-enamine moieties that can catalyze the NADH-dependent reduction of O_2 to superoxide. Such redox-cycling action can result in site-specific damage at the protein loci of sugar attachment, as exemplified by the fact that glycated proteins invariably contain oxidative modifications. In addition, glycation of membrane lipids can be a mechanism for initiating lipid peroxidation, which adds to that resulting from oxidative stress. Evidence for such a scenario is that malondialdehyde, a major product of lipid peroxidation whose concentration is increased in neurofibrillary tangle (NFT)-containing neurons, is induced by AGE-modification of cultured cell lines.[172] Overall, modification of proteins by oxidation, glycation, and products of lipid peroxidation, can occur in both additive and synergistic mechanisms.

Neuronal membranes contain a high proportion of free-radical-susceptible polyunsaturated fatty acids compared with normal plasma membrane, making them more susceptible to peroxidative damage. Moreover, any oxidative stress in neurons would undoubtedly be exacerbated by the very high consumption of oxygen that is required for the brain. Compared with other cellular proteins, neurofilaments and cytoskeletal proteins such as tau have relatively long half-lives, and contain high proportions of lysine residues.[173] Therefore, while the lysine residues have been implicated in supramolecular cytoskeletal organization, the presence of these multiple lysine-derived amino groups is probably the single most important feature that renders them particularly susceptible to cumulative oxidative damage, as well as modification by AGE and lipid-peroxidation products.

References

1. Hirano A. Cytopathology of amyotrophic lateral sclerosis. Adv Neurol 1991; 56: 91-101.
2. Trojanowski J, Schmidt M et al. Altered tau and neurofilament proteins in neurodegenerative diseases: Diagnostic implications for Alzheimer's disease and Lewy body dementias. Brain Pathol 1993; 3:45-54.
3. Goldman JE, Yen SH. Cytoskeletal protein abnormalities in neurodegenerative diseases. Ann Neurol 1986; 19:209-223.
4. Halliwell B, Gutteridge JMC. Free Radicals in Biology and Medicine. 2nd edn, Oxford University Press, Oxford, 1989.
5. McCord JM, Fridovich I. Superoxide dismutase an enzymic function for erythrocuprein (hemocuprein). J Biol Chem 1969; 244:6059-6055.
6. Moncada S, Higgs EA. Endogenous nitric oxide: physiology, pathology and clinical relevance. Eur J Clin Invest 1991; 21:361-374.
7. Beckman JS, Beckman TW, Chen J et al. Apparent hydroxyl radical production by peroxynitrite: Implications for endothelial injury from nitric oxide and superoxide. Proc Natl Acad Sci USA 1990; 87:1620-1624.
8. Fenton HJH. Oxidation of tartaric acid in the presence of iron. J Chem Soc (Lond) 1894; 65:899-903.
9. Halliwell B. Reactive oxygen species in living systems: source, biochemistry and role in human disease. Am J Med 1991; 91:14-21.
10. Freeman BA, Crapo JD. Biology of disease: free radicals and tissue injury. Lab Invest 1982; 47:412-427.
11. Boveris A, Chance B. The mitochondrial generation of hydrogen peroxide. Biochem J 1973 ; 134:707-16.
12. Turrens JF, Boveris A. Generation of superoxide anion by the NADH dehydrogenase of bovine heart mitochondria. Biochem J 1980; 191:421-27.
13. Evans PH. Free radicals in brain metabolism and pathology. British Medical Bull 1993; 49:577-587.
14. Reiter JR. Oxidative processes and antioxidative defense mechanisms in the aging brain. FASEB J 1995; 9:526-533.
15. Ceballos I, Javoy-Agid F, Delacourte A et al. Parkinson's disease and Alzheimer's disease: neurodegenerative disorders due to brain antioxidant system deficiency? In: Emerit et al, eds. Antioxidants in Therapy and Preventive Medicine, Plenum Press:New York, 1990:493-498.
16. Halliwell B. Reactive oxygen species and the central nervous system. J Neurochem 1992; 59:1609-1623.
17. Sokoloff L. The metabolism of the central nervous system in vivo. In: Field J, Magoun HW, Hall VE, eds. Handbook of Physiology-Neurophysiology American Physiological Society, Washington, DC, 1960; 3:1843-1864.
18. Rice-Evans C, Burdon R. Free radical-lipid interactions and their pathological consequences. Prog Lipid Res 1993; 32:71-110.
19. Harman D, Hendricks S, Eddy DE et al. Prolongation of life: role of free radical reactions in aging. J Am Geriatr Soc 1976; 17:721-725.
20. Sadrzadek SMH, Eaton JW. Hemoglobin related oxidant damage to the central nervous system requires exogenous ascorbate. J Clin Invest 1988; 82: 1510-1516.
21. Sadrzadek SM, Anderson DK, Porter SS et al. Hemoglobin potentiates central nervous system damage. J Clin Invest 1987; 79:662-664.
22. Dugan LL, Choi DW. Excitotoxicity, free radicals, and cell membrane changes. Ann Neurol 1994; 35:517-521.
23. Choi DW. Excitotoxic cell death. J Neurobiol 1992; 23:1261-1276.

24. Dawson VL, Dawson TM, London ED et al. Nitric oxide mediates glutamate neurotoxicity in primary cortical cultures. Proc Natl Acad Sci USA 1991; 88: 6368-6371.
25. McCord J, Ormar BA. Sources of free radicals. Toxicol Indust Health 1993; 9:23-37.
26. Winterbourn CC. Comparison of superoxide with other reducing agents in the biological production of hydroxyl radicals. Biochem J 1979; 182:625-628.
27. Spina MB, Cohen G. Dopamine turnover and glutathione oxidation: implications for Parkinson's disease. Proc Natl Acad Sci USA 1989; 86:1398-1400.
28. Graham DG. Oxidative pathways for catecholamines in the genesis of neuromelanin and cytotoxic quinones. Molec Pharmac 1978; 14:633-643.
29. Colton CA, Gilbert DL. Production of superoxide anions by a CNS macrophage, the microglia. FEBS Lett 1987; 223:284-288.
30. Lafon-Cazal M, Pietri S, Culcasi M et al. NMDA-dependent superoxide production and neurotoxicity. Nature 1993; 364:535-537.
31. Glenner GG. Alzheimer's disease: its proteins and genes. Cell 1988; 52:307-308.
32. Loo DT, Copani A, Pike CJ et al. Apoptosis is induced by β-amyloid in cultured nervous system neurons. Proc Natl Acad Sci USA 1993; 90:7951-7955.
33. Behl C, Davis JB, Lesley R et al. Hydrogen peroxide mediates amyloid β protein toxicity. Cell 1994; 77:817-827.
34. Thery C, Chamak B, Mallat M. Cytotoxic effect of brain macrophages on developing neurons. Eur J Neurosci 1991; 3: 1155-1164.
35. Jaffrey SR, Snyder SH. Nitric oxide: a neural messenger. Annu Rev Cell Dev Biol 1995; 11:417-440.
36. Itzhak Y, Ali SF. The neuronal nitric oxide synthase inhibitor, 7-nitroindazole, protects against methamphetamine-induced neurotoxicity in vivo. J Neurochem 1996; 67:1770-1773.
37. Dawson VL, Kizushi VM, Huang PL et al. Resistance to neurotoxicity in cortical cultures from neuronal nitric oxide synthase deficient mice. J Neuroscience 1996; 16:2479-2482.
38. Hantraye P, Brouillet E, Ferrante R et al. Inhibition of neuronal nitric oxide synthase prevents MPTP-induced parkinsonism in baboons. Nature Med 1996; 2:1017-1021.
39. Schultz JB, Matthews RT, Muqit MMR et al. Inhibition of neuronal nitric oxide synthase by 7-nitroindazole protects against MPTP-induced neurotoxicity in mice. J Neurochem 1995; 64:936-939.
40. Przedborski S, Jackson-Lewis V, Yokoyama R et al. Role of nitric oxide in MPTP-induced dopaminergic neurotoxicity. Proc Natl Acad Sci USA 1996; 93:4565-4571.
41. Ischiropoulos H, Chen J, Tsai HM et al. Peroxynitrite (ONOO-) reacts with superoxide dismutase to give the reactive nitronium ion. Free Rad Biol Med 1990; 9(Suppl.1):131.
42. Ischiropoulos H, Zhu L, Chen J et al. Peroxynitrite-mediated tyrosine nitration catalyzed by superoxide dismutase. Arch Biochem Biophys 1992a ; 298:431-437.
43. Beckman JS, Ischiropoulos H, Zhu L et al. Kinetics of superoxide dismutase and iron catalyzed nitration of phenolics by peroxynitrite. Arch Biochem Biophys 1992; 298 :438-445.
44. Dawson TM, Dawson VL, Snyder SH. A novel neuronal messenger molecule in brain: the free radical, nitric oxide. Ann Neurol 1992; 32:297-311.
45. Garthwaite J. Nitric oxide signaling in the nervous system. Semin Neurosci 1993; 5:171-80.
46. Shuman EM, Madison DV. The intercellular messenger nitric oxide is required for long-term potentiation. Science 1991; 254:1503-6.
47. Shibuki K, Okada D. Endogenous nitric oxide release required for long-term synaptic depression in the cerebellum. Nature 1991; 349:326-8.
48. Bredt DS, Hwang PM, Snyder SH. Localization of nitric oxide synthase indicating a neural role for nitric oxide. Nature 1990; 347:768-70.
49. Karanth S, Lyson K, McCann SM. Role of nitric oxide in interleukin 2-induced corticotropin-releasing factor release from incubated hypothalami. Proc Natl Acad Sci USA 1993; 90:3383-7.

50. Garthwaite J, Charles SL, Chess-Williams R. Endothelium-derived relaxing factor release on activation of NMDA receptors suggests role as intercellular messenger in the brain. Nature 1988; 336:385-8.
51. Knowles RG, Palacios M, Palmer RMJ et al. Formation of nitric oxide from L-arginine in the central nervous system : a transduction mechanism for stimulation of the soluble guanylate cyclase. Proc Natl Acad Sci USA 1989; 86:5159-62.
52. Huang Z et al. Effects of cerebral ischemia in mice deficient in neuronal nitric oxide synthase. Science 1994; 265:1883-1885.
53. Lipton SA et al. A redox-based mechanism in the neuroprotective and neurodestructive effect of nitric oxide and related nitroso-compounds. Nature 1993; 364:626-632.
54. Zhang J, Dawson VL, Dawson TM and Snyder SH. Nitric oxide activation of polyADP-ribose synthetase in neurotoxicity. Science, 1994; 263:687-690.
55. Bolanos JP, Simon JR, Heales SJR, Land JM, Clark JB. Effect of peroxynitrite on the mitochondrial respiratory chain: differential susceptibility of neurones and astrocytes in primary culture. J Neurochem 1995; 64:1965-1972.
56. Brown GC. Nitric oxide regulates mitochondrial respiration and cell functions by inhibiting cytochrome oxidase. FEBS Lett 1995; 369:136-139.
57. Solodkin A, Traub RJ, Gebhart GF. Unilateral hindpaw inflammation produces a bilateral increase in NADPH-diaphorase hisotchemical staining in the rat lumbar spinal cord. Neuroscience 1992; 51:495-499.
58. Moncada S, Palmer RMJ, Higgs EA. Nitric oxide: physiology, pathophysiology, and pharmacology. Pharmacol Rev 1991; 43:109-142.
59. Boje KM, Arora PK. Microglial-produced nitric oxide and reactive nitrogen oxides mediate neuronal cell death. Brain Res 1992; 587:250-256.
60. Murphy S, Minor RL Jr, Wel G et al. Evidence for an astrocyte-derived vasorelaxing factor with properties similar to nitric oxide. J Neurochem 1990; 55:349-351.
61. Galea E, Feinstein D, Reis D. Induction of calcium-independent nitric oxide synthase activity in primary rat glial cultures. Proc Natl Acad Sci USA 1992; 89:10945-10949.
62. Nakane M, Schmidt H, Pollock J et al. Cloned human brain nitric oxide synthase is highly expressed in skeletal muscle. FEBS Lett 1993; 316:175-180.
63. Ignarro LJ. Signal transduction mechanisms involving nitric oxide. Biochem Pharmacol 1991; 41:485-90.
64. Schmidt HHHW, Lohmann WU. The nitric oxide and cGMP signal transduction system: regulation and mechanism of action. Biochim Biophys Acta 1993; 1178:153-75.
65. Hibbs JR Jr, Taintor RR, Vavrin Z et al. Synthesis of nitric oxide from a terminal guanidino nitrogen atom of L-arginine: a molecular mechanism regulating cellular proliferation that targets intracellular iron. In: Moncada S, Higgs EA, eds. Nitric Oxide from L-arginine: a Bio-regulatory System. Amsterdam: Elsevier, 1990: 189-224.
66. Lepoivre M, Fieschi F, Coves J et al. Inactivation of ribonucleotide reductase by nitric oxide. Biochem Biophys Res Commun 1991; 179:442-8.
67. Wink DA, Kasprzak KS, Maragos CM et al. DNA deaminating ability and genotoxicity of nitric oxide and its progenitors. Science 1991; 254:1001-3.
68. Stamler JS, Simon DI, Osborne JA et al. S-nitrosylation of proteins with nitric oxide: Synthesis and characterization of biologically active compounds. Proc Natl Acad Sci USA 1992; 89:444-8.
69. Gryglewski RJ, Palmer RMJ, Moncada S. Superoxide ion is involved in the breakdown of endothelium-derived vascular relaxing factor. Nature 1986; 320:454-6.
70. Beckman J, Tsai JH. Reactions and diffusion of nitric oxide and peroxynitrite. Biochemist 1994; 16:8-10.
71. Marklund SL, Holme E, Hellner L. Superoxide dismutase in extracellular fluids. Clin Chim Acta 1982; 126:41-51.
72. Sinet PM, Heikkila RE, Cohen G. Hydrogen peroxide production in rat brain in vivo. J Neurochem 1980; 34:1421-1428.

73. Nakajima M, Kashiwagi K, Hayashi Y et al. Alpha-tocopherol supports survival and neurite extension of neurons cultured from various regions of fetal rat brain. Neurosci Lett 1991; 133:49-52.
74. Raps SP, Lai JCK, Hertz L et al. Glutathione is present in high concentrations in cultured astrocytes but not in cultured neurons. Brain Res 1989; 493:398-401.
75. Gutteridge JM. Antioxidant properties of the proteins ceruloplasmin, albumin and transferrin. A study of their activity in serum and synovial fluid from patients with rheumatoid arthritis. Biochem Biophys Acta 1986; 869:119-127.
76. Thorburne SK, Jurlink BHJ. Low glutathione and high iron govern the susceptibility of oligodendroglial precursors to oxidative stress. J Neurochem 1996; 67:1014-1022.
77. Makar TK, Nedergaard M, Preuss A et al. Vitamin E, ascorbate, glutathione, glutathione disulfide, and enzymes of glutathione metabolism in cultures of chick astrocytes and neurons: evidence that astrocytes play an important role in antioxidative processes in the brain. J Neurochem 1994; 62:45-53.
78. Sagara J, Miura K, Bannai S. Cystine uptake and glutathione level in fetal brain cells in primary culture and suspension. J Neurochem 1993; 61:1667-1671.
79. Sagara J, Miura K, Bannai S. Maintenance of neuronal glutathione by glial cells. J Neurochem 1993; 61:1672-1676.
80. Devesa A, O'Connor JE, Garcia C et al. Glutathione metabolism in primary astrocyte cultures: flow cytometric evidence of heterogeneous distribution of GSH content. Brain Res 1993; 618:181-189.
81. Janssen YW, Van Houten B, Borm PJA et al. Biology of the disease: cell and tissue responses to oxidative damage. Lab Invest 1993; 69:261-274.
82. Spitz DR, Dewey WC, Li GC. Hydrogen peroxide or heat shock induces resistance to hydrogen peroxide in Chinese hamster fibroblasts. J Cell Physiol 1987; 131:364-373.
83. Damier P, Hirsch EC, Zhang Y et al. Glutathione peroxidase, glial cells and Parkinson's disease. Neuroscience 1993; 52:1-6.
84. Takizawa S, Matsushima K, Shinohara Y et al. Immunohistochemical localization of glutathione peroxidase in infarcted human brain. J Neurol Sci 1994; 122: 66-73.
85. Muller DPR, Goss-Sampson MA. Neurochemical, neurophysiological and neuropathological studies in vitamin E deficiency. Crit Rev Neurobiol 1990; 5:239-255.
86. Spector R, Eells J. Deoxynucleoside and vitamin transport into the central nervous system. Federation Proc 1984; 43:196-200.
87. Tan DX, Chen LD, Poeggeler B et al. Melatonin: a potent endogenous hydroxyl radical scavenger. Endocrine J 1993; 1:57-60.
88. Pieri C, Marra M, Moroni F et al. Melatonin: a peroxyl radical scavenger more effective than vitamin E. Life Sci 1994; 55:271-276.
89. Sinet PM, Ceballos-Picot I. Role of free radicals in Alzheimer's disease and Down's syndrome. In: L Packer, L Prilipko, Y Christen, eds. Free Radicals in the Brain: Aging, Neurological and Mental Disorders, Springer-Verlag, Berlin, Heidelberg, 1992:91-98.
90. Marklund SL, Adolfsson R, Gottfries CG et al. Superoxide dismutase isoenzymes in normal brains and in brains from patients with dementia of Alzheimer type. J Neurol Sci 1985; 7:319-325.
91. Ceballos I, Javoy-Agid F, Delacourte A et al. Neuronal localization of copper-zinc superoxide dismutase protein and mRNA within the human hippocampus from control and Alzheimer's disease brains. Free Rad Res Comms 1991; 12-13:571-580.
92. Delacourte A, Defossez A, Ceballos I et al. Preferential expression of copper-zinc superoxide dismutase in the vulnerable cortical neurons in Alzheimer's disease. Neurosci Lett 1988; 92:247-253.
93. Ceballos I, Lafon M, Javoy-Agid F et al. Superoxide dismutase and Parkinson's disease. The Lancet 1990; 335:1035-1036.
94. Zhang P, Damier P, Hirsch EC et al. Preferential expression of superoxide dismutase messenger RNA in melanized neurons in human mesencephalon. Neurosci 1993; 55:167-175.

95. Bergeron C, Muntasser S, Somerville MJ et al. Copper/zinc superoxide dismutase mRNA levels are increased in sporadic amyotrophic lateral sclerosis motorneurons. Brain Res 1994; 659: 272-276.
96. Bergeron C, Petrunka C, Weyer L. Copper/zinc superoxide dismutase expression in the human central nervous system. Am J Pathol 1996; 148:273-279.
97. Ceballos I, Javoy-Agid F, Hirsch EC et al. Localization of copper-zinc superoxide dismutase mRNA in human hippocampus by in situ hybridization. Neurosci Lett 1989; 105:41-46.
98. Elroy-Stein O, Bernstein Y, Groner Y. Overproduction of human CuZn-superoxide dismutase in transfected cells; extenuation of paraquat-mediated cytotoxicity and enhancement of lipid peroxidation. EMBO J 1986; 5:615-622.
99. Ceballos I, Delabar JM, Nicole A et al. Expression of transfected human CuZn superoxide dismutase gene in mouse L cells and NS20Y neuroblastoma cells induces enhancement of glutathione peroxidase activity. Biochem Biophys Acta 1988; 949:58-64.
100. Elroy-Stein O, Groner Y. Impaired neurotransmitter uptake in PC12 cells overexpressing human Cu/Zn-superoxide dismutase. Implication for gene dosage effects in Down syndrome. Cell 1988; 52:259-267.
101. Bahmanyar S, Higgins GA, Goldgaber D et al. Localization of amyloid beta protein messenger RNA in brains from patients with Alzheimer's disease. Science 1987; 237:77-80.
102. Goedert TM. Neuronal localization of amyloid beta protein precursor mRNA in normal human brain and in Alzheimer's disease. EMBO J 1987; 6:3627-3632.
103. Blanquet V, Goldgaber D, Turleau C et al. The β amyloid protein (AD-AP) cDNA hybridizes in normal and Alzheimer individuals near the interface of 21q21 and q22.1. Ann Genet 1987; 30:68-69.
104. Wakai M, Mokuno K, Hashizume Y, Kato K. An immunohistochemical study of the neuronal expression of superoxide dismutase in sporadic amyotrophic lateral sclerosis. Acta Neuropathol 1994; 88:151-158.
105. Ceballos-Picot I, Nicole A, Briand P et al. Neuronal-specific expression of human copper-zinc superoxide dismutase gene in transgenic mice: animal model of gene dosage effect in Down's syndrome. Brain Res 1991; 552:198-214.
106. Furuta A, Price DL, Pardo CA et al. Localization of superoxide dismutases in Alzheimer's disease and Down's syndrome neocortex and hippocampus. Am J Pathol 1995; 146:357-367.
107. Somerville MJ, Percy ME, Bergeron C et al. Localization and quantitation of 68 kDa neurofilament and superoxide dismutase-1 mRNA in Alzheimer brains. Mol Brain Res 1991; 9:1-8.
108. Pardo CA, Xu Z, Borchelt DR et al. Superoxide dismutase is an abundant component in cell bodies, dendrites, and axons of motor neurons and in a subset of other neurons. Proc Natl Acad Sci 1995; 92: 954-958.
109. Tyrrel RM, Applegate LA, Tromvoukis Y. The proximal promotor region of the human heme oxygenase gene contains elements involved in stimulation of transcriptional activity by a variety of agents including oxidants. Carcinogenesis 1993; 14:761-765.
110. Smith MA, Kutty RK, Richey PL et al. Heme oxygenase-1 is associated with the neurofibrillary pathology of Alzheimer's disease. Am J Pathol 1994; 145:42-47.
111. Maines MD. Heme oxygenase: function, multiplicity, regulatory mechanisms, and clinical applications. FASEB J 1988; 2: 2557-2568.
112. Stocker R, Yamamoto Y, McDonagh AF et al. Bilirubin is an antioxidant of possible physiological importance. Science 1987; 235:1043-1046.
113. Verma A, Hirsch DJ, Glatt CE et al. Carbon monoxide: a putative neural messenger. Science 1993; 259:381-384.
114. Smith MA, Taneda S, Richey PL et al. Immunocytochemical demonstration of advanced Maillard reaction products in Alzheimer disease pathology. Proc Natl Acad Sci USA 1994; 91: 5710-5714.
115. Selkoe DJ, Ihara Y, Salazar FJ. Alzheimer's disease: insolubility of

partially purified paired helical filaments in sodium dodecyl sulfate and urea. Science 1982; 215:1243-1245.
116. Corder EH, Saunders AM, Strittmatter WJ et al. Gene dose of apolipoprotein E type 4 allele and the risk of Alzheimer's disease in late onset families. Science 1993; 261:921-923.
117. Strittmatter WJ, Weisgraber KH, Huang DY et al. Binding of human apolipoprotein E to synthetic amyloid beta peptide: isoform-specific effects and implications for late-onset Alzheimer's disease. Proc Natl Acad Sci USA 1993; 90:8098-8102.
118. Nakamura S, Kawamata T, Akiguchi I et al. Expression of monoamine oxidase B activity in astrocytes of senile plaques. Acta Neuropathol 1990; 80:419-425.
119. Asoh KI, Watanabe Y, Mizoguchi H et al. Induction of manganese superoxide dismutase by tumor necrosis factor in human breast cancer MCF-7 cell line and its TNF-resistant variant. Biochem Biophys Res Commun 1989; 162:794-801.
120. Ho YS, Howard AJ, Crapo JD. Molecular structure of a functional rat gene for manganese-containing superoxide dismutase. Am J Respir Cell Molec Biol 1991; 4:278-286.
121. Liu X-H, Kato H, Araki T et al. An immunohistochemical observation of manganese superoxide dismutase in rat substantia nigra after occlusion of middle cerebral artery. Neurosc Lett 1994; 173: 103-106.
122. Akai F, Maeda M, Suzuki K et al. Immunocytochemical localization of manganese superoxide dismutase (MnSOD) in the hippocampus of the rat. Neurosc Lett 1990; 115:19-23.
123. Inagaki S, Takagi H, Suzuki K et al. Intense immunoreactivity for Mn-superoxide dismutase (MnSOD) in cholinergic and non-cholinergic neurons in the rat basal forebrain. Brain Res 1991; 541: 354-357.
124. Inagaki S, Suzuki K, Taniguchi N et al. Localization of Mn-superoxide dismutase (MnSOD) in cholinergic and somatostatin-containing neurons in the rat neostriatum. Brain Res 1991; 549:174-177.
125. Zhang P, Anglade P, Hirsch EC et al. Distribution of manganese-dependent superoxide dismutase in the human brain. Neuroscience 1994; 61:317-330.
126. Ransmayr G, Cervera P, Hirsch E et al. Choline acetyltransferase-like immunoreactivity in the hippocampal formation of control subjects and patients with Alzheimer's disease. Neuroscience 1989; 32:701-714.
127. Pappolla MA, Omar RA, Kim KS et al. Immunohistochemical evidence of antioxidant stress in Alzheimer's disease. Am J Pathol 1992; 140:621-628.
128. Fleming JE, Miquel J, Cottrell SF et al. Is cell aging caused by respiration-dependent injury to the mitochondrial genome? Gerontology 1982; 28:44-53.
129. Shapira AHV, Mann VM, Cooper JM et al. Anatomical and disease specificity of NADH CoQ1 reductase (Complex 1) deficiency in Parkinson's disease. J Neurochem 1990; 55:2142-2145.
130. Wallace DC. Mitochondrial genetics: a paradigm for aging and degenerative diseases? Science 1992; 256:628-632.
131. Norrgren G, Ebendal T, Belew M et al. Release of nerve growth factor by human glial cells in culture. Expl Cell Res 1980; 130:31-39.
132. Mattson MP, Rychlik B, Chu C et al. Evidence for calcium-reducing and excito-protective roles for the calcium-binding protein calbindin-D_{28K} in cultured hippocampal neurons. Neuron 1991; 6:41-51.
133. Dexter DT, Wells FR, Lees A et al. Increased nigral iron content and alterations in other metal ions occurring in brain in Parkinson's disease. J Neurochem 1989; 52:1830-1836.
134. Dringen R, Hamprecht B. Glutathione content as an indicator for the presence of metabolic pathways of amino acids in astroglial cultures. J Neurochem 1996; 67:1375-1382.
135. Bolanos JP, Heales SJR, Land JM et al. Effect of peroxynitrite on the mitochondrial respiratory chain: differential susceptibility of neurons and astrocytes in primary cultures. J Neurochem 1995; 64:1965-1972.

136. Han S-K, Mytilineou C, Cohen G. L-Dopa upregulates glutathione and protects mesencephalic cultures against oxidative stress. J Neurochem 1996; 66: 501-510.
137. Slivka A, Mytilineou C, Cohen G. Histochemical evaluation of glutathione in brain. Brain Res 1987; 409:275-284.
138. Philbert MA, Beiswanger CM, Waters DK et al. Cellular and regional distribution of reduced glutathione in the nervous system of the rat: histochemical localization by mercury orange and O-phtaldialdehyde-induced histofluorescence. Toxicol and Applied Pharmacol 1991; 107:215-227.
139. Whittemore ER, Loo DT, Cotman CW. Exposure to hydrogen peroxide induces cell death via apoptosis in cultured rat cortical neurons. NeuroReport 1994; 5:1485-1488.
140. Desagher S, Glowinski J, Premont J. Astrocytes protect neurons from hydrogen peroxide toxicity. J Neuroscience 1996; 16:2553-2556.
141. Coyle JT, Puttfarken P. Oxidative stress, glutamate, and neurodegenerative disorders. Science 1993; 262:689-69.
142. Hartley DM, Kurth MC, Bjerkness L et al. Glutamate receptor-induced $^{45}Ca^{2+}$ accumulation in cortical cell culture correlates with subsequent neuronal degeneration. J Neurosci 1993; 13:1993-2000.
143. Tymianski M, Wallace MC, Spigelman I et al. Cell-permeant Ca^{2+} chelators reduce early excitotoxic and ischemic neuronal injury in vitro and in vivo. Neuron 1993; 11:221-35.
144. Tymianski M, Charlton MP, Carlen PL et al. Source specificity of early calcium neurotoxicity in cultured embryonic spinal neurons. J Neurosci 1993; 13: 2085-2104.
145. Mattson MP, Lovell MA, Furukawa K et al. Neurotrophic factors attenuate glutamate-induced accumulation of peroxides, elevation of intracellular Ca^{2+} concentration, and neurotoxicity and increase antioxidant enzyme activities in hippocampal neurons. J Neurochem 1995; 65:1740-1751.
146. Dykens JA. Isolated cerebral and cerebellar mitochondria produce free radicals when exposed to elevated Ca^{2+} and Na^{+}: implications for neurodegeneration. J Neurochem 1994; 63:584-91.
147. Monyer H, Hartley DM, Choi DW. 21-aminosteroids attenuate excitotoxic neuronal injury in cortical cell cultures. Neuron 1990; 5:121-126.
148. Chan PH, Chu L, Chen SF et al. Reduced neurotoxicity in transgenic mice overexpressing human copper-zinc-superoxide dismutase. Stroke 1990; 21: 11180-11182.
149. Verity MA. Mechanisms of phospholipases A2 activation and neuronal injury. Ann NY Acad Sci 1993; 679:110-120.
150. Mattson MP, Barger SW, Begley JG et al. Calcium, free radicals, and excitotoxic neuronal death in primary cell culture. Methods Cell Biol 1995; 46:187-216.
151. Dugan LL, Sensi SL, Canzoniero LMT et al. Imaging of mitochondrial oxygen radical production in cortical neurons exposed to NMDA. Soc Neurosci Abstr 1994 ; 20:1532 (Abstract).
152. Beckman JS. The double-edged role of nitric oxide in brain function and superoxide-mediated injury. J Dev Physiol 1991; 15:53-59.
153. Beckman JS, Crow JP. Pathological implications of nitric oxide, superoxide and peroxynitrite formation. Biochem Soc Trans 1993; 21:330-334.
154. Williams JH, Errington ML, Lynch MA et al. Arachidonic acid induces a long-term activity-dependent enhancement of synaptic transmission in the hippocampus. Nature 1989; 341:739-742.
155. Gally JA, Montague PR, Reeke GN Jr et al. The NO hypothesis: possible effects of a short-lived, rapidly diffusible signal in the development and function of the nervous system. Proc Natl Acad Sci USA 1990; 87:3547-3551.
156. Garthwaite J. Glutamate, nitric oxide and cell-cell signaling in the nervous system. Trends Neurosci 1991; 14:75-82.
157. Farinelli SE, Park DS, Greene LA. Nitric oxide delays the death of trophic factor-deprived PC12 cells and sympathetic neurons by a cGMP-mediated mechanism. J Neurosci 1996; 16:2325-2334.
158. De Jong AMP, Albracht SPJ. Ubisemiquinones as obligatory intermediates in the electron transfer from NADH to

ubiquinone. Eur J Biochem 1994; 222: 975-82.
159. Patole MS, Swaroop A, Ramasarma T. Generation of H_2O_2 in brain mitochondria. J Neurochem 1986; 47:1-8.
160. Beal MF. Aging, energy and oxidative stress in neurodegenerative diseases. Ann Neurol 1995; 38:357-66.
161. Werth JL, Thayer SA. Mitochondria buffer physiological calcium loads in cultured rat dorsal root ganglion neurons. J Neurosci 1994; 14:348-56.
162. Novelli A, Reilly JA, Lysko PG et al. Glutamate becomes neurotoxic via the *N*-methyl-D-aspartate receptor when intracellular energy levels are reduced. Brain Res 1988; 451:205-12.
163. Polla BS, Kantengwa S, Francois D et al. Mitochondria are selective targets for the protective effects of heat shock against oxidative injury Proc Natl Acad Sci USA 1996; 93:6458-6463.
164. Pacifici RE, Doviesk JA. Protein, lipid and free-radical theory of aging revisited. Gerontology 1991; 37:166-180.
165. Stadtman ER. Oxidation of proteins by mixed-function oxidation system: Implication in protein turnover, ageing and neutrophil function. Trends Biochem Sci 1986; 11:11-12.
166. Teebor GW, Boorstein RJ, Cadet J. The repairability of oxidative free radical mediated damaged to DNA: a review. Int J Radiat Biol 1988; 54:131-150.
167. Thomas JP, Maiorino M, Ursini F et al. Protective action of phospholipid hydroperoxide glutathione peroxidase against membrane-damaging lipid peroxidation. J Biol Chem 1990; 265:454-461.
168. Niki E, Noguchi N, Gotoh N. Dynamics of lipid peroxidation and its inhibition by antioxidants. Biochem Soc Trans 1993; 21:313-317.
169. Stadtman ER. Protein oxidation and aging. Science 1992; 257:1220-1224.
170. Smith CD, Carney JM, Starke-Reed PE et al. Excess brain protein oxidation and enzyme dysfunction in normal aging and in Alzheimer disease. Proc Natl Acad Sci USA 1991; 88:10540-10543.
171. Yan SD, Schmidt AM, Anderson GM et al. Enhanced cellular oxidant stress by the interaction of advanced glycation end products with their receptors/binding protein. Proc Natl Acad Sci USA 1994; 91:7787-7791.
172. Smith MA, Sayre LM, Monnier VM et al. Radical AGEing in Alzheimer's disease. Treds Neurosci 1995; 18:172-176.
173. Wang J-Z, Grundke-Iqbal I, Iqbal K. Glycosylation of microtubule-associated protein tau: an abnormal posttranslational modification in Alzheimer's disease. Nature Med 1996; 2:871-875.

CHAPTER 3

Oxidative Stress in Neuronal Death and Apoptosis

INTRODUCTION

Oxidative stress, resulting from a deregulation of the balance of pro-oxidants/antioxidants, results in a number of phenomena from activation of signaling mechanisms, deregulation of Ca^{2+} homeostasis, alterations of mitochondrial structure and function and alterations in cellular structure to alterations in gene expression, all of which contribute to or play a critical role in neuronal death. However the sequence of events leading to neuronal death from an initial insult, as well as the causal role of a particular event guiding the progression of changes from initiation (reversible) to maintenance (reversible-irreversible) phases and finally to cell death is not actually known.

There are two general pathways of neuronal death that can be activated in response to oxidative insult: apoptosis and necrosis.[1,2] The term apoptosis, which literally means a "falling away," has been widely equated with programmed cell death (the term programmed was initially taken to imply the need for new RNA and protein synthesis) because it occurs in cells destined for removal in a variety of normal situations. Neuronal loss is an integral part in the normal development of a functional integrated nervous system, and 50%, or more, of all neurons die before adulthood.[3] Loss of neurons via apoptosis during development is beneficial, but apoptosis that occurs in the mature brain, as may be occurring in neurodegenerative illnesses, is harmful.[4] The mechanism for apoptosis appears to proceed via entry into an abortive cycle. Thus the same machinery that contributes to the generation of cells through cell division also can contribute to the death of cells through apoptosis. Apoptosis is a form of cell death in which an individual cell undergoes an internally controlled transition from an intact metabolically active state into a number of shrunken remnants retaining their membrane bound integrity.[5] Lysis of internal organelles apparently does not occur during this process, and little external leakage of the contents of the dying cell can be detected. As a consequence, apoptotic cells do not induce an inflammatory response in vivo. Instead the shrunken apoptotic bodies are phagocytosed and their contents presumably recycled.[5] During apoptosis there is a dispersion of polyribosomes, condensation of the cytoplasm as the cell undergoes shrinkage, and condensation and ultimately degradation of the nuclear DNA.[6] In contrast, necrosis reflects an acute loss of cellular regulation that results in cell lysis and inflammation.[3] Apoptosis has both prelethal and lethal stages. The prelethal changes are particularly characteristic and interestingly, the signals that initiate apoptosis appear to be similar to signals that occur in neurodegenerative disorders. Calcium and reactive oxygen species, in particular, are able to induce proteins that are involved

The Role of Oxidative Stress in Neuronal Death, by Irène Ceballos-Picot.

in apoptotic pathways, such as p53. Other proteins, such as Bcl-2, inhibit apoptosis. Thus, cells normally exist perched in an equilibrium between forces that induce apoptosis and forces that prevent apoptosis.

The role of oxidants as inducers of apoptosis has been highlighted by the discovery of the oncogene *bcl-2*, which may act in an antioxidant pathway to protect cells against free radical damage.[7] Overexpression of *bcl-2* protects neuronal cells from undergoing apoptosis induced by many different stimuli including hydrogen peroxide and other agents that cause free radical damage, such as radiation.[8–14]

Because the signals for apoptosis are the same molecules that are produced during oxidative stress, it is increasingly evident that oxidative stress is a common inducer of apoptosis. In this chapter we discuss the potential for oxidative stress and for changes in intracellular redox state to be involved in the apoptotic process. First, several examples are presented demonstrating that direct oxidative stress or agents that stimulate the production of free radicals through cellular metabolism or a decrease of antioxidant potential can induce apoptosis. We then discuss evidence that apoptotic cells accumulate oxidatively damaged products irrespective of the nature of the inducing stimulus. We also describe some recent experimental findings concerning mechanisms of neuronal apoptosis depending on changes in glutathione or intracellular reducing potential.

All these data strongly suggest that a loss of intracellular reducing potential is a common component of the downstream events of neuronal apoptosis and that reactive oxygen species serve as an early signal and/or a toxic agent to mediate neuronal apoptosis. The involvement of similar mediators in neurodegenerative disorders suggests that apoptosis may also be important in neurodegenerative diseases.

REACTIVE OXYGEN SPECIES AS COMMON MEDIATORS OF NEURONAL APOPTOSIS

Several recent studies suggest that oxidative stress may be an important cause of neuronal death and have implicated ROS as mediators of apoptosis.

ROS and the overall redox status of the cell play a pivotal role in the progression of neuronal apoptosis in response to divergent apoptotic stimuli.[7,12,15,16] ROS have been reported to be required during apoptosis resulting from growth factor withdrawal in neuronal cells,[17,18] UV-irradiation-induced apoptosis,[7] chemical hypoxia-induced (rotenone and MPP^+) apoptosis,[19] hydrogen peroxide induced apoptosis[20] and hypoxia-induced cell death in hypothalamic neurons.[21] ROS are downstream mediators of p53-dependent apoptosis.[22]

Antioxidants reduce ROS production and restore cell survival. The *bcl-2* gene product, an inhibitor of neural apoptosis, is reported to function as an antioxidant because it is associated with a marked decrease in the net cellular generation of ROS and prevention of lipid peroxidation.[7,12] Furthermore, overexpression of CuZnSOD in a neural cell line or glutathione peroxidase in a lymphoid cell line inhibits apoptosis.[7,12,17] In cortical neurons, 2-mercaptoethanol, recombinant thioredoxin (thiol donors) and N-acetylcysteine (a ROS scavenger or precursor of reduced glutathione) effectively delay, whereas buthionine sulfoximine (an inhibitor of glutathione synthetase) enhances, the serum deprivation-induced apoptotic neuronal death.[23,24] Addition of nerve growth factor (NGF) to PC12 cells increases the intracellular level of catalase and decreases the sensitivity of PC12 cells to exogenous hydrogen peroxide.[25] The catalase inhibitor 3-aminotriazole counters the protective effect of NGF against hydrogen peroxide. Ultrastructurally, early degenerative changes of mitochondria occur, and protection against cytotoxicity is afforded by antioxidants such as ascorbate, and by the iron ion chelator desferrioxamine. NGF-withdrawal triggered a transient increase of ROS in the early phase of sympathetic neuronal death and a microinjection of CuZnSOD delays death,[17,26] indicating that ROS production is integrated into the signal transduction pathway of physiologically-occurring apoptosis.

Evidence from both in vitro and in vivo studies has suggested a role for the excitatory amino acid glutamate and ROS in several neurological disorders ranging from acute insults such as stroke, hypoglycemia and epilepsy to chronic degenerative states. Cortical neurons from the Down's syndrome brain have been shown to generate significantly increased levels of ROS, leading to apoptotic cell death in culture.[27] Oxidative stress is now associated with other pathological processes of the CNS, such as Parkinson's disease,[28,29,30] Alzheimer's disease,[31,32] amyotrophic lateral sclerosis[33,34] and glutamate toxicity.[35,36] Overstimulation of glutamate receptors leads to sustained Ca^{2+} influx, predominantly through the NMDA receptor subtype.[37] The resulting Ca^{2+} overload activates Ca^{2+} downstream mechanisms and eventually neuronal death.[38,39] In addition Ca^{2+} overload can directly stimulate free radical production[40] and activate endogenous NO generation.[41,42] The mode of excitotoxic neuronal death in glutamate treated neurons has been studied in primary cultures in vitro[2] and both necrotic and apoptotic cell death could take place. The duration and extent of the insult elicited by glutamate or oxidative stress may determine whether neurons die by apoptosis or undergo necrotic lysis.[2] For instance, stimulation of cortical neurons with high concentrations of NMDA results in necrosis, whereas exposure to lower concentrations causes apoptosis.[1] Correspondingly, neuronal death in experimental stroke models is necrotic in the ischemic core, but delayed and apoptotic in the less severely compromised penumbra or border regions.[43] The sensors that switch neurons towards one or the other fate may be multiple. However, there is reason to believe that some would be located in the mitochondria.[2] Cellular stress following Ca^{2+} overload or free radical accumulation may elicit mitochondrial permeability transition[44] and a consequent release of mitochondrial Ca^{2+} plus a proteinaceous factor related to apoptosis.[45,46] The latter may be required as a cofactor in the activation of the proteolytic cascade involving a family of enzymes known as ICE-like proteases. A complete deenergization of the cell however, may not allow the ordered sequence of changes required for apoptotic demise. In such a case, the combination of multiple random processes would result in rapid uncontrolled cell lysis/necrosis. Therefore, it seems likely that apoptosis ensues under conditions where there remains sufficient energy production to execute an internal death program. Accordingly, ATP levels are maintained in PC12 cells or hippocampal neurons undergoing apoptosis.[2,47]

These findings argue that ROS serve as mediators of neuronal apoptosis since they satisfy the following criteria of a mediator: (1) demonstrate a change during apoptosis, prior to cell death; (2) modulation of the putative mediator modulates apoptosis accordingly; (3) expression of anti-apoptotic genes affect the putative mediator, unless the effects of the anti-apoptotic genes are downstream from the putative mediator; (4) the effect of anti-apoptotic genes should be overriden by independent modulation of the putative mediator, unless the effects of the antiapoptotic genes are downstream from the putative mediator.

The development of therapeutic and preventive approaches would benefit from understanding of the sensor, effectors and modulators of cell death. In particular, since apoptotic cell death takes place with a delayed time-course compared to necrotic lysis, the understanding of relevant modulatory factors may be used for the development of therapeutic strategies that would limit neuronal loss.

SOURCES OF REACTIVE OXYGEN SPECIES IN NEURONAL APOPTOSIS

What is/are the source(s) and chemical structure(s) of the ROS that mediate apoptosis? The brain and neural tissue contain sources of oxidative stress unique to this tissue. The amino acid neurotransmitters glutamate and aspartate, and many of their derivatives, can trigger a series of events leading to neuronal damage and death. Oxidative processes are part of these events. Activation of the N-methyl-D-aspartate

(NMDA) receptor, considered one of the most sensitive receptors to excitatory amino acids, leads to elevation of lipid peroxidation in the hippocampus.[48] Stimulation of NMDA receptors can also lead to increases of superoxide[49] or in intracellular nitric oxide,[41] which rapidly reacts with superoxide to form peroxynitrite, which can be cytotoxic, and which is involved in tyrosine nitration and formation of hydroxyl radicals.[41]

One event is an increase in intracellular Ca^{2+}, which can lead to production of free radicals from many sources, including microglial activation, arachidonic acid release, conversion of xanthine dehydrogenase to xanthine oxidase and subsequent release of ROS. In addition, high levels of intracellular calcium are taken up by mitochondria, where elevated calcium levels lead to production of free radicals, hydroxyl radicals and carbon-centered radicals.[50]

The metabolism of dopamine and other neurotransmitters can produce a variety of toxic compounds, many of which are ROS. Monoamine oxidase catabolizes dopamine and produces hydrogen peroxide as a product[51] and nonenzymatic reactions of dopamine also produce hydrogen peroxide and other ROS.[52] Autooxidation of dopamine produces neuromelanin, and hydrogen peroxide and oxyradicals are formed in the process.[30] The autooxidation of epinephrine and norepinephrine produces superoxide radicals and neurotoxic products. In addition, reaction of serotonin with hydroxyl radicals results in the formation of the neurotoxin tryptamine-4,5-dione.[53] A decrease of thiol donors' uptake may cause the increase of ROS production. Studies of tumor necrosis factor cytotoxicity implicate superoxide[54] and studies of amyloid peptide neurotoxicity implicate H_2O_2 or NO and more recently 4-hydroxynonenal, a product of lipid peroxidation (see chapter V). The potential sources of Aβ induced H_2O_2 production was tested pharmacologically with reagents inhibiting superoxide and H_2O_2 production with varying degrees of specificity.[31] Only diphenylene iodinum (DPI) and neopterin, reagents originally developed as inhibitors of neutrophil NADPH oxidase, inhibit Aβ toxicity in B12 cells and CNS neurons. Since DPI and neopterin may inhibit the activity of other flavin-containing enzymes, it can be concluded that flavin oxidases such as NADPH oxidase are involved in H_2O_2 generation. Impairment in energy homeostasis may be a more significant early factor in the neurodegenerative process.[55] Because the intracellular free calcium concentration rises during the first 24 h exposure to Aβ,[56] it would be expected that ATP turnover would be enhanced following Aβ exposure owing to increased Ca^{2+}-ATPase activity.

Even from these few examples, it is clear that more than one source of ROS generation is likely. Excitotoxic mechanisms, neurotransmitter metabolism, decrease in scavenging and other antioxidant defense systems and mitochondrial dysfunction can occur together and reinforce each other in ROS elevation leading to apoptosis. Any model of apoptosis that includes ROS as mediators must address the question of how cells derive the remarkable order of the apoptotic program, with the regular DNA fragmentation and appropriate upregulation and downregulation of specific sets of genes, depending on cell type, that distinguish apoptosis from the relative disorder of ROS diffusion and high reactivity. One possibility is that coordinated functional changes might be brought about in structural proteins by sequential and redox changes.[57] To account for such changes, and for the finding that Bcl-2 inhibits both apoptosis and necrosis, a model was suggested in which ROS act as effectors in necrosis but as signaling molecules in apoptosis[58] leading to the transcriptional changes of apoptosis via redox-sensitive transcription factors such as AP-1 and NF-κB.[59,60] This model would explain why similar insults may lead to apoptosis or necrosis, depending on severity or cell type.

OXIDANTS/ANTIOXIDANTS IN THE MODULATION OF NEURONAL APOPTOSIS

Superoxide Dismutase Delays Neuronal Apoptosis

Given the constant threat to cell viability imposed by ROS derived from normal cellular metabolism, a down regulation of antioxidant defense may provide an efficient mechanism by which physiologic and pathologic cell deletion can be achieved. This may have a role in the mechanisms of neurodegeneration and apoptosis in the central nervous system. A means to analyze the role played by a particular antioxidant in the central nervous system is to selectively modulate it by genetic or metabolic manipulations.[61,62] This is exemplified by copper-zinc SOD. A down regulation in PC12 cells by exposure to an appropriate antisense oligonucleotide causes their apoptotic death.[63] In this model, NO generators expedite cell death while inhibitors of NOS abolish cell death, suggesting that formation of peroxynitrite is necessary for this process.[64] However, motor neurons in CuZnSOD-deficient mice develop normally but exhibit enhanced cell death after axotomy.[65] FALS-associated SOD mutants promote neural apoptosis in a dominant fashion, despite the fact that these mutants retain enzymatic activity in neural cells[66] and mice expressing a human CuZnSOD mutation develop motor neuron degeneration.[67] Overexpression in neurons of the superoxide-metabolizing enzyme superoxide dismutase was shown to delay apoptosis.[17,26,66] Increased CuZnSOD promotes cell survival and neuronal process development in post-natal midbrain neurons of transgenic mice, probably by providing more efficient detoxification of free radicals.[68]

These results demonstrated that increased CuZnSOD activity attenuates apoptosis, whereas decreased CuZnSOD activity stimulates it. Which mechanisms are involved in the apoptosis modulation by CuZnSOD?

Evidence for participation of superoxide anions in neurotoxicity elicited by NMDA, the calcium ionophore, as well as for NO is evident from the protective effect of SOD.[49] The importance of superoxide anions in NMDA neurotoxicity is emphasized by findings that cortical cultures from transgenic mice overexpressing SOD are resistant to NMDA neurotoxicity[69] and these same animals are relatively resistant to focal ischemia.[70] SOD protection might relate to superoxides directly generated by NOS, as purified NOS directly forms superoxide anions.[71] Both NO[72] and the superoxide anion[73] block both NMDA receptor currents and the associated increase in intracellular calcium. Thus both potential toxic agents might exert feedback inhibition on NMDA receptors under physiologic conditions. With excessive release of both NO and the superoxide anion, normal regulatory mechanisms might be overwhelmed, leading to cell death. Superoxide dismutase ameliorates the formation of peroxynitrite and histone-associated DNA fragments in cortical cultures.[74] In the models of trophic factor deprivation-induced death,[75] NO generators promote survival, and inhibition of NOS has no effect. It is conceivable that NO becomes cytotoxic only under conditions of increased superoxide. NO is necessary but not sufficient for neuronal injury and is toxic only in the presence of other factors, such as the superoxide anions. It appears likely that the oxidative state of the target cell can determine whether NO is cytoprotective or cytotoxic. NO can combine with the superoxide anion to form peroxynitrite, which decomposes into the hydroxyl free radical (OH°) and the nitrogen dioxide free radical (NO_2°), both of which are highly reactive and potentially toxic.[76] Conceivably NOS neurons are enriched in SOD, which would account in part for their resistance to toxicity. It has been known for some time that SOD prolongs the effect of exogenous NO, presumably by reducing the level of superoxide and thus diminishing its reaction with NO to form peroxynitrite. SOD protects from NO neurotoxicity in primary brain cultures for the same reasons cited above.[41,42] With regard to the finding by Greenlund et al,[17]who reported that microinjection of

CuZnSOD or SOD expression vector, delayed the death of sympathetic neurons deprived of NGF, it is conceivable that this treatment results in an increase in intracellular levels of NO by the mechanisms mentioned above, and consequently leads to cGMP accumulation and prolonged survival.

The nature of the original insult as well as the decision to enter into the necrotic versus the apoptotic pathway might have therapeutic consequences in terms of the possible effectiveness of SOD. Some findings predict that signals in addition to NO that result in elevation or maintenance of intracellular cGMP may contribute to cell survival.

Glutathione as Modulator of Apoptosis in Neuronal Cells

Glutathione is a tripeptide consisting of glycine, cysteine, and glutamic acid moieties is a major antioxidant and functions directly in the elimination of ROS. It is also a substrate for scavenging ROS through the enzymes GSHPx and GSSG-Rd. A major role of GSH is to maintain the intracellular redox potential. Because it is the most prevalent constituent in the cellular pool of reducing equivalents, even modest variations in GSH concentration can strongly modulate redox state. Thus, aside from its antioxidant properties, GSH acts as a cellular redox buffer. GSH is also involved in cellular metabolism, participating in thiol-disulfide exchange reactions with cellular components, and in transport of amino acid sulfur.[77] GSH may also be directly involved in intercellular communication in the central nervous system. Evidence has been obtained of calcium-dependent GSH release from brain slices[78] and of GSH binding sites in rat synaptic membranes,[79,80] suggesting a possible additional role for GSH as a neuromodulator/neurotransmitter. The binding of certain glutamate agonists and antagonists to rat synaptic membranes is inhibited by GSH.[81] Glutathione may also affect the redox site of the N-methyl-D-Aspartate (NMDA) receptor.[80] It is thus possible that GSH may exert important modulatory effects on glutamaergic neurotransmission and glutamate neurotoxicity. In rat brain, biotinylated glutathione binds to sites predominantly within the white matter, suggesting that there are GSH receptors on neuroglial cells.[82] Immunocytochemical and radioligand receptor essays of GSH binding on cultured rat cortical astrocytes have confirmed the presence of GSH receptors in cortex.[82] In addition, GSH stimulation of GSH receptors on cultured rat brain astrocytes produces an increase in the level of inositol 1,4,5-triphosphate (IP_3), an action in keeping with the view that GSH may stimulate specific receptors, leading to changes in second messenger activity. IP_3 activation is part of a cascade leading to Ca^{2+} release from internal stores and is also associated with diacylglycerol production. As GSH receptors also appear to be regulated by PKc, abnormal kinase/phosphatase action leads to abnormal receptor regulation (NMDA or GSH) and that may be one of key steps in the pathogenesis leading to neurodegeneration. It could influence the activity of a number of redox-regulated transcription factors including NF-κB and AP-1 and therefore could modulate some gene expression.[59,60,83,84]

In the brain GSH has been shown to protect against toxic effects of glutamate and N-methyl-D-Aspartate;[81] furthermore, GSH has recently been implicated in protection against the induction of apoptotic and necrotic cell death in a variety of cell types.[85-88] Reduced levels of GSH have been associated with mitochondrial brain damage [89] and neuronal cells may be particularly susceptible to the effects of GSH depletion, since these cells, in contrast to glia, have relatively low concentrations of GSH.[90-94] Altering the intracellular redox balance by decreasing GSH levels profoundly affects early signal transduction events.[60] It was also shown that GSH could modify the activity of redox sensitive enzymes including protein tyrosine kinases (PTKs) and protein tyrosine phosphatases (PTPs) and could have an impact on signal transduction.[95] A thiol-mediated redox regulation of apoptosis is suggested in human cells implicated in immune function.[96,97]

However, no data exist for neuronal cells and the nervous system; particularly, it is not known whether GSH depletion represents a primary cause of neurodegeneration and neuronal death. Therefore, the role played by glutathione deficiency in neuronal loss and the mode of cell death remains undefined. The mechanisms may come to be understood by creating GSH depletion in neuronal cell cultures by specifically inhibiting its biosynthesis with L-buthionine-(S,R)-sulfoximine (BSO),[98] a specific inhibitor of γ-glutamylcysteine synthetase. This treatment led to a progressive decrease of intracellular GSH concentration during the first 24 hours.

A GSH depletion of 50 percent after one hour and 98 percent after 24 hours was evidenced (Fig. 3.1). BSO presumably depletes both the cytosolic and mitochondrial GSH pools, as supported by the nearly complete loss of GSH in treated cells.

The effects of BSO on cell morphology and on cell number were studied using phase contrast microscopy. As shown on Figure 3.2, incubation of NS20Y cells with BSO for 48 hours did not cause morphological changes as compared to controls. The effects of GSH depletion on cell morphology was detected after 5 days of BSO treatment. The morphological changes were a loss of cell connections and neurites, a

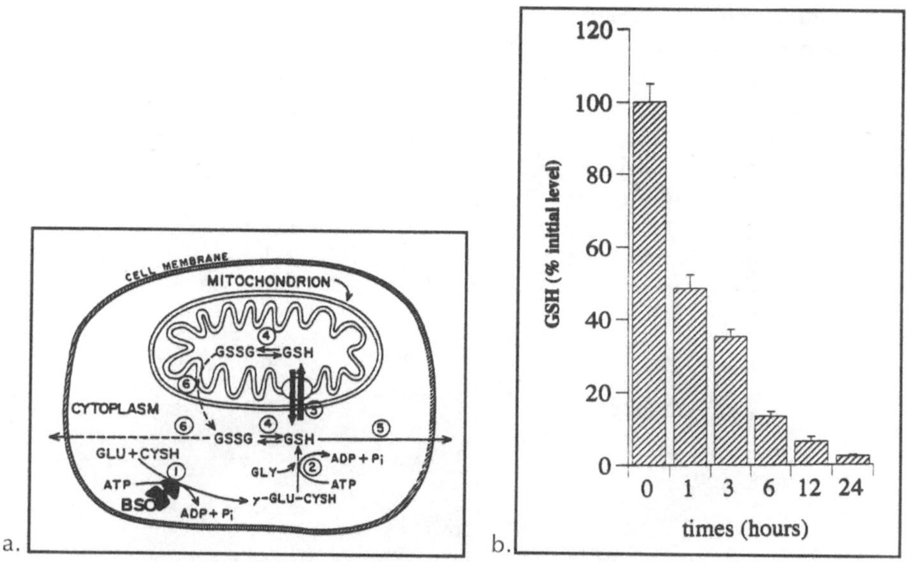

Fig. 3.1. (a) Scheme for synthesis and transport of GSH in mitochondria and cytoplasm. GSH is synthetized only in the cytoplasm (reactions 1 and 2). L-buthionine-S,R-sulfoximine (BSO) is an inhibitor of γ-glutamylcysteine synthetase (reaction 1). BSO, after phosphorylation by ATP on the enzyme, binds tighly to its active site thus inhibiting it irreversibly. The absence of synthetase from mitochondria showed that mitochondrial GSH must arise from the cytosol. Extramitochondrial GSH promotes mitochondrial uptake and exchange, and the intermembranous space appears to function as a recovery zone that facilitates efficient cycling of matrix glutathione. GSH is transported into mitochondria by a system characterized by slow net transport and more rapid exchange transport (reaction 3). Reversible oxidation of GSH to GSSG occurs in both cytoplasm and mitochondria (reaction 4). GSH is exported across the membrane of cells (reaction 5). Under conditions of oxidative stress, GSSG is transported out of the mitochondria and across cell membranes (reaction 6) Adapted from Griffith and Meister.[288]
(b) Time course of glutathione depletion induced by L-buthionine-(S,R) sulfoximine on NS20Y cultures. A nearly complete depletion is observed after 24 hours with BSO. Data shown are expressed as percentages of control values and are the mean of quadruplicate cultures ± SD (bars). The experiment was performed in duplicate.

rounding of cell soma and a decrease in cell number (Fig. 3.2). This was followed by extensive detachment of the cell lawn. Our experiments demonstrate that simply lowering intracellular GSH levels is sufficient to induce a decrease in cellular proliferation and/or to induce a cellular death after 5 days of BSO treatment.

We used the TUNEL technique to identify the nuclei of cells undergoing the DNA fragmentation characteristic of apoptosis at different times after BSO exposure. This technique permits the specific labeling of high concentrations of 3'-OH ends of DNA in apoptotic cells. After 5 days of BSO exposure, the type of cellular mortality was characterized as apoptosis, as demonstrated by the *in situ* staining of internucleosomal DNA fragmentation at the cellular level. Nuclei demonstrated chromatin condensation and nuclear fragmentation after 5 days of BSO treatment (Fig. 3.2F), while no labeling was observed after 2 days (Fig. 3.2E), as compared to control cultures (Fig. 3.2D).

To clarify the role of GSH depletion in the apoptotic process in this model, it was important to know if other enzymes implicated in free radical detoxification were affected during the course of BSO treatment. GSH depletion led to secondary changes in the antioxidant system; particularly, a down regulation of GSH peroxidase in living NS20Y after 2 days of BSO treatment was observed (-40%; $p < 0.001$; Fig. 3.3). This decrease precedes nuclear fragmentation since no labeling of DNA fragmentation was observed at the cellular level.Thus GSH depletion being the initial insult, a down regulation of GSH peroxidase and consequent oxidative stress could constitute early stages in the apoptotic process.

After 5 days of BSO exposure, GSH-depleted NS20Y cells still living showed a significant increase in CuZnSOD and GSSG reductase activities ($p < 0.001$; Fig. 3.3) and the significant decrease in GSH peroxidase activity still exist ($p < 0.01$). Whatever the time of BSO exposure, MnSOD activity was unchanged as compared to control cells.

Our study indicate that GSH depletion induce by itself a death characterized as apoptosis of a clonal cell line of neuronal origin. This result adds to the evidence that direct oxidative damage [1,20,99] or indirect oxi-

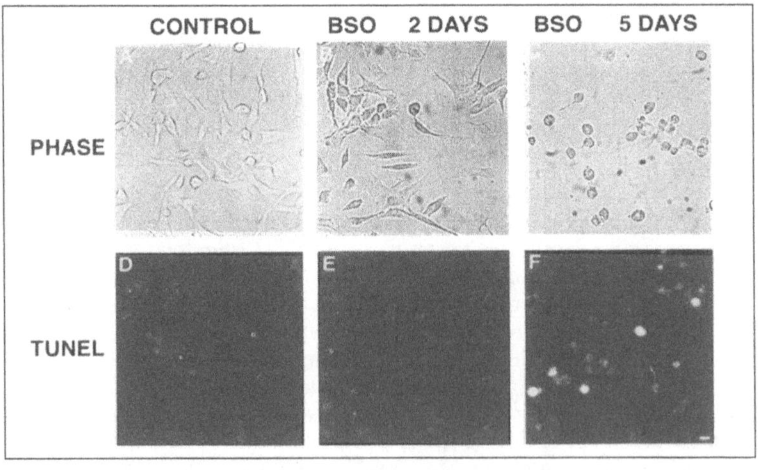

Fig. 3.2. Morphological changes of NS20Y neuroblastoma in culture and in situ demonstration of DNA fragmentation induced by GSH depletion. Morphology of NS20Y cells and cellular death by phase-contrast microscopy (A, B, C) and fragmentation of nuclei characteristic of apoptosis evidenced *in situ* by the FITC-based TUNEL technique (D, E ,F). A, D: untreated cells. B, E: 2 days of BSO treatment demonstrated neither morphological changes (B) nor chromatin fragmentation (E). C, F: 5 days of BSO treatment induced cellular death and morphological changes (C) and a strong fluorescent staining of fragmented chromatin (F). Scale Bar= 5 μM.

dative stress[15,16,61,100] may cause neuronal death that proceeds through an apoptotic pathway. The glutathione redox cycle is an important component of the antioxidant machinery; mRNA levels for redox-related genes (CuZnSOD, MnSOD and catalase) fall during apoptosis[17] and depletion of reduced glutathione has been reported to induce or increase sensitivity to apoptosis in different systems.[85] In particular, glutamate-induced glutathione depletion in embryonic cortical neurons[16] and in neuroblastoma and hippocampal neurons[100] results in oxidative stress and apoptotic cell death. Diethylmaleate reduces cellular stores of GSH by forming a thioether conjugate in a reaction catalyzed by glutathione S-transferase and as a consequence ROS accumulating in the cells induce cell death.[12] Thus, redox modulation seems to play a major role in the regulation of cell death and apoptosis.

Any new model for the biochemical events defining apoptosis must distinguish between an inducer of apoptosis, of which there are many, (the list of inducers of apoptosis is large including growth factor withdrawal, ischemia, many toxins, amyloid peptides, calcium ionophores, and many others)[101] and a mediator of apoptosis. Such a mediator should satisfy some criteria: (1) demonstrate a change during apoptosis, prior to cell death; (2) modulation of the putative mediator should modulate apoptosis accordingly; and (3) expression of anti-apoptotic genes should affect the putative mediator, unless the effect of the anti-apoptotic genes are downstream from the putative mediator. GSH fulfill all of the criteria of a mediator of apoptosis. A decrease of GSH concentration was demonstrated in preapoptotic cells[16,102] and its depletion generated cell death via apoptosis in neuronal clones. In addition, the protooncogene *bcl-2*, which inhibits many forms of apoptotic cell death, has been suggested to mediate an antioxidant pathway to prevent apoptosis. Particularly, cells expressing Bcl-2 have two times more GSH than controls.[88] Moreover an increase of intracellular levels of GSH by N-acetylcysteine (NAC), approximately 10-fold in PC12 cells, prevents cell death caused by withdrawal of trophic support.[103] Other results showed that the upregulation of cellular GSH evoked by autooxidable agents such as L-DOPA is associated with significant protection of cells.[86] The

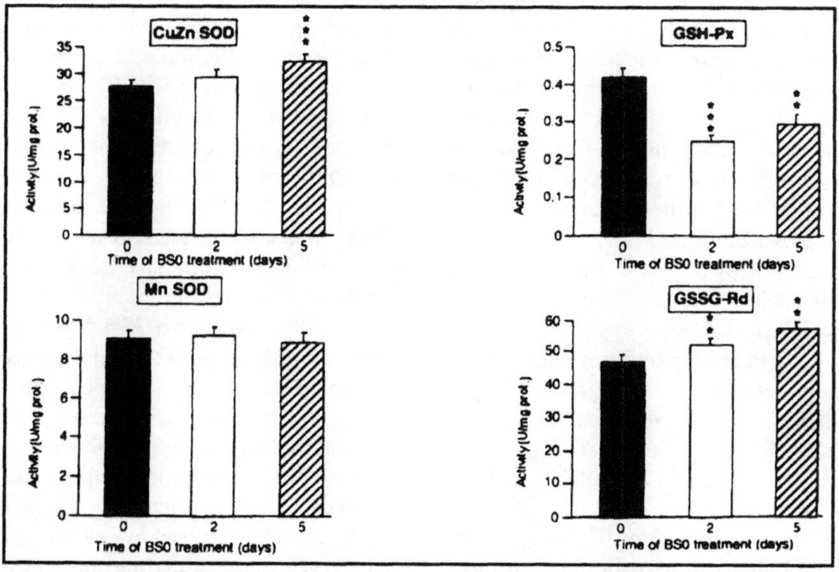

Fig. 3.3. Basal antioxidant enzyme activities of NS20Y cells before and during the course of glutathione depletion. Data are means ± SD (n=6). The experiment was duplicated. * $P<0.05$; ** $P<0.01$; *** $P<0.001$.

induction of apoptosis by any given factor offers insight neither on the toxic biochemistry of the given factor nor into the mechanism of apoptosis itself. In contrast, the identification of a mediator of apoptosis such as glutathione in neural cells offers some insight into the biochemistry of apoptosis. Thus GSH could be the mediator of a common pathway triggered by both physiologic and pathologic inducers of neuronal apoptosis and could play a key role in inducing naturally occuring neuronal death during development and in pathologic conditions.

To explain the mechanism of neurotoxic action of GSH depletion several hypotheses may be put forward. GSH depletion could intervene in the complex redox chemistry of nitric oxide (NO) because a simple change in the ambient redox milieu could allow NO interconversion from neuroprotective to neurotoxic species.[74] Glutathione depletion could also modify NMDA receptor activity by affecting the receptor's redox modulatory site and could modulate glutamate neurotransmission and neurotoxicity. The neuronal cells have a relative paucity of GSH as compared to astrocytes[104] and hence may be particularly vulnerable to the effects of GSH depletion. Decreased levels of GSH would be expected to cause an increase in the production of hydroxyl radicals from the Fenton reaction because the competing GSH peroxidase reaction would be diminished. Consequently, one indirect way in which GSH depletion can kill cells is via the oxidation of membrane polyunsaturated fatty acids to form fatty acid hydroperoxides. Mammalian cells usually avoid the toxic effects of hydrogen peroxide and hydroperoxy fatty acids by converting them in a reaction requiring reduced glutathione and catalyzed by the antioxidant enzyme GSH peroxidase. GSH depletion led to secondary changes in the antioxidant system; particularly, a down regulation of GSH peroxidase appeared before nuclear fragmentation. GSH peroxidase is involved in hydroperoxide detoxification, and its down regulation could potentiate oxidative stress and cause an increased sensitivity to ROS produced. These results support the view that glutathione depletion leading to oxidative stress may serve as a general trigger for apoptosis in neuronal cells and that oxidative stress is a key component of the apoptotic pathway. These events could constitute early stages in the apoptotic process. Presumably obligate and specific oxidation is required at some step between initial insult and DNA fragmentation. ROS production, rather than a toxic agent, could serve as an early signal to mediate apoptosis. The decision as to whether a cell will undergo programmed cell death or not may be expected to be regulated by switches that, once activated, trigger self amplificatory metabolic pathways. One of these switches may reside in a perturbation of antioxidant protection implicated in hydrogen or lipoperoxide detoxification.[17] A reduction in growth rate and increase in detaching, trypan blue staining cells can also be brought about by exposure of BHK cells to mercaptosuccinate, an inhibitor of GSH peroxidase.[105] It appears that molecules that participate in apoptotic decision-making also exert functions that are vital for normal cell proliferation since it was also demonstrated an endogenously generated ROS and modification of cellular glutathione levels during cell proliferation.[105] It is possible that specific intracellular levels of GSH are critical in determining whether proliferation is enhanced or diminished, although the higher levels of hydrogen peroxide may also be significant, in as much as hydrogen peroxide itself may directly affect growth regulatory molecules.[106] These results underline the possibility of some special relationships between proliferation, viability and intracellular GSH levels. This GSH depletion may cause an imbalance of the thiol status and, thereby, could affect Ca^{2+} homeostasis. Oxygen radical production and thiol depletion are required for Ca^{2+}-mediated endogenous endonuclease activation in apoptotic thymocytes.[107] These phenomena have been reported to be involved in oxidative stress-related cell death[37] and thus may contribute to the cytocidal activity of GSH depletion.

Insight into the mechanism by which GSH depletion leads to cellular death may also be gleaned from the observation that GSH depletion is accompanied by increased neuronal nitric oxide synthase activity and NO production.[108] This finding may have implications for the understanding of the pathogenesis of neurodegenerative disorders such as Parkinson's disease, in which loss of brain glutathione is considered to be an early event. Thus, factors which directly or indirectly influence GSH availability may therefore influence brain NOS activity, NO production and also induce alterations of mitochondrial function.[109] The exact mechanism for NO-mediated cytotoxicity is not known but may arise as a result of energy depletion due to inactivation of the mitochondrial respiratory chain.[110,111] In this context, it is particularly interesting that mitochondrial damage and glutathione depletion is associated with Parkinson's disease.[112] Cells that are deprived of GSH typically suffer severe oxidative damage associated with mitochondrial degeneration.[89,113] One of the essentials of GSH is the mitochondrial function.[114] Thus, in the model of neuronal apoptosis induced by GSH depletion, mitochondrial dysfunction could be a critical step in the apoptotic cascade. Recent demonstration of the mitochondrial control of nuclear apoptosis argues in favor of this hypothesis.[115]

Nitric Oxide Produces or Delays the Neuronal Death

Recent evidence implicates NO as a major synaptic signaling agent in the nervous system.[116,117] Apart from this role under physiological conditions, NO appears to mediate neuronal death when produced in excess. Many of the actions of NO in neurons are linked to glutamate, the principal excitatory neurotransmitter in the brain.[118] When released in excess, glutamate stimulation of NMDA receptors in nNOS neurons causes a massive release of NO, which kills adjacent neurons in neuronal cultures.[119] Evidence for this includes blockade of neurotoxicity by NOS inhibitors and the lack of such toxicity in cultures from mice in which the gene for nNOS has been deleted ($nNOS^-$). Because inhibitors of NOS reduce the extent of brain damage caused by experimentally induced cerebral ischemia, NO appears to be intimately involved in the death process.[120] Neural damage following vascular stroke elicited by occlusion of the middle cerebral artery is diminished by treatment with NOS inhibitors or in $nNOS^-$ mice.[121] This is further supported by observations that NOS inhibitors as well as other agents that reduce NO levels protect cultured neurons from glutamate toxicity.[41,122] Inhibition of NO synthesis in baboons prevents the loss of dopamine neurons in the MPTP (1-methyl,4-phenyl-1,2,3,6-tetrahydropyridine) model, which faithfully mimics clinical parkinsonism.[123] In mice, NOS inhibitors prevent loss of dopaminergic neurons,[124] while Dawson et al reported that MPTP effects are obliterated in $nNOS^-$ mice.[125] $nNOS^-$ mice are resistant to excitotoxic lesions produced by malonate, and the resistance is associated with reduced generation of 3-nitrosotyrosine.[126] Application of the NO synthase inhibitor nitroarginine after, but not during, the NMDA exposure inhibited NO formation and protected against delayed neuronal death.[127] Thus, glutamate receptors and NO operate interdependently and sequentially to cause neurodegeneration. At the core of the mechanism is a vicious cycle in which NMDA receptor stimulation causes glutamate release and further NMDA receptor stimulation. The output of the cycle is an ongoing production of NO from neuronal sources, and this is responsible for delayed neuronal death. The mechanism by which excessive NO leads to neuronal death is unclear but could include direct activation or inhibition of signaling or metabolism[128,129] or reaction with superoxide to form the highly reactive peroxynitrite anion,[64,74,130–132] which in turn may degenerate into the even more noxious hydroxyl free radicals. Numerous sequelae are possible including lipid and protein oxidation and protein nitration.[74] There is evidence that neurotoxic cell death derives primarily from DNA damage. Damaged DNA fragments activate poly-ADP-ribose synthetase (PARS), which attempts to repair DNA by attaching chains of ADP-ribose to nuclear proteins such as histones and PARS itself. Massive activation of PARS by excess amounts of NO depletes NAD, the PARS substrate, and subsequently ATP, leading to cell death by energy deprivation. PARS inhibitors block glutamate-NO-mediated neurotoxicity.[129]

NO has also been reported to be protective in certain circumtances. NO generators retard the death of PC12 cells and sympathetic neurons after withdrawal of serum and/or NGF.[75] The mechanism of this effect is not dependent on inhibition of cellular proliferation, but rather on the activation of guanylate cyclase and generation of intracellular cGMP since inhibition of guanylate cyclase blocks the ability of NO to promote survival. This survival promoting effect of NO is mediated by elevation of intracellular cGMP, implicating either or both of these messenger molecules as regulators of survival or apoptosis. Copper-zinc superoxide dismutase could be the modulator of the role played by NO (cytotoxic or cytoprotector) since microinjection of CuZnSOD or a SOD expression vector delays the death of sympathetic neurons induced by NGF deprivation.[17] The mechanism of cytoprotection induced by SOD in this model could be to maintain a high level of NO and to avoid peroxynitrite formation. It has also been reported that SOD itself is capable of catalyzing the reversible conversion of nitroxyl anion to NO, a reaction that would serve to prolong the action of NO.[133] Moreover Barger et al demonstrated that secreted forms of the β-amyloid precursor protein protected cultured hippocampal neurons against glucose deprivation and glutamate toxicity by increasing cGMP levels.[134] Taken together, these data strongly suggest that cGMP may be a regulator of neuronal survival. It was reported recently that PC12 cells express a constitutively active form of NOS and that this is upregulated after NGF treatment.[135] NO has previously been shown to be cytostatic for a variety of cell types, possibly through a mechanism involving inhibition of ribonucleotide reductase.[136,137] Because it has been demonstrated that NGF treatment of PC12 cells leads to the induction of a constituvely active form of NOS and that the production or application of NO initiates cessation of proliferation of these cells,[135] the cytoprotective effect of NO could be causally related to inhibition of cellular proliferation. Effectively, blockade of the cell cycle by molecular or pharmacological means is sufficient to prevent death of PC12 cells and sympathetic neurons after withdrawal of trophic elements.[138–140] The effects of NO on neuronal survival are short-lived and cannot be prolonged by readdition of NO generators, suggesting that the protective response is somehow downregulated. The observations that inhibition of cGMP phosphodiesterase prolongs the protective action of NO indicate that the downregulation occurs at the level of guanylate cyclase.[140]

Bcl-2 Functions in an Antioxidant Pathway to Prevent Apoptosis

Expression of *bcl-2* inhibits neural cell death induced by serum and growth factor withdrawal, calcium ionophores, glucose withdrawal, free radical-inducing agents such as menadione, membrane-peroxidizing agents such as tert-butyl hydroperoxide, glutamate.[8–12] These findings suggest that neural cell death from a wide variety of insults is mediated by a common biochemical pathway. Somewhat surprisingly, although *bcl-2* was shown previously to be an antiapoptotic gene, it inhibited nonapoptotic neural cell death more effectively in some paradigms.[141] Thus the label "death suppressor gene" may be more accurate than "antiapoptotic gene" for *bcl-2*. The clinical relevance of the death suppressive effect of *bcl-2* expression has been stressed by the recent finding that *bcl-2* overexpression in the brains of transgenic mice decreases experimental infarct size.[13]

The mechanism by which the *bcl-2* gene product inhibits neural cell death in such a wide range of circumstances has been investigated. Some studies suggested targeting of the 26-kd protein to the mitochondrial membrane,[142] but subsequent reports favored a less restricted localization, including perinuclear endoplasmic reticulum, outer mitochondrial membrane, and other membranous structures.[143,144] There is no signal sequence to suggest initial targeting into the endoplasmic reticulum; rather, membrane targeting is a result of a hydrophobic carboxyl terminus. Deletion of the hydrophobic C terminus results in a shift of Bcl-2 protein into the cytosol, but approximately half of its antiapoptotic activity is retained.[7]

Expression of *bcl-2* in neural cells does not alter the rate of cell division (in undifferentiated neuroblasts), neural differentiation, oxygen consumption, ATP levels, intacellular free calcium, secretion of trophic factors, protein synthesis, or phosphoprotein patterns on two-dimensional gels.[11] The failure to find a difference in *bcl-2*-expressing cells with respect to intracellular free calcium suggests that the effect of Bcl-2 on neural cell death is not mediated through inhibition of a rise in intracellular free calcium.

One mechanism by which Bcl-2 has been proposed to act is by decreasing the net cellular generation of reactive oxygen species.[7,12] Overexpression of *bcl-2* in neuronal cells results in decreased generation of reactive oxygen species and decreased lipid peroxidation. The nature of the reactive oxygen species affected by *bcl-2* expression has been evaluated preliminarily; expression of *bcl-2* has no direct effect on NO, but rather appears to have its predominant effect on molecular species involved in Fenton chemistry: hydrogen peroxide, metals such as iron and copper and hydroxyl radical.[12,145] It is interesting that studies looking at the effect of expression of *bcl-2* in *Escherichia coli* strains deficient in SOD led to a recent proposal that Bcl-2 exerts its physiological effect by acting as a pro-oxidant to induce the expression of antioxidant enzymes.[146] The concentration of GSH is elevated in some but not all cells expressing *bcl-2*.[12,147] Buthionine sulfoximine, a specific and essentially irreversible inhibitor of γ-glutamylcysteine synthetase, decreases glutathione concentration in the presence or absence of *bcl-2* expression, yet toxicity of this inhibitor is virtually confined to cells that do not express *bcl-2*.[12] Cell death across cell types does not correlate with glutathione levels, suggesting that the effect of *bcl-2* expression is not simply to increase glutathione concentration within the cell. In the brain of transgenic mice expressing the human *bcl-2* gene, partially protected from cerebral ischemia,[13] the concentration of glutathione and related enzymes as well as of superoxide dismutase were unmodified (unpublished data). Overexpression of *bcl-2* in two different neural cell lines, rat pheochromocytoma PC12 and the hypothalamic cell line GT1-7, shifts the cellular redox potential to a more reduced state, without consistently affecting the major cellular antioxidant enzymes.[147] This reductive shift includes a decrease in the (GSSG)/(GSSG+GSH) ratio, which is the major determinant of the cellular thiol redox status. Furthermore, oxygen consumption experiments have been reported previously[148] and the respiratory characteristics of the controls are identical to those of the *bcl-2* transfectants in GT1-7 cells.

It remains controversial whether the antioxidant-like effect of *bcl-2* expression on cells is a requisite property for its antiapoptotic effect. In addition to *bcl-2*, *bcl-2* homologues including *bax*, $bcl\text{-}x_L$, $bcl\text{-}x_S$, and *mcl-1* are expressed in central neurons.[149,150] Interactions of Bcl-2 with a proapoptotic homologue, Bax, as well as R-ras and Raf-1 kinase have been demonstrated.[151,152] The ratio of Bax to Bcl-2 sets a cellular "apostart" that determines the propensity of the cell to undergo apoptosis: an excess of Bax favors apoptosis, whereas an excess of Bcl-2 favors survival.[150] The transcription factor p53 induces neuronal apoptosis,[153] and ROS are downstream mediators of p53-mediated apoptosis.[22] Transcription factor p53 binds to the Bax promoter, and leads to an increase in the Bax/Bcl-2 ratio, thereby enhancing apoptosis;[154] and its DNA binding activity is redox-regulated.[155]

How can we rationalize these findings? One possibility is that Bcl-2 blocks several intracellular pathways, one of which involves the generation of ROS and utilization of reducing equivalents. This possibility would allow for reactive oxygen species to mediate a physiological signal into and out of the death pathway but would not exclude the involvement of other key signaling pathways that do not require a physiological signal from ROS. Given the central role of the ICE family of cysteine proteases in apoptosis, one possibility is that Bcl-2 inhibits this pathway, and this result is an alteration of the redox status of the cell. Perhaps one or more of the substrates are involved in the cellular redox status, or the activation of ICE family members is influenced by the redox status of the cell. The combination of a reducing shift and a slight decrease in ATP/ADP ratios suggests the possibility of a shunt of reducing equivalents from energy production to cellular reduction, i.e., to cellular oxidant protection. Theoretically, such a shift would seem to be an appropriate cellular strategy to prevent damage (and the resultant cell death) from many different insults.

OXIDATIVE STRESS AND CALCIUM HOMEOSTASIS IN THE DEVELOPMENT OF NEURONAL APOPTOSIS

Oxidant challenge has been shown to perturb intracellular Ca^{2+} homeostasis. Such an effect may exacerbate free radical reactions, activate endonucleases, and contribute to apoptosis. Oxidative stress may affect calcium homeostasis at various levels:

(1) Receptor-mediated calcium mobilization is perturbed. Oxidative stress impairs muscarinic, serotoninergic, alpha-adreno and NMDA receptor function and binding sites in rat brain synaptosomes and cortical membrane lipids due to increased lipid peroxidation.[73,156,157] These receptor types are Ca^{2+}-mobilizing receptors and the decrease in receptor function, as well as a disturbance in IP_3-dependent Ca^{2+} release after oxidative stress, suggest that Ca^{2+} mobilization is impaired.

(2) Intracellular calcium sequestration is also modified by oxidative stress. Normally, intracellular concentration is kept low by extrusion of Ca^{2+} from the cell by Na^+-Ca^{2+}-exchange or by Ca^{2+}-ATPases. Ca^{2+} is however also sequestered in mitochondria and endoplasmic and sarcoplasmic reticulum by Ca^{2+}-ATPases.[40,158] In several reports, oxidative stress has been shown to promote Ca^{2+} release from mitochondria[159] and from the sarcoplasmic reticulum, which is generally explained by an inactivation of Ca^{2+}-ATPases, thereby impairing Ca^{2+} sequestration.[160] This will contribute to the observed increase in intracellular Ca^{2+} level after oxidative stress.

(3) An increased membrane permeability to Ca^{2+} is demonstrated to be an earlier event than inactivation of Ca^{2+}-ATPase after oxidative stress. The increase of Ca^{2+} release from the above-mentionned intracellular organelles is probably promoted by peroxidation of membrane lipids, which promotes membrane permeability to Ca^{2+}. Moreover, Na^+-Ca^{2+} exchange activation might be induced by a promoted thiol/disulfide interchange in the carrier protein.[161] The increase of ROS levels is upstream of the increased Ca^{2+} level in the apoptotic cascade[162] and thus, ROS could be the mediators of the downstream events characteristic of apoptosis.

Elevated Ca^{2+} initiates a cascade of signaling leading to activation of hydrolytic enzymes such as phospholipases A_2 (PLA_2) and C (PL_C), endonucleases, or proteases, and the expression of several early genes including *c-fos, c-myc, c-jun and egr-1*. Elevated Ca^{2+} modulates signaling pathways; this includes ligand-activated kinases as well as intracellular messengers, e.g., IP_3, cADPR and nitric oxide. Moreover, elevated intracellular Ca^{2+} is an early and possibly ubiquitous event in the initiation and progression of the reparative response to injury, in that it comprises an integral part of signal transduction through growth factors and other agents. Increased Ca^{2+} exerts a number of effects on the cytoskeleton, including actin, tubulin, and intermediate filaments, separation of gap junctions leading to decreased or absent cell-cell communications, nuclear changes such as DNA strand breaks, gene activation such as immediate early genes and stress genes.[163]

OXIDATIVE STRESS, IMMEDIATE EARLY GENES AND NEURONAL APOPTOSIS

Regulation of gene expression can be significantly affected by intra- and extra-cellular ROS levels leading to modification of intracellular redox (reduction/oxidation) state. Several results indicate that ROS could be involved in gene regulation also through indirect mechanisms, for example by modulating the activity of enzymes playing key roles in membrane signal transduction such as PLA_2 and PL_C, protein tyrosine kinases (PTK) and the protein tyrosine phosphatases (PTP)[164–167] or by modulation of redox-regulated transcriptional factors/activators such as p53, AP-1 and NF-κB.[168] This fine regulation of gene expression is of major importance to mitogenesis, cell adhesion, cell differentiation, oncogenic cell transformation and apoptosis.

ROS and elevated Ca^{2+} may initiate a cascade of signaling leading to activation PLA_2 and PL_C, endonucleases and proteases, and the expression of several immediate-early genes (IEGs) including *c-fos, c-jun, c-myc.* These IEGs may be important in the initiation and progression of both cell death and in the regeneration response in sublethally injured cells that will undergo recovery and regeneration. Elevated ROS could be an early and possibly ubiquitous event in the initiation and progression of the reparative response to injury, in that it comprizes an integral part of signal transduction through growth factors and other agents. ROS could link key characteristics of cell injury and death with proliferation, regeneration, and differentiation.

Apoptosis has been shown to occur via a program that is dependent on macromolecular synthesis in some paradigms[3,5] suggesting that this process involves the initiation of a set of genes that regulate events leading to or associated with cell death. Such genes would presumably mediate the controlled destruction of the cell, possibly regulating DNA fragmentation or maintaining membrane integrity as a part of the apoptotic process. At present there is little information available on gene induction during apoptosis in primary neurons. The level of expression of over 60 messages was quantitated in superior cervical ganglion cells induced to undergo apoptosis by the withdrawal of NGF.[169] Expression of *c-jun* was shown to be required for apoptosis in these cells.[170] Cyclin D1 expression was also increased, suggesting that apoptosis and mitosis may share some features. Morphologically, some features of mitosis, such as the rounding up of the cell, are shared by apoptosis. However, expression of other cell cycle genes was not increased during apoptosis, DNA replication did not occur, and cyclin D1 was neither required nor sufficient. Thus, the relationship between mitosis and apoptosis, although intriguing, remains poorly understood. Apoptosis has been referred to as cellular suicide, since the morphological, genetic and biochemical changes that occur are orderly and reproducible, and inhibitors of transcription and translation inhibit the process, suggesting an active role for the cell in its own demise.

One group of genes that has been associated with apoptosis in a number of paradigms are the IEGs. Studies in nonneuronal and neuronal systems have suggested that the induction of a number IEGs may be related to the initiation of apoptosis, including several members of the *c-jun* family. In this context, a relationship between the initiation of apoptosis and the induction of *c-jun* has been described in neuronal cells treated with amyloid peptides[171,172] suggesting a role of *c-jun* in the apoptotic cascade. In Alzheimer's brain, an intensification of both Jun and Fos immunoreactivity was observed as compared to controls[173] and a large increase in *c-jun* mRNA occurred in the dorsal and ventral horns of the SLA spinal cord.[174] A direct role for *c-jun* in mediating apoptosis has been described in growth factor-dependent lymphoid cells[175] and in nerve growth factor-deprived primary sympathetic neurons.[170] These studies show that *c-jun* is induced during apoptosis, is correlated with DNA fragmentation in cells undergoing apoptosis, and can directly mediate apoptosis.Thus, the rapid induction and transcriptional regulatory capability of IEGs may provide an early marker for the signal transduction pathways initiated during apoptosis.

The product of the IEG c-*jun*, c-Jun, is a transcriptional regulatory protein and is a member of a larger family of c-Jun-related transcriptional regulatory proteins that includes JunB and JunD. c-Jun functions as a trans-activator for various promoters by forming heterodimers with itself or heterodimers with the products of other IEGs such as JunD and c-Fos.[171] Promoter consensus sequence specificity/affinity is dictated by the composition of the c-Jun dimer. The cooperative interaction of different sets of IEG protein products allows the regulation of individual IEGs by specific second messenger systems and stimuli to interact and produce a specialized, or combinatorial response. Furthermore, in addition to their transcriptional activation characteristics, IEGs can mediate transcriptional repressor activity, as in the case of c-Fos/Jun B heterodimers at the AP-1 site. IEGs have been shown to play an important role in this regard in neurons potentially regulating multiple processes such as proliferation, differentiation, and apoptosis, within a single cell. IEGs are induced in a stimulus and a cell type-specific fashion, and with differential kinetics.This temporal patterning of responses, in conjunction with the combinatorial nature of IEG dimerization, illustrates an added level of specificity in target gene regulation. This combined properties are critical for the role of IEGs in signal transduction. Although *c-jun* and *c-fos* related genes appear to play a role in apoptosis in some systems, it is clear that the cellular mechanism of apoptosis differs among cell types, and that these IEGs do not necessarily play the same role in all systems. Furthermore, a variety of additional immediate early and other genes have been linked to the initiation or regulation of mammalian apoptosis; these include *c-myc, bcl-2, bcl-x, bax*, RP-8, p53, p35, *nur77*, cyclin D1, and interleukin 1-β converting enzyme (ICE). A detailed understanding of these interactions will be critical to understanding any potential role for IEGs in cell death and defining the gene targets regulated by these transcription complexes.

Other apoptosis-inducing stimuli such as oxidative stress induced by H_20_2 and intracellular thiol levels appear to influence the expression of IEGs, notably *c-fos* and *c-jun*, in non-neuronal cells[172,173] or neuronal cells.[176] The mechanism of induction of *c-jun* protooncogene expression by H_20_2 is through hydroxyl radical generation and $p60^{SRC}$ tyrosine kisase activation.[168] Despite the need to clarify this issue, it is possible to speculate that there may be some similarities between the cellular signaling pathways associated with Aβ-mediated and oxidative or redox-mediated cell death. This include the induction of a specific set of genes that are associated with or participate in apoptosis. Future studies will help to clarify this issue, definitively identify specific apoptosis-related genes from these models and define the role of these genes. The products of the *jun* and *fos* genes form heterodimers that constitute the transcription factor AP-1 which is involved in several cellular processes, including cell proliferation and differentiation. Recently, numerous studies reported that AP-1 plays a prominent role in the pathway(s) leading to apoptosis since increased AP-1 activity and/or increased gene expression of *c-fos* and/or *c-jun* and apoptosis are causally related.[170,177] In Parkinson's disease, dopaminergic cell death in substantia nigra was associated with a marked decrease of the powerful antioxidant GSH. Apoptosis induced by GSH depletion was created in a neuroblastoma cell line by using an inhibitor of its biosynthesis: L-buthionine sulfoximine. Redox disequilibrium induced by GSH depletion led to an increase in ROS production and to apoptotic cell death (see in this chapter paragraph 4). To understand the early events by which redox-disequilibrium could affect genetic responses, the possible redox modulation of *c-jun, c-fos, jun B, c-myb* and *c-myc* was examined by Northern blot in the course of apoptosis. The *c-jun* mRNA was increased as early as 15 min of BSO treatment, before a strong decrease in GSH content. This induction continued at least until 6 h of BSO-treatment with an ongoing decrease in GSH (Fig. 3.4). The protooncogenes *jun B* and *c-fos* were coinduced transiently at one hour of BSO treatment.

Thus, the upregulation of *c-jun* protooncogenes seems to be one of the first events which occurs at the beginning of the GSH depletion in BSO-treated neuroblastoma in culture. This increase is similar to the magnitude of *c-jun* induction in another neuronal apoptosis model induced by amyloid peptide.[171] By contrast, *c-myb* and *c-myc* expression are not modified. This indicates that the regulation of *c-jun*, *c-fos* and *jun B* protooncogenes observed here does not reflect a nonspecific deregulation of cellular signaling pathways and that IEG expression is differentially regulated in response to apoptotic inducers. The induction of (*c-jun*, *junB* and *c-fos*) IEGs might play a role in this neuronal apoptotic process due to the GSH depletion.

Different schemes concerning the role of IEGs in neuronal apoptosis could be possible. First, IEGs induction is only a response to the initial insult. The *c-jun* protooncogene could be considered as a sensitive sensor of thiol variations in neuronal cells since its induction takes place while a slight decrease in GSH concentration is evidenced. Other investigators have reported that the redox status of non-neuronal cells may be important in the regulation of *c-jun* and *c-fos* expression. In particular, the induction of AP-1 activity by chemical agents is prevented by thiol compounds and is enhanced by lowering the cellular GSH levels.[178] Pretreatment of rat pleural mesothelial cells with BSO increases the magnitude of induction of *c-fos* and *c-jun* mRNA by asbestos fibers.[177] Second, IEGs induction was also

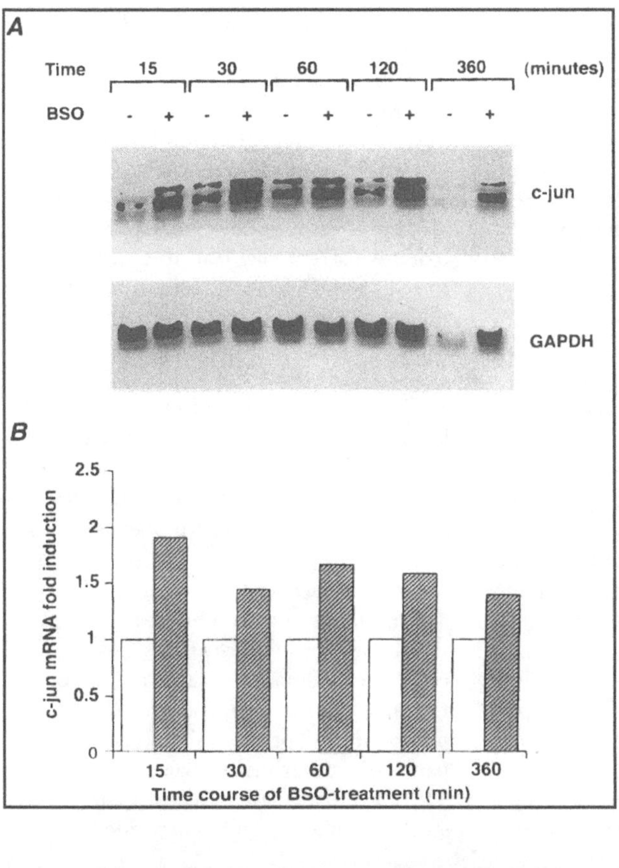

Fig. 3.4. Expression of *c-jun* protooncogene in BSO-treated NS20Y. NS20Y cells were cultivated in medium without (-) or with (+) L-buthionine-(S,R)-sulfoximine (BSO). (A) Total RNA was extracted at various times (15-360 min) and examined for gene expression by Northern blot analysis. The same blot has been serially hybridized to *c-jun* and a GAPDH probe. (B) The hybridization signals were quantified by scanning the aotoradiograms. The *c-jun* values were normalized to GAPDH values obtained on the same blot.

associated with an apoptotic process in neurons after hypoxia-ischemia or status epilepticus[179] and in primary hippocampal cultures treated with amyloid peptide.[171] This last study clearly demonstrated that Jun proteins are not induced in cells that are resistant to amyloid peptide-induced cell death and that expression of these proteins is restricted to cells that undergo Aβ-mediated apoptosis.

This last scheme implies the existence of a direct relationship between IEGs induction and apoptosis. We have shown that GSH depletion in a murine neuroblastoma cell line led to cellular death via an apoptotic pathway and that GSH may act as the mediator of this process. In this model, the rapid inducibility of *c-jun* might play a role in the progression of the apoptotic process. Recently, different studies have evidenced the role of c-fos and *c-jun* in the programmed cell death both in non-neuronal[171] and neuronal cells.[167,172] Furthermore, functional inhibition of *c-fos or c-jun* inhibits in vitro cell death. C-jun and at least one of Fos family member are essential for neuronal death in response to NGF-deprivation in vitro.[177] Overexpression of *c-jun* alone can trigger cell death in neurons and that expression of a *c-jun* dominant negative mutation protects sympathetic neurons from cell death induced by NGF deprivation.[177] In primary rat hippocampal neurons, the inhibition of *jun B* expression with antisense oligonucleotide results in a reduction in neuronal survival. By contrast, hippocampal neurons treated with anti-*c-jun* oligonucleotide display strongly increased survival compared to control neurons.[180] All these data strongly support the role of these genes in activating the genetic program responsible for cell death in neuronal cells. Nevertheless, the occurence of apoptosis in animals lacking functional genes for *c-fos* and/or *c-jun* clearly demonstrated that these genes are not essential for apoptosis to take place normally in certain tissues in vivo [181,182] suggesting that the control of apoptosis could be achieved by the heterodimerization of other Jun and Fos family members.

Consequently, it is tempting to speculate that the IEGs induction evidenced in the models of neuronal death induced by GSH depletion or amyloid peptide, might play a major role in this apoptotic process. Further experiment of functional inhibition of c-*jun* or *c-fos* genes should emphazise the involvement of these protooncogenes in the neuronal apoptotic process. Moreover, the identification of the target genes (transactivated by AP-1) leading to neuronal death should contribute to the further understanding of the molecular mechanisms underlying apoptosis in neurodegenerative disorders.

ALTERATIONS OF CELL SIGNALING PROCESSES BY OXIDANTS: IMPLICATION IN NEURONAL DEATH

ROS metabolites affect binding of ligands to membrane receptors and also coupling of receptors to G-proteins and effectors enzymes. Peroxidation of membranes lipids may lead to a lowered receptor density and also will alter the viscosity of the plasma membrane, which affects receptor coupling.[157] ROS may also interact with thiol/disulfide moieties on receptor proteins or on other factors in the receptors system, which is responsible for alterations in receptor binding or coupling. Moreover, lipid peroxidation is associated with the phospholipase A_2 pathway, which might indirectly affect receptor function. Moreover, oxidative stress may lead to a disturbance in cellular Ca^{2+} homeostasis. This might be related to an effect on Ca^{2+}-mobilizing receptors, but there is also evidence for a decreased Ca^{2+}-sequestration by ATPases. In addition, peroxidation of membrane lipids increases membrane permeability to Ca^{2+}. ROS interfere with action of NO, thus affecting another messenger system.[183] In fact, there is now a growing body of evidence implicating ROS, sulfhydryl oxidants, redox metals and the reactive nitrogen species, NO, as modulators of protein tyrosine kinases (PTK) and protein tyrosine phosphatases (PTP) activities. ROS play a major role as modulators/mediators of tyrosine phophorylation-dependent signal transduction pathways operating in eukaryotic cells.[166]

The neuroprotective and neurodestructive effect of NO is dependent on redox-based mechanisms.[74] Owing to its redox and additive chemistry, NO is able to stimulate signaling through protein kinases as well as through its reactions with ROS and redox metals. Under physiological conditions, NO reacts with O_2, superoxide anions $O_2^{\circ -}$, and transition metals generating peroxynitrite and metal-NO adducts. Any of these products is highly reactive towards sulfhydryl groups yielding S-nitrosothiols.[128] Thus, the major intracellular targets sites for NO are the sulfhydryl- and metal-containing proteins: protein kinases, phosphatases and transcription factors that contain either redox metals or sulfhydryls located at either allosteric or active sites.[128] Nitration of tyrosine residues inhibits the activity of glutamine synthetase,[184] and it disables the ability of tyrosine kinase to phosphorylate the tyrosine[185] which is a critical event in cell-signaling pathways and in cell cycle control.

H_2O_2 may function in triggering apoptosis. Responses of some types of cells to growth factor, which include tyrosine phosphorylation, mitogen-activated protein kinase (MAPK) stimulation, DNA synthesis and chemotaxis, were inhibited when the growth factor-stimulated rise in H_2O_2 concentration was blocked.[164] ROS have a part in the control of cell cycle progression as a consequence of growth signal transduction by activating mitogen-activated protein kinase (MAPK)[165] and in turn by inducing the cell cycle regulator WAF1/CIP1.[186] The MAPK family is comprized of key regulatory proteins that control the cellular response to both proliferation and stress signals. Their stimulation results in phosphorylation and activation of the SRF:SRE p62 complex leading to gene expression.[187] In accordance with these observations, H_2O_2 increased the mRNA expression of MAPK-dependent genes *c-jun*, *c-fos* and MAPK phosphatase-1.[168]

In Alzheimer's disease, the phosphorylation events critical for the conversion of Tau to paired helical filaments constituted by the highly phosphorylated forms of Tau protein could be due to activation of kinases in the signal transduction pathways mediated by ROS and/or by Aβ peptide.[188] Pkc and F-actin are essential for stimulation of neuronal FAK-tyrosine phosphorylation by G proteins and amyloid peptide.[189] The activity of Tau is regulated by phosphorylation, and tau can be phosphorylated at up to 14 different sites by a plethora of kinases, including protein kinase C, casein kinase, protein kinase A, tau kinase, calcium/calmodulin-dependent protein kinase, p^{cdc34} and MAP kinase. MAP kinase, p^{cdc34} and casein kinase are capable of converting Tau to paired helical filaments.[190] Potent activation of extracellular signal-regulated protein kinase (ERK_2) occurred within a few minutes of H_2O_2 treatment, whereupon rapid inactivation ensued. This rapid and transient nature of ERK_2 activation by H_2O_2 highlights the reversible and direct nature of alterations stimulated by H_2O_2 and underscores a role for phosphatases in regulating this response. This H_2O_2-stimulated activation of ERK_2 was abolished in PC12 cells by inducible and constitutive expression of the dominant negative Ras-N-17 allele.[165] Interestingly, PC12/Ras-N-17 cells were more sensitive than wild-type PC12 cells to H_2O_2 toxicity. Moreover, NIH 3T3 cells expressing constitutively active MAPK kinase (MEK, the immediate upstream regulator of ERK) were more resistant to H_2O_2 toxicity, while those expressing kinase-defective MEK were more sensitive than cells expressing wild-type MEK. Taken together, these studies provide insight into mechanisms of MAPK regulation and suggest that ERK plays a critical role in cell survival following oxidant injury.

These results suggest that ROS and particularly H_2O_2 and NO may act as signal transducing molecules. Furthermore by analysis of genes whose expression is induced by ROS through the mRNA differential display technique, it was found that one of these genes encodes a *ras* regulator[187] indicating that ROS could interplay with the *ras* pathway at the level of the GTP/GDP exchange, or on the kinases acting downstream. Thus ROS could play a role as physiological mediators of gene regulation and membrane signal transduction. The functions of numerous cellular proteins, including transcription factors, calcium regulatory proteins, and other cell and organelle surface molecules, is subject to redox regulation. Oxidation-reduction mechanisms are in fact a likely physiological means for reversible regulation of protein function and provide a likely target through which exogenous oxidants can usurp normal signal transduction pathways. These studies provide strong support of a crucial role for ROS in controlling the MAPK pathway to regulate cellular protection and proliferation in response to oxidative stress.

ROLE OF APOPTOSIS IN DISEASE-RELATED NEURAL CELL DEATH

There is no irrefutable evidence that apoptosis plays an important role in neurological diseases. However, circumstantial evidence favoring such a role is growing rapidly. For example, recently it was shown that β-amyloid peptide induces apoptosis in neurons in primary culture.[191] A peptide fragment of a prion protein, including amino acids 106 to 126, also was shown to induce apoptosis.[192] The oncogene *bcl-2* inhibits the death of neural cell lines induced with multiple agents. Conversely, expression of the proapoptotic gene $p75^{NTR}$ enhances the sensitivity of cells to β-amyloid peptide toxicity.[193] Cellular signaling pathways that are associated with apoptosis in response to amyloid peptide provide support for apoptosis as a mechanism of cell death for other neurodegeneration-inducing stimuli (e.g., oxidative injury).[194] The expression of *sgp-2*, a gene expressed at increased levels during apoptosis in prostatic cells, lymphocytes, and other cell types, is increased in the brains of patients with Alzheimer's disease.[195] Approximately 10% of patients with amyotrophic lateral sclerosis (ALS) have a familial form of the disease. Pathologically, FALS is indistinguishable from sporadic ALS. In 1993, Rosen and colleagues[196] reported that a subset of patients with FALS have mutations in *sod1*, the gene that encodes CuZnSOD. Approximately 30 different mutations in *sod1* have now been associated with FALS, and about 25% of the FALS patients demonstrate such mutations. The mutations cluster at turns of the β-barrels of the protein, and at the dimerization interface.[197] Initially, it was believed that FALS in these patients resulted from insufficient activity of CuZnSOD, since most of the mutants show at least some reduction in enzymatic activity.[198] However, subsequent studies in transgenic mice that have normal or supranormal SOD activity, yet develop motor neuron degeneration in association with the expression of mutant CuZnSODs, suggested that the FALS-associated mutations act as dominant, gain-of-function mutations.[199] Supporting this conclusion is the finding that the wild-type CuZnSOD acts as a neuronal antiapoptotic gene in cultured neural cells,[17,66] whereas the FALS-associated mutants act as dominant proapoptotic genes, despite the retention of significant SOD activity by some of the mutants.[66] Moreover, spinal muscular atrophy has been associated with mutations in the *naip* (neuronal apoptosis inhibitor protein) gene.[200] The *naip* gene product is likely to be a transcription factor, based on its structure. The genes controling *naip* that result in apoptosis inhibition are as yet undetermined. However, the finding of mutations in *naip* in patients with spinal muscular atrophy suggests that the antiapoptotic activity of the Naip protein plays an important role in the survival of motor neurons.

The connection of multiple stimuli that induce neuronal death to an apoptotic mechanism suggests that apoptosis could play a central role in neurodegeneration in the brain.

OXIDATIVE STRESS, TRANSCRIPTION FACTORS AND REDOX REGULATION OF GENE TRANSCRIPTION: ROLE IN NEURONAL DEATH

A delicate balance between intracellular oxidants and antioxidants likely influences neuronal degeneration, death or health.

ROS may be cytotoxic when produced in excess; however, moderate concentrations of intracellular ROS influence gene expression as well as posttranslational modification of proteins. To develop a better understanding of the exact mechanisms that underlie ROS-dependent disorders in biological systems, recent studies have investigated the regulation of gene expression by oxidants, antioxidants, and other determinants of the intracellular reduction-oxidation (redox) state.[201]

At least three well-defined transcription factors, nuclear factor (NF)-κB, activator protein AP-1, and p53 are regulated by the intracellular redox state.[167] NF-κB and AP-1 are implicated in the inducible expression of a wide variety of genes involved in oxidative stress and cellular response mechanisms (Fig. 3.5).

The regulation of these transcriptional activators occurs mainly at the postranslational level by redox molecules and a redox modulation of target genes could ensue. Many basic events of cell regulation such as protein phosphorylation and binding of transcription factors to consensus sites of DNA are driven by physiological oxidant-antioxidant homeostasis, especially by the thiol-disulfide balance. Binding sites of these redox-regulated transcription factors are located in the promoter region of a large variety of genes that are directly involved in pathogenesis of the diseases (AIDS, cancer, atherosclerosis) and more recently in neuronal apoptosis and Alzheimer's disease. The efficacy of different antioxidants to favorably influence the molecular mechanisms implicated in neuronal death should be a critical determinant of its selection for clinical studies or for prevention of neurodegeneration.

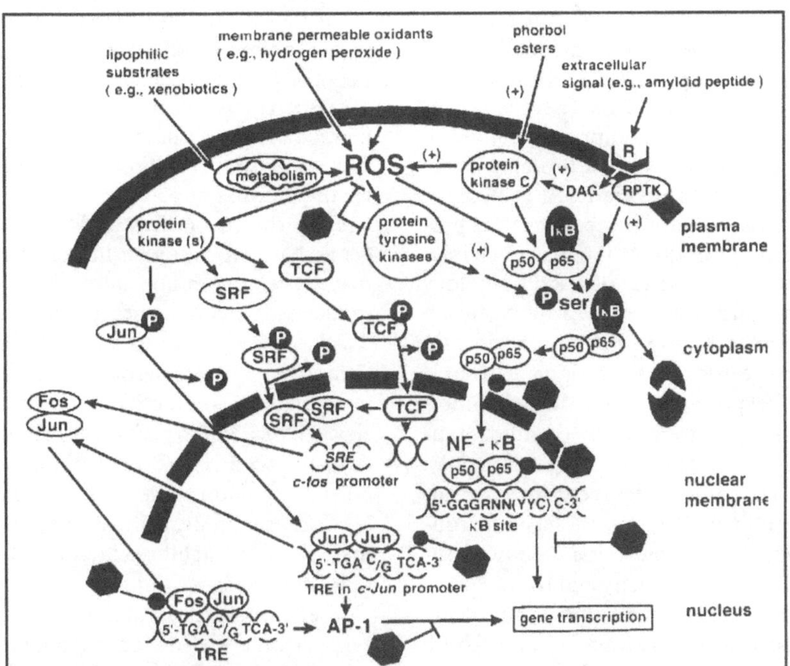

Fig. 3.5. Hypothetical scheme illustrating the steps in cellular NF-κB and AP-1 activation that may be influenced by oxidants and antioxidants. Filled hexagons indicate the antioxidant-regulated steps. DAG, diacyl glycerol; RPTK, receptor-associated protein tyrosine kinases; ROS, reactive oxygen species; TCF, ternary complex factor; SRF, serum response factor; SRE, serum response element; TRE, tetra decanoylphorbolacetate (TPA) response element/ Adapted with permission from Sen and Packer, FASEB J 1996; 10:709.[201]

NF-κB: An Oxidative Stress-Responsive Transcription Factor Implicated in Neuronal Death

The eukaryotic transcription factor NF-κB is a member of the Rel family proteins that are present in the cytosol in an inactive form.This transcription factor increases the expression of genes involved in viral activation, such as human immunodeficiency virus type 1, and those involved in adhesion and inflammatory responses[202] and is regulated through redox mechanisms.[203,204] Target genes that are regulated by NF-κB activity include those that affect acute inflammation responses, cytokine production, and nitric oxide synthase expression and other genes wich contain a κB *cis*-regulatory element in their enhancer regions, such as MnSOD and CuZnSOD.[205] Presently NF-κB is regarded as a central regulator of inflammatory, immune and acute phase reactions.[206] The Rel protein family comprizes two groups classified on the basis of their structure, function, and mode of synthesis. Group I includes p50 and p52, which are synthesized as precursor proteins of 105 (p105) and 100 (p100) kDa, respectively. Group II comprizes p65, Rel or c-RelB, and the *Drosophila* proteins dorsal and Dif. Rel proteins undergo proteolytic processing to be transformed to mature proteins. Group I mature proteins are characterized by the presence of a Rel homology domain, which is about 300 amino acids in length, and includes a DNA binding domain and a nuclear localization signal. Group II mature Rel proteins possess the Rel homology domain as well as one or more transcriptional activator domain (or domains). Rel/NF-κB family proteins can form homo- or heterodimers. Dimerization is functionally important because it is required for DNA binding of the activated factor.[206]

In the dormant form, two types of Rel protein complexes are found in the cytosol: (1) a Rel homo- or heterodimer (e.g., p50-p65) bound to a member of the IκB family of inhibitor proteins; and (2) a complex between a mature Rel group II protein (e.g., p65) and a group I precursor protein (e.g., p105). The IκB protein family is characterized by the presence of: (1) a partially conserved domain that docks between five to seven ankyrin repeat motifs 33 amino acid long that allow these proteins to interact with the Rel proteins; and (2) a carboxy-terminal protein destabilizing sequence rich in the amino acids proline, glutamate/aspartate, and serine/threonine, also known as the PEST sequence. The presence of this PEST sequence facilitates proteolytic degradation of IκB proteins.[202,203]

Cytosolic NF-κB activation can be brought about by a variety of pathogenic or pathogen-elicited stimuli including amyloid peptides,[31,207] cytokines, mitogens, bacteria and related products, virus and viral products, physical stress such as UV and ionizing radiation,[205] oxidants such as H_2O_2,[206] and a variety of chemical agents such as phorbol esters, certain phosphatase inhibitors, etc.[198,208] These stimuli trigger reactions that lead to the degradation of the bound IκB protein. Phosphorylation of IκB is a prerequisite for the activation of NF-κB. Loss of the inhibitory IκB protein bound to the Rel protein complex p50-p65 is followed by a rapid translocation of the p50-p65 heterodimer to the nucleus. Under favorable conditions, transcription is switched on after the binding of p50-p65 to *cis*-acting κB sites in the promoters and enhancers of genes. Numerous studies indicate that ROS may serve as common intracellular agents that contribute to the process of NF-κB activation in response to a diverse range of stimuli. Hydrogen peroxide is a sensitive inducer of NF-κB activation. A sustained elevated $[Ca^{2+}]_i$ also play a crucial role in oxidant-induced NF-κB activation.[207,208] The NF-κB regulatory effect of α-lipoate and NAC appear to be, at least in part, due to their ability to stabilize elevation of $[Ca^{2+}]_i$ following oxidant challenge.[208]

Cells overexpressing antioxidant enzymes with a well-defined activity for ROS removal are a valuable tool to test the specific form of ROS that are involved in triggering the activation of NF-κB. Cells overexpressing catalase were unable to activate NF-κB in response to TNF-α or okadaic acid. The catalase inhibitor aminotriazole restored the NF-κB response. Overexpression of cytosolic superoxide dismutase, which causes cytosolic hydrogen peroxide accumulation, potentiated such NF-κB responses. These results from different cells using a different overexpression systems support the hypothesis that hydrogen peroxide and/or its derivatives in certain cell types is involved in the activation of NF-κB.[209] Not only ROS themselves but biological derivatives such as oxidatively damaged DNA and advanced glycation end products (AGE) have also been identified as inducers of NF-κB activation.[207,210,213]

Intracellular thiol redox status appears to be a critical determinant of NF-κB activation, an intermediate optimal level of intracellular GSSG is required for effective NF-κB activation. Another physiologically relevant thiol that plays a crucial role in the regulation of NF-κB function is reduced thioredoxin, an important cellular protein oxidoreductase with antioxidant functions.[214] Other reducing thiol agents are also known to influence NF-κB activation. For example, pyrrolidinedithiocarbamate (PDTC) inhibits the induced degradation of IκB-α[212,213] and N-acetylcysteine, a progluta-thione antioxidant, inhibits activation of NF-κB induced by UV light.[215]

The binding of activated NF-κB to its cognate DNA site is redox-sensitive and a cysteine residue in a short amino acid stretch of Rel proteins regulates NF-κB DNA binding. Reduced thiols, including dithiothreitol, cysteine, dihydrolipoate, and reduced thioredoxin, enhance the DNA binding of activated NF-κB.[198,216] Recently the DNA binding activity of NF-κB was found to be enhanced by dihydrolipoate and inhibited by α-lipoic acid. Inhibition of NF-kB DNA binding induced by a nonreducing environment or by exposure to the thiol-oxiding agent diamide, was reversed by dihydrolipoate.[217] The highly conserved Rel homology domain is responsible for DNA binding of NF-κB. A short stretch of amino acids (the RXXRXRXXC motif, R = arginine, C = cysteine, X = other amino acid) at the beginning of the domain is essential to contact DNA.[218,219] The cysteine residue in the motif is critical and must be maintained in a reduced state to allow DNA binding because oxidation of this cysteine residue interferes with DNA binding of NF-κB. Perhaps reduced thiols enhance DNA binding of NF-κB by maintaining this critical cysteine residue in a reduced state.[221] Another cellular redox mechanism that may independently or in synergism with the thioredoxin system help to enhance NF-κB DNA binding is the apurinin/apyrimidinic endonuclease system or APEX nuclease,[219] also known as redox factor-1 or Ref-1.[222] The nuclease activity of Ref-1 is specific for the initial repair of DNA templates damaged by various noxious stimuli: ROS, UV light, and ionizing radiation nucleases.

NFκ-B Activation in Neuronal Pathologies

Recently, it has been reported that rat and mouse neurons have constitutive NF-κB activity.[223–225] Moreover, NF-κB activity was found to change dramatically during development of the cerebellum.[226] Enhancement of immunoreactivity for NF-κB in the the hippocampal formation and cerebral cortex of AD brains was also reported.[227] The distribution of dystrophic neurites and neurofibrillary tangles stained for NF-κB is similar to that of the general neuropathological damages. Advanced glycation end products (AGEs) on neuronal paired helical filaments (PHF-tau) in AD brain may generate ROS, thereby activating transcription via NF-κB, increasing amyloid β-protein precursor and release of amyloid peptide.[228] β-Amyloid peptides also activate NF-κB.[31,207] The different cytotoxic effects of Aβ, i.e., induction of oxidant stress, activation of NF-κB, cytotoxicity and microglial activation occur after the interaction of Aβ with a cell surface receptor RAGE (receptor for glycated end product).[207] Microglia express scavenger receptors (SR) that mediate adhesion of microglia to β-amyloid fibrils leading to secretion of ROS.[229] Further support for the relationship between RAGE and ROS is the presence of two functional NF-κB sites in the RAGE promoter whose activity can be triggered by oxidants. These data suggest a possible feedback loop whereby binding of Aβ to RAGE induces oxidant stress, which further upregulates RAGE, thereby augmenting further Aβ-RAGE interaction.

Tumor necrosis factor α and β (TNF α and β) protects hippocampal neurons against Aβ toxicity by suppressing accumulation of ROS and Ca^{2+}, and that κB-dependent transcription is sufficient to mediate these effects.[230] NF-κB is also one of the most ubiquitously conserved components of TNF signaling. In addition the genes for TNF-α and TNF-β are themselves activated by NF-κB. Therefore, it is possible that tissues damaged by an oxidative attack from Aβ respond with an increase in the levels of cytokine that offer protection against the insult. Alternatively, because NF-κB modulates the expression of a variety of immune and inflammatory molecules such as cytokines and complement proteins, the expression of many of those molecules might possibly be induced by activation of NF-κB. Moreover, evidence is accumulating that these inflammatory components play a significant role in exacerbating neuronal damage in AD. As in the monocytes/macrophages activated during an inflammatory process, activation of NF-κB in microglial cells is probably responsible for the secretion of inflammatory cytokines (TNF-α, IL-6) induced by Aβ. Recently, it has been demonstrated that one of the anti-inflammatory mechanisms of salicylates is the inactivation of NF-κB by preventing the degradation of IκB.[231,232] At concentrations compatible with amounts in plasma during chronic anti-inflammatory therapy, acetylsalicylic acid and its metabolite sodium salicylate were found to be protective against neurotoxicity elicited by glutamate in rat primary neuronal cultures and hippocampal slices.[210] The site of action of the drugs appeared to be downstream of glutamate receptors and to involve specific inhibition of glutamate-mediated induction of NF-κB,[210] suggesting a link between neuroprotection and the nuclear event. NO has similar effects for the NF-κB/IκB system; the induction and stabilization of IκBα by nitric oxide mediate inhibition of NF-κB activity in human vascular endothelial cells.[231] NO can possibly be used as a NF-κB inhibitor in neurons since it has been shown that neurons containing nitric oxide synthase resist neuronal cell loss in advancing AD.[233–235]

These results suggest a major role of NF-κB activation in glutamate and amyloid-induced neurodegeneration and also in the progression of Alzheimer's disease. Effective strategies to regulate the activity of NF-κB or to block RAGE and SR receptors should be of preventive or therapeutic importance.

Induction of Activator Protein AP-1 by Reactive Oxygen Species

Interaction between cJun and cFos proteins, products of c-*jun* and c-*fos* protooncogenes, form Fos-Jun or Jun-Jun complexes, or AP-1. Via a "leucine-zipper" domain AP-1 has been observed to have crucial regulatory implications on the expression of a wide variety of genes, especially growth factor-inducible genes.[167,236] cJun allows cFos to regulate gene expression by serving as an anchor that allows the Fos-Jun heterodimer to bind to a cognate DNA site. In addition, cJun contains three short regions in its amino-terminal half that are important for transcriptional activation in vivo. Dimerization of Fos-Jun or Jun-Jun is a prerequisite for DNA binding; however, such interaction is not limited only to cFos and cJun. Other members of the Fos and Jun family of proteins are also known to interact with each other. Members of the Fos family identified so far include vFos, cFos, FosB, Fra-1 (Fra, Fos-related antigen), and possibly Fra-2. At present, the Jun family is comprized of vJun, cJun, JunB, and JunD. All the Jun family proteins are capable of forming homo- and heterodimers, which are capable of binding to a cognate DNA site. Fos proteins do not associate with each other but are capable of associating with any member of the Jun family to form stable dimers that have higher DNA binding activity than Jun dimers. Despite the high degree of sequence conservation of Fos and Jun proteins, certain Fos-Jun combinations appear to be more active in DNA binding than others.[167,236]

DNA binding of Fos-Jun or Jun-Jun complexes, or AP-1, is governed by a "basic region" found immediately upstream of the "leucine-zipper" domain. Binding of the dimers to a 12-*0*-tetradecanoylphorbol 13-acetate (TPA) response element (TRE) of several cellular and viral genes is a prerequisite for AP-1-induced transcriptional regulation. TPA is a remarkable activator of protein kinase C and other agents that lead to protein kinase C activation such as serum and growth hormones. cFos proteins and mRNA have shorter half-lives than cJun proteins and mRNA. As a result, before induction most of the AP-1 complex is present in the form of Jun homo-and heterodimers. After induction, Fos proteins are expressed and most of the induced AP-1 are Fos-Jun heterodimers. The differential regulation of the *c-fos* and c-*jun* genes supports the contention that the function of cFos induction is to generate an initial spike of transcriptional activation, whereas cJun is responsible for a steady sustained AP-1 activity. The *cis*-acting element that mediates cFos expression in response to activation by serum, certain growth factors, UV irradation, and TPA is a 20 bp dyad symmetry or serum response element (SRE) located in the c-*fos* promoter. A 67 kDa transcription factor serum response factor (SRF) is phosphorylated and binds to the SRE as a homodimer. This DNA binding of SRF is enhanced by the physical interaction of a phosphorylated 62 kDa ternary complex factor (TCF) protein that binds to a second DNA binding site adjacent to SRE. cJun expression is driven by the binding of preexisting AP-1 proteins to a TRE in the c-*jun* promoter.

Evidence suggesting the direct involvement of ROS in AP-1 activation has been obtained mostly by using defined oxidative stress generating systems to challenge cultured cells. Studies from a number of laboratories have demonstrated that superoxide produced by a xanthine/xanthine oxidase system and hydrogen peroxide induces the expression of several early response genes including c-*fos* and c-*jun*, the major components of AP-1.[237,238] Mitogen-induced AP-1 activation has been observed to be antioxidant inhibitable and has thus been suggested to be ROS mediated.[239,240] Oxidative stress caused by ionizing radiation and hydrogen peroxide challenge was a potent inducer of c-*jun* expression. This effect was observed to be protein kinase C dependent.[240]

Not much is known about the exact mechanisms underlying ROS-dependent AP-1 activation. Perturbation of cellular thiol redox status has been suggested to be a signal that may be implicated in the induction of c-*fos* and c-*jun* expression. In support of this, high intracellular GSSG has been shown to be implicated in AP-1 activation.[241] It has been also suggested that AP-1 activation under oxidative conditions may be, at least in part, mediated by phosphorylation of Jun proteins.[168]

Information available regarding the pathway (or pathways) involved in ROS-induced AP-1 activation suggests that SRF binding to SRE, and TCF phosphorylation, are involved in such processes.[201] Both SRE and AP-1 sites in the c-*fos* promoter have been shown to be implicated in upregulation cFos expression in response to hydrogen peroxide challenge (Fig. 3.5). Consistent with the notion of the presence of a ROS sensitive kinase cascade (or cascades), oxidative stress-sensitive mitogen-activated protein (MAP) kinase and MAP kinase phosphatase have been evident.[166]

A role for c-*fos* and c-*jun* in the specific activation of gene expression, mediated by a complex site called the electrophile response element (EpRE) found in the mouse glutathione S-transferase Ya-subunit gene, has been suggested.[242] This EpRE includes a smaller site called the antioxidant response element (ARE), which was originally identified in the rat homolog of the same gene and in the NADP(H) quinone reductase gene. The ARE consensus sequence contains a part that is similar to the AP-1 consensus. Moreover, EpRE and ARE are known to be bound by cFos-cJun and c-Jun, respectively, indicating that AP-1 may be implicated in the electrophile and antioxidant responses.

AP-1 function is responsive to thiol antioxidants. N-acetylcysteine inhibited a tyrosine-kinase-dependent induction of cJun expression. Regulation of Fos and Jun expression via interaction of oxidants and antioxidants with specific cysteine groups of these polypeptides had been evident previously.[168,243] One, or most, proteins that might interact with SRE during AP-1 activation may be sensitive to intracellular changes in thiol redox milieu caused by antioxidants. It has been consistently found that AP-1 transactivation is influenced by intracellular thioredoxin and glutathione status.[214,243-244] However, some results available from different experimental systems indicate that the mode of AP-1 regulation by reduced thiols is opposite from the regulation of NF-κB. These studies also showed that other reducing thiol agents (PDTC and NAC) enhance DNA binding activity transactivation of AP-1. Since AP-1 activation by antioxidants is mediated by induction of *c-jun/c-fos* expression, AP-1 has been classified as a secondary antioxidant responsive factor.[167]

Fos and Jun DNA binding in vitro is regulated by the reduction-oxidation of a single conserved cysteine residue (Lys-Cys-Arg) in the DNA binding domains of the two proteins. It was suggested that conversion of the cysteine to reversible oxidation products such as sulfenic (RSOH) or sulfinic (RSO_2H) acids could contribute to the regulation of DNA binding.[59] Ref-1, the protein product of the *ref-1* gene, and other chemical reducing agents stimulate AP-1 DNA binding in vitro by acting on the regulatory cysteine residue of Lys-Cys-Arg. Ref-1 (for redox factor) is a nuclear factor initially identified to activate AP-1 DNA binding. Ref-1 was found to be a previously characterized DNA repair enzyme. Replacement of the critical cysteine residue of a truncated Fos protein by serine resulted in a 3-fold increase in AP-1 DNA binding activity that was no longer redox regulated. Oxidized thioredoxin and GSSG inhibit AP-1 DNA binding in vitro, the effect being most pronounced with the former.

AP-1 Activation in Neurological Disorders

Apart from being implicated in tumor promotion, AP-1 activation is involved in disorders of the nervous system. Most neuronal tissues express basal levels of AP-1, which increase dramatically in response to certain stimuli. Electric stimulation or drug-induced seizure induces AP-1 activation in the hippocampus, which in turn enhances growth factor mRNA expression. Stimulation or blocking of dopamine receptors is another signal that activates AP-1 DNA binding activity, depending on which type of dopamine receptor is expressed in the responding cell.

NO activates transcription from AP-1-responsive promoters. NO has been implicated in the regulation of many biological processes, including smooth muscle relaxation, neurotransmission, immune regulation and cellular differenciation.[245] NO produces its effects through a variety of mechanisms including nitrosylation of (or redox reaction with) metal-and thiol-containing proteins.[246] Although the number of newly discovered potential targets of NO continues to increase, the effect of NO in many systems appears to be mediated principally through the stimulation of cytosolic guanylate cyclase activity and thereby activates the cGMP signal transduction pathway. Little is known about transcriptional regulation of gene expression by NO/cGMP. NO releasing agents and cGMP analogs change expression of differentiation-associated genes and induce cellular differentiation and activate TRE-regulated but not CRE-regulated reporter genes. Activation of TRE-regulated genes by NO-releasing agents and cGMP analogs appeared to be mediated by the AP-1 (Jun/Fos) transcription factor complex because increased DNA binding of AP-1 and increased JunB and c-Fos mRNA was observed in cells treated with these agents.[247,248] Strategies to pharmacologically manipulate AP-1 transcriptional activity are expected to be part of future research on neurological disorders.

Conclusions

ROS are central mediators of apoptosis. This changes the perception of the biological roles of these reactive intermediates. For the most part, ROS have been considered to be an unavoidable and undesirable consequence of life in the presence of molecular oxygen. ROS have been implicated as the primary destructive intermediates in a wide range of environmental toxins, as well as an increasing number of human pathological conditions. Their biological utility has until recently been limited to their generation by neutrophils for use in microbial killing. However, a signaling role for NO has been established in endothelia and in the nervous system. Modulation of signaling systems by ROS indicate that they are utilized in messenger functions. The accumulating evidence for a central role of ROS in apoptosis suggests that virtually all eukaryotic cells possess the means of active, regulated generation of such species and for selective channeling of these intermediates to produce the intricately orchestrated process of apoptosis.

The regulation of gene expression by oxidants, antioxidants and redox state has emerged as a novel subdiscipline in molecular biology that has promising therapeutic implications in neurodegenerative disorders. Binding sites of the redox-regulated transcription factors NF-κB and AP-1 are located in the promotor region of a large variety of genes that are probably involved in neuropathogenesis. Moreover, many of the basic events of cell regulation such as protein phosphorylation, critical steps in the signal transduction cascade and binding of transcription factors to consensus sites of the DNA, are driven by the physiological oxidant-antioxidant homeostasis, especially by the thiol disulfide balance. Thus, the efficacy of different compounds that modulate intracellular redox state to favorably influence the molecular mechanisms known to be implicated in the etiology of neurodegenerative diseases should be a critical determinant of their selection for prevention or therapy.

OXIDATIVE STRESS AND MITOCHONDRIA IN NEURONAL DEATH

Mitochondria and Oxidative Stress

The mitochondrial electron transport system is responsible for major ROS production (see chapter 2). Glutamate toxicity to cortical neurons depends of the intracellular content of ROS, and an uncoupler of aerobic metabolism effectively abolishes ROS production.[249] Addition of either ascorbate or catalase to the cultures primarily attenuates the loss of cell viability induced by cyanide, and decreases the incidence of apoptotic cells.[250,251] Defects in the mitochondrial electron transport chain can directly contribute to free radical production and components of the mitochondrial respiratory chain have been shown to be susceptible to damage by free radicals.[252–254] Antioxidant depletion leads to mitochondrial damage to the brain.[89,109] Mitochondria are known to be the most important physiological source of superoxide radicals in mammals. Furthermore, an increase in $[Ca^{2+}]_i$ leads to increased free radical generation, and to activation of NOS, which is dependent of calmodulin.[252,253] The interaction of NO with superoxide radical leads to the formation of peroxynitrite. This reaction is extremely fast, being 3-fold faster than the rate of dismutation od superoxide by SOD. It was reported a dramatic decrease in cytochrome oxidase activity by an induction of NO synthase or by peroxynitrite in primary astroglial cultures[110] and neurons.[111,251] Recent studies have demonstrated a link between glutamate-induced increases in intracellular $[Ca^{2+}]_i$ and mitochondrial free-radical production.[251–258] A direct role for superoxide and inactivation of aconitase in excitotoxicity is also reported.[259–260]

Oxidative Stress and Alterations of Mitochondrial Functions Are Early Events of Apoptosis

The scientific community is paying growing attention to the analysis of mitochondrial function during apoptosis. Apoptosis has been implicated as a final mediator of cell death in a variety of hypoxic and ischemic insults resulting in neurodegenerative pathology, including vessel occlusion and mitochondrial inhibition by 1-methyl-4-phenylpyridinium (MPP^+), the neurotoxic metabolite of 1-methyl-4-phenyl-1,2,3,4,6-tetrahydropyridine (MPTP).[261] Inhibition of mitochondrial respiration by MPP^+ and rotenone, both complex I inhibitors, induced apoptosis in PC12 cells. MPP^+ treatment in primates induced a Parkinson-like syndrome associated with increases in lipid peroxidation in the substantia nigra and inhibition of neuronal nitric oxide synthase prevents MPTP-induced parkinsonism in baboons[262] suggesting that ROS play a role in MPTP neurotoxicity and associated mitochondrial dysfunction. Cyanide has been reported to mediate apoptosis in neuronal cells[261] and cells were protected by transfection with the antiapoptotic/antioxidant *bcl-2*.[263] It has become increasingly evident during recent years that gene expression of the enzymatic complexes involved in the mitochondrial respiratory chain and oxidative phosphorylation process is regulated by neuronal functional activity.[264,265] It is still unresolved whether the mitochondrial or the nuclear genome is the main target of this regulation.

A number of recent data suggest that mitochondria play a complex role during the different phases of apoptotic process. Their functional integrity appears to be important during the early phases of apoptosis when a variety of energy-requiring intracellular processes occur.[266] Cells undergoing apoptosis in vivo showed increased release of cytochrome c to their cytosol, suggesting that mitochondria may function in apoptosis by releasing cytochrome c.[267] Since cytochrome c is encoded by a nuclear gene and the translocation of apocytochrome c into mitochondria does not require membrane potential and general protein translocation machinery,[268] it may be totally functional in apoptosis in cells lacking mitochondrial DNA.[266] A functional impairment of mitochondria follows, and this event likely contributes to the drop of ATP required for apoptosis to occur. It seems that alterations in the cellular energy charge play a major role in the cellular decision to die following an apoptotic or necrotic pathway.[269–272]

Moreover there is increasing evidence that defects in mitochondrial energy metabolism may be involved in the pathogenesis of neurodegenerative diseases.[253–256] Mitochondrial defects that result in a decrease in energy production lead to a number of deleterious consequences. In neurons, this decrease in energy production causes partial depolarization, which is followed by activation of excitatory amino acid receptors by ambient concentrations of glutamate.[258] Impaired mitochondrial function also impairs intracellular $[Ca^{2+}]_i$ buffering.[259]

Only a few data exist on the role and the function of mitochondria in the apoptotic process in neuronal cells. Recently we described an alteration of mitochondrial function in a neuronal model of apoptosis induced by GSH (see above). A profound decrease in complex II and complex IV activities appeared before the DNA fragmentation, suggesting that mitochondrial alteration and energy depletion could play a functional role in the progression of apoptosis. In neuronal death induced by glutamate, a succession of necrosis and apoptosis occurred depending on mitochondrial function.[2]

Oxidative stress and reduction of mitochondrial transmembrane potential seem to be early irreversible steps of early programmed cell death.[273,275] When stimulted to undergo apoptosis, cells hyperproduce superoxide anion, as determined by fluorochrome dihydroethidine.[162,257] In addition, cell membranes undergo peroxidation, as can been shown with the fluorescent dye cis-parinaric acid.[7] Both processes are inhibited by hyperexpression of *bcl-2*. Alterations in mitochondrial permeability transition linked to membrane potential dysruption precede nuclear and plasma membrane changes.[162,273–276] In vitro induction of permeability transition in isolated mitochondria provokes the release of a protein factor capable of inducing nuclear chromatin condensation and fragmentation.[276] This permeability transition is regulated by multiple endogenous effectors, including members of the *bcl-2* gene family. The capacity of drugs such as NAC to inhibit lymphocyte apoptosis in vivo correlates well with its potential to stabilize mitochondrial permeability transition.[277] NAC is a well-established thiol antioxidant, which, after uptake, deacylation, and conversion to glutathione, functions as both a redox buffer and a ROS scavenger. NAC has been reported to prevent or postpone the death of cell lines deprived from essential growth factor,[7] as well as the antigen-driven apoptosis of hybridomas.[277] Similarly to NAC, overexpression of the protooncogene product Bcl-2, a protein that is thought to protect against apoptosis via an antioxidant pathways,[7] has been shown to actively enhance mitochondrial permeability transition in transfected cells and prevents its loss induced by TNF-α in vitro.[278] Short-term exposure of cells to H_2O_2 for several minutes causes an immediate reduction in permeability transition. These data suggest that ROS may be involved in this disruption. Bcl-2 was initially thought to be located in the mitochondrial membrane, suggesting that it might primarily directly affect mitochondrial function. Recent models have proposed that Bcl-2 has antioxidant properties and inhibits PCD by suppressing the formation or effects of ROS.[7,12,279] This fits well with the location of Bcl-2 in organelles known to participate in redox reactions and the formation of ROS. High expression of *bcl-2* also correlates with enhanced ATP generation by mitochondria,[275] but in the brain of transgenic mice overexpressing *bcl-2*[13] the activities of the respiratory chain complexes were found to be normal (unpublished data). To integrate these data, it is tempting to speculate that both events, loss in mitochondrial permeability transition and accumulation/action of ROS, could result in a self-amplification loop in which disruption of the mitochondrial compartments favors the accumulation of radicals and radical-mediated damage, as well as the breakdown of respiratory metabolism, and that, together, they participate in the execution of apoptosis. Bcl-2 may thus function to protect cells against cytoplasmic oxidants involved in mediating apoptotic cell death and/or from ROS generated as normal by-products of the normal mitochondrial reactions during ATP synthesis. Cells devoid of mtDNA have been selected for their ability to grow in the presence of ethidium bromide (which inhibits mtDNA replication).[280] These cells generate ATP by glycolysis and maintain a normal mitochondrial membrane potential, probably by using the ADP/ATP translocator to import glycolytic ATP in the mitochondria.[280,281] Thus it remains that a mitochondrial function, not directly linked to respiration but to mitochondrial membrane potential, is involved in apoptosis. A crucial common step in apoptosis is the opening of the mitochondrial megachannel or permeability transition pores in their inner membranes.[115,276] Increase in the level of products of the one-electron reduction of O_2 (ROS) is known to induce the mitochondrial pores.[281] Thus the pore formation and apoptosis are involved in the organization of a defense system preventing ROS formation. It is proposed that ROS-induced pore opening lowers ROS production due to: (1) maximal stimulation of mitochondrial O_2 consumption and hence, intracellular O_2 lowering; and (2) complete dissipation of mitochondrial membrane potential and, as a consequence, maximal oxidation of respiratory chain carriers such as $CoQ^{\circ-}$ which serve as one-electron O_2 reductants. ROS decrease allows pore closure. If, nevertheless, ROS are still accumulating in a mitochondrion, long-lived pores cause degradation of the organelle which cannot import and synthesize proteins due to the absence of the membrane potential. In this way, ROS-producing mitochondria can be eliminated (mitochondrial selection). Another result of the long-lived pores is mitochondrial swelling. This disrupts the outer mitochondrial membrane and releases the apoptosis-inducing protein. Apoptosis eliminates ROS producing cells (cell selection).

ROS elevation, deregulation of Ca^{2+} homeostasis and energic changes during apoptosis appeared as key elements in the induction of the apoptotic program. A better knowledge of the mechanisms underlying the induction of apoptosis in neuropathologies in which deregulation of mitochondrial function exists, such as Parkinson's and Alzheimer's disease, can have a beneficial impact in establishing strategies of pharmacological intervention at the mitochondrial level.

REACTIVE OXYGEN SPECIES IN NEURONAL DIFFERENTIATION

The underlying nature of the mechanisms by which selected genes become repressed or derepressed at critical times in development is presently unknown. Many embryologists believe that the cellular environment interacts with the genetic control mechanisms that direct developmental events and that cellular oxidation can influence genetic control of development.[282] Oxidation gradients resulting from differential nutrient and oxygen supplies were implicated as major environmental factors that could influence the development. Free radical generation by various metabolic pathways governs some developmental changes.[283]

Nitric oxide triggers a switch to growth arrest during differentiation of neuronal cells.[284] Arrest of cell division is a prerequisite for cells to enter a program of terminal differentiation. Mitogenesis and cytostasis of neuronal cell precursors can be induced by the same or by different growth or trophic factors. Response of PC12 cells to nerve growth factor (NGF) involves the induction of different forms of NOS in neuronal cells accompanied by a proliferative phase that is followed by growth arrest and differentiation. The cytostatic effect of NGF is mediated by NO, a second messenger molecule with both para- and autocrine properties that can diffuse freely and act within a restricted volume. Inhibition of NOS leads to reversal of NGF-induced cytostasis and thereby prevents full differentiation. To test if NO can act as a anti-mitogenic agent in PC12 cells, cells were treated with different chemical NO donors. These compounds suppressed DNA synthesis in a concentration-dependent manner implying that NO is an active suppressor of DNA replication in PC12. Both DNA synthesis and cell proliferation were restored once NO sources were washed away, indicating that the decrease in DNA synthesis that is seen with NO donors is caused by a decrease in DNA synthesis and cell proliferation (cytostasis) and is not due to cytotoxicity.The results of this important study suggest a model for NGF action in PC12 cells in which at least three stages can be outlined. In the first, proliferative stage NGF activates a cascade of genes, eventually leading to the induction of the NOS gene. In the second stage, the accumulated NOS enzyme produces enough NO to inhibit DNA synthesis and probably to alert further checkpoints, thereby blocking further progression of the cell cycle, completing the proliferative phase of NGF action, and switching the cells to the cytostatic phase; NO probably also directly induces some of the later differentiation markers by promoting gene activity. NOS-dependent cytostasis could explain the action of other growth factors during terminal differentiation in which the initial mitogenic response is replaced by a cytostatic phase. It is possible that NO diffuses from the cell that produces it to promote cessation of growth in adjacent cells. If NO molecules, produced by a group of adjacent cells, act additively within a limited volume, this could contribute to the synchronization of development of precursor cells. Finally because NO is also implicated in neuronal plasticity, induction of NOS could provide a link between the activity of a cell and the development fate of it and its nearest neighbors.

There is now a growing body of evidence to suggest that superoxide and hydrogen peroxide may play a crucial role in the mechanisms underlying proliferative responses.[105,285]

One of the most intensive investigative efforts of cell and molecular biology is devoted to the elucidation of the molecular switches that regulate mammalian cell proliferation. The control of cell proliferation involves a variety of genetic and biochemical factors that still remain poorly understood. Control and regulation of cell proliferation is of extreme importance during development and for the maintenance of various organ systems. A general outline is now emerging of the mechanisms underlying the regulation of cell proliferation. Basically growth factors, cytokines or hormones interact with specific cell surface receptors that set in motion intracellular signal or second-messenger systems that, in turn, activate specific genes, including protooncogenes, and initiate a variety of key biochemical timing devices. Whereas there are considerable complexities involved, superoxide and hydrogen peroxide at low concentration can each stimulate growth and growth responses in a considerable variety of cultured mammalian cell types. Such observations led to suggestions that superoxide and hydrogen peroxide might function as mitogenic stimuli through biochemical processes common to natural growth factors. In relation to the growth promoting effects of low concentrations of hydrogen peroxide, recent studies have shown that following exposure, the early growth response 1 transcription factor gene (*egr1*) in human cells is activated and also stimulates *c-jun* as well as *c-fos* activation. In relation to *c-fos* induction it has been argued that ROS would induce DNA damage and particularly DNA breaks. Another suggestion is that hydrogen peroxide-activated c-Fos expression was mediated through PLA_2-dependent release of arachidonic acid. Studies have shown that cis-unsaturated free fatty acids can stimulate protein kinase C in several tissues. This enzyme can be activated in vitro by low concentration of oxidized glutathione, suggesting that the oxidant can modulate this important kinase through the alteration of the thiol/disulfide balance of the enzyme.

Although protein kinases and protein phosphatases play a pivotal role in intracellular signal transduction pathways, another group of molecules central to the regulation of signal induced gene responses are the nuclear transcription factors since their activity is highly dependent on their redox status. For example, the specific DNA-binding activity in vitro of the Fos/Jun heterodimer to the AP-1 promotor site is modulated by reduction-oxidation of a highly conserved cysteine residue in the DNA-binding domain of the protein. The DNA-binding in vitro of other transcription factors such as p53 and NF-κB seems also to be favored by reduction and appears particularly to involve the redox state of specific cysteine residues. In view of the ability of NAC to counteract the toxic effect of hydrogen peroxide, it would be appropriate to consider the relationship between cellular GSH levels and proliferation responses. An elevation of intracellular GSH was associated with the mitogenic stimulation of quiescent 3T3 fibroblasts.[286] Fluorescence-activated cell sorting analysis reveals that low-level oxidative stress leads to an arrest in both G1 and G2/M of the cell cycle. Cells in G1 arrest in G1, while S-phase cells complete DNA synthesis and subsequently arrest in G2/M. Both the G1 and G2/M arrest appear to be mediated by events occurring at the cell surface.[287]

A wide variety of normal and malignant cell types generate and release superoxide or hydrogen peroxide in vitro either in response to specific cytokine/growth factor stimulus or constitutively in the case of tumor cells. These species at submicromolar levels appear to act as novel intra and intercellular "messengers" capable of promoting growth responses in culture. The mechanisms may involve direct interaction with specific receptors or oxidation of growth signal transduction molecules such as protein kinases, protein phosphatases, transcription factors, or transcription factor inhibitors. It is also reported that hydrogen peroxide may modulate the redox state and activity of these important signal transduction proteins indirectely through changes in cellular levels of GSH and GSSG. Critical balance may exist in relation to cell proliferation on one hand and lipid peroxidation and cell death on the other. Progression to a more pro-oxidant state whilst initially leading to enhanced proliferative responses results subsequently in increased cell death.

REFERENCES

1. Bonfoco E, Krainc D, Ankarcrona M et al. Apoptosis and necrosis: two distinct events induced respectively by mild and intense insults with NMDA or nitric oxide/superoxide in cortical cells in cultures. Proc Natl Acad Sci USA 1995; 92:72162-72166.
2. Ankarcrona M, Dypbukt JM, Bonfonco E et al. Glutamate-induced neuronal death: a succession of necrosis or apoptosis depending on mitochondrial function. Neuron 1995; 15:961-973.
3. Oppenheim R. Cell death during development of the nervous system. Annu Rev Neurosci 1991; 14:453-501.
4. Raff M. Programmed cell death and the control of cell survival: lessons for the nervous system. Science 1993; 262: 695-700.
5. Willie AH, Kerr JF, Currie AR. Cell death: the significance of apoptosis. Int Rev Cytol 1980; 68:251-306.
6. Batistatou A, Greene L. Internucleosomal DNA cleavage and neuronal cell survival/death. J Cell Biol 1993; 122:523-532.
7 Hockenbury DM, Oltvai ZN, Yin XM et al. Bcl-2 function in an antioxidant pathway to prevent apoptosis. Cell 1993; 75:241-251.
8. Garcia I, Martinou I, Tsujimoto Y et al. Prevention of programmed cell death of sympathetic neurons by the bcl-2 protooncogene. Science 1992; 258:302-304.
9. Mah SP, Zhong L-T, Liu Y et al. The proto-oncogene *bcl-2* inhibits apoptosis in PC12 cells. J Neurochem 1993; 60:1183-86.
10. Zhong L-T, Kane D, Bredesen D. Bcl-2 blocks glutamate toxicity in neural cell lines. Mol Brain Res 1993; 19:353-55.
11. Zhong L-T, Sarafian T, Kane DJ et al. Bcl-2 inhibits death of central neural cells induced by multiple agents. Proc Natl Acad Sci USA 1993; 90:4533-37.
12. Kane DJ, Sarafian TA, Anton R et al. Bcl-2 inhibition of neural death: decreased generation of reactive oxygen species. Science 1993; 262:1274-77.
13. Martinou J-C, Dubois-Dauphin M, Staple JK et al. Overexpression of BCL-2 in transgenic mice protects neurons from naturally occurring cell death and experimental ischemia. Neuron 1994; 13: 1017-30.
14. Wood K, Dipasquale B et al. *In situ* labeling of granule cells for apoptosis-associated DNA fragmentation reveals different mechanisms of cell loss in developing cerebellum. Neuron 1993; 11: 621-632.
15. Ratan RR, Murphy TH, Baraban JM. Oxidative stress induced apoptosis in embryonic cortical neurons. J Neurochem 1994; 62:376-3379.
16. Ratan RR, Murphy TH, Baraban JM. Macromolecular synthesis inhibitors prevents oxidative stress-induced apoptosis in embryonic cortical neurons by shunting cysteine from protein synthesis to glutathione. J Neurosci 1994; 14: 4358-4392.
17. Greenlund LJS, Deckwerth TL, Johnson EM Jr. Superoxide dismutase delays neuronal apoptosis: a role for reactive oxygen species in programmed neuronal death. Neuron 1995; 14:303-315.
18. Satoh T, Sakai N, Enokido Y et al. Survival factor-insensitive generation of reactive oxygen species induced by serum deprivation in neuronal cells. Brain Res 1996; 733:9-14.
19. Hartley A, Stone JM, Heron C et al. Complex I inhibitor induce dose-dependent apoptosis in PC12 cells: relevance to Parkinson's disease. J Neurochem 1994; 63:1987-1990.
20. Whittemore ER, Loo DT, Cotman CW. Hydrogen peroxide induces apoptosis in cultured neurons. Neuroreport 1994; 5:1485-1488.
21. Myers KM, Fiskum G, Liu Y et al. Bcl-2 protects neural cells from cyanide/aglycemia-induced lipid oxidation, mitochondrial injury, and loss of viability. J Neurochem 1995; 65:2432-2440.
22. Johnson TM, Yu Zu-Yu, Ferrans VJ et al. Reactive oxygen species are downstream mediators of p53-dependent apoptosis. Proc Natl Acad Sci 1996; 93:11848-11852.

23. Hori K, Katayama M, Sato N et al. Neurotrophin by glial cells through adult T cell leukemia-derived factor/human thioredoxin. Brain Res 1994; 652: 304-310.
24. Ferrari G, Yan CYI, Greene LA. N-Acetylcysteine (D- and L-stereoisomers) prevents apoptotic death of neuronal cells. J Neurosci 1995; 15:2857-2866.
25. Jackson GR, Apffel L, Werrbach-Perez K, Perez-Polo JR. Role of nerve growth factor in oxidant-antioxidant balance and neuronal injury. Stimulation of hydrogen peroxide resisitance. J Neurosci Res 1990; 25:360-368.
26. Jordan J, Ghadge GD, Phren JHM et al. Expression of human copper-zinc superoxide dismutase inhibits the death of rat sympathetic neurons caused by withdrawal of nerve growth factor. Mol Pharmacol 1995; 47:1095-1100.
27. Busciglio J, Yankner BA. Apoptosis and increased generation of reactive oxygen species in Down's syndrome neurons in vitro. Nature 1995; 378:776-779.
28. Ceballos I, Agid F, Delacourte A et al. Parkinson's disease and Alzheimer's disease: neurodegenerative disorders due to brain antioxidant system deficiency? In: I Emerit, L Packer, C Auclair, eds. Antioxidants in Therapy and Preventive medecine. Adv Exp Biol Med. Plenum Press 1990:493-498.
29. Ceballos I, Agid F, Lafon M et al. Superoxide dismutase and Parkinson's disease. Lancet 1990; 335:1035 -1036.
30. Fahn S, Cohen G. The oxidant stress hypothesis in Parkinson's disease: evidence supporting it. Ann Neurol 1992; 32: 804-812.
31. Behl C, Davis JB, Lesley R, Schubert D. Hydrogen peroxide mediate amyloid b protein toxicity. Cell 1994; 77:817-827.
32. Sinet PM, Ceballos-Picot I. Role of free radicals in Alzheimer's disease and Down syndrome. In: L Packer, A Philippo, Y Christen, eds. Free Radicals in the brain; WHO-Fondation IPSEN. Springer Verlag, 1992:91-98.
33. Bowling AC, Schultz JB, Brown Jr, RH et al. Superoxide dismutase activity, oxidative damage, and mitochondrial energy metabolism in familial and sporadic amyotrophic lateral sclerosis. J Neurochem 1993; 61:2322-2325.
34. Rosen DR, Siddique T, Patterson D et al. Mutations in CuZn superoxide dismutase gene are associated with familial amyotrophic lateral sclerosis. Nature 1993; 362:59-62.
35. Dugan LL, Sensi SL, Canzoniero LMT et al. Mitochondrial production of reactive oxygen species in cortical neurons following exposure to N-methyl-D-aspartate. J Neurosci 1995; 15:6377-6388.
36. Reynolds IJ, Hastings TG. Glutamate induces the production of reactive oxygen species in cultured forebrain neurons following NMDA receptors action. J Neurosci 1995; 15:3318-3327.
37. Coyle JT, Puttfarken P. Oxidative stress, glutamate and neurodegenerative disorders. Science 1993; 262:804-812.
38. Choi DW. Bench to beside: the glutamate connection. Science 1992; 258:241-243.
39. Lipton SA, Rosenberg PA. Excitatory amino acids as a final common pathway for neurologic disorders. New England J Med 1994; 330:613-622.
40. Richter C. Prooxidant aand mitochondrial Ca^{2+}; their relationship to apoptosis and oncogenesis. FEBS Lett 1993; 325: 104-107.
41. Dawson VL, Dawson TM, London ED et al. Nitric oxide mediates glutamate neurotoxicity in primary cortical cultures. Proc Natl Acad Sci USA 1991; 88: 6368-6371.
42. Dawson VL. Nitric oxide: role in neurotoxicity. Clin Exp Pharmacol Physiol 1995; 22:305-308.
43. Dykens JA. Isolated cerebral and cerebellar mitochondria produce free radicals when exposed to elevated Ca^{2+} and Na^{+}: implications for neurodegeneration. J Neurochem 1994; 63:584-591.
44. Zamzani N, Rubin SA, Marchetti P et al. Mitochondrial control of nuclear apoptosis. J Exp Med. 1996; 183: 1533-1544.
45. Liu X, Kim CN, Yang J et al. Induction of apoptotic program in cell-free extracts: requirement for ATP and cytochrome c. Cell 1996; 86:147-157.
46. Mills JC, Nelson D, Brecinska M, Pittman RN. Metabolic and energetic changes during apoptosis in neural cells. J Neurochem 1995; 65:1721-1730.

47. Mills EM, Gunasekar PG, Pavlakovic G, Isom GE. Cyanide-induced apoptosis and oxidative stress in differentiated PC12 cells. J Neurochem 1996; 67:1039-1046.
48. Haba K, Ogawa K, Mizukawa K, Mori A. Time course of changes in lipid peroxidation, pre and post synaptic cholinergic indices, NMDA receptor binding and neuronal death in the gerbil hippocampus following transient ischemia. Brain Res 1990; 40:116-122.
49. Lafon-Cazal M, Pietri S, Culcasi M et al. Nitric oxide, superoxide production and neurotoxicity. Nature 1993; 364:535-537.
50. Patel M, Day BJ, Crapo DJ et al. Requirement for superoxide in excitotoxic cell death. Neuron 1996; 16:345-355.
51. Cohen G. The pathobiology of Parkinson's disease: biochemical aspects of dopamine neuron senescence. J Neural Transm Supp 1983; 19:89-103.
52. Graham DG. Oxidative pathways for catecholamines in the genesis of neuromelanin and cytotoxic quinones. 1978; 14:633-643.
53. Fishman JB, Rubins JB, Chen JC et al. Modification of brain guanine nucleotide-binding regulatory proteins by tryptamine-4,5-dione, a neurotoxic derivative of serotonin. J Neurochem 1991; 56:1851-1854.
54. Schulze-Osthoff K, Bakker AC, Vanhaesebroeck B et al. Cytotoxic activity of tumor necrosis factor is mediated by early damage of mitochondrial functions. J Biol Chem 1991; 267:5317-5323.
55. Zhang Z, Rydel RE, Drzewiechi GJ et al. Amyloid β-mediated oxidative and metabolic stress in rat cortical neurons: no direct evidence for a role for H_2O_2 generation. J Neurochem 1996; 67: 1595-1606.
56. Mattson MP, Rydell RE, Liederburg I et al. Altered calcium signaling and neuronal injury: stroke and Alzheimer's disease as examples. Ann NY Acad Sci 1993; 679:1-21.
57. Mirabelli A, Salis A, Perotti M et al. Alterations of surface morphology caused by the metabolism of menadione in mammalian cells are associated with the oxidation of critical sulfhydryl groups in cytoskeletal proteins. Biochem Pharmacol 1988; 37:3423-3427.
58. Sarafian T, Bredesen DE. Is apoptosis mediated by reactive oxygen species? Free Rad Res 1994; 21:1-8.
59. Abate C, Patel L, Rauscher FJ et al. Redox regulation of Fos and Jun DNA-binding activity in vitro. Science 1990; 249:1157-1161.
60. Stall FJT, Anderson MT, Staal GEJ et al. Redox regulation of signal transduction: tyrosine phosphorylation and calcium influx. Proc Natl Acad Sci USA 1994; 65: 3619-3622.
61. Buckman T, Sutphin MS, Mitrovic B. Oxidative stress in a clonal cell line of neuronal origin:effects of antioxidant enzyme modulation. J Neurochem 1993, 60:2046-2058.
62. Merad-Boudia M, Fideler L, Nicole A et al. Selective modulation of brain antioxidant capacity by genetic or metabolic manipulation. In: Oxidative stress in cell activation and viral infection. Birkhaüser Verlag 1994:173-192.
63. Troy CM, Shelanski ML. Down regulation of copper-zinc superoxide dismutase causes apoptotic death in PC12 neuronal cells. Proc Natl Acad Sci USA 1994; 91:6384-6387.
64. Troy CM, Derossi D, Prochiantz A et al. Downregulation of SOD1 leads to cell death by the NO-peroxynitrite pathway. J Neurosci 1996; 16:253-261.
65. Reaume AG, Elliot JL, Hoffman EK et al. Motor neurons in CuZn superoxide dismutase-deficient mice develop normally but exhibit enhanced cell death after axonal injury. Nature Genetics 1996; 13:43-47.
66. Rabizadeh S, Gralla EB, Borchelt DR et al. Mutations associated with amyotrophic lateral sclerosis convert superoxide dismutase from an antiapoptotic gene to a proapoptotic gene: studies in yeast and neural cells. Proc Natl Acad Sci USA 1995; 92:3024-3028.
67. Gurney ME, Pu H, Chiu AY et al. Motor neuron degeneration in mice expressing a human CuZn superoxide dismutase mutation. Science 1994; 264:1772-1775.
68. Przedborski S, Khan U, Kostic V et al. Increased superoxide dismutase activity improves survival of cultured postnatal midbrain neurons. J Neurochem 1996; 67:1383-1392.

69. Chan PH, Yang GY, Chen SF et al. Cold-induced brain edema and infarction are reduced in transgenic mice overexpressing copper-zinc superoxide dismutase. Ann Neurol 1991; 29:482-486.
70. Kinouchi H, Eipstein CJ, Mizui T et al. Attenuation of focal cerebral ischemic injury in transgenic mice overexpressing CuZn superoxide dismutase. Proc Natl Acad Sci USA 1991; 88:11158-11162.
71. Pou S, Pou WS, Bredt DS et al. Generation of superoxide by purified brain nitric oxide synthase. J Biol Chem 1992; 267:24173-24176.
72. Lei SZ, Pan Z-H, Aggarval SK et al. Effect of nitric oxide production on the redox modulatory site of the NMDA receptor-channel complex. Neuron 1992; 8:1087-1099.
73. Aizenman E, Lipton SA, Loring RH. selective modulation of NMDA responses by reduction and oxidation. Neuron 1989; 2:1257-1263.
74. Lipton SA, Choi Y-B, Pan Z-H et al. A redox-based mechanism for the neuroprotective and neurodestructive effects of nitric oxide and related nitroso-compounds. Nature 1993; 364:626-632.
75. Farinelli SE, Park DS, Greene LA. Nitric oxide delays the death of trophic factor-deprived PC12 cells and sympathetic neurons by a cGMP-mediated mechanism. J Neurosci 1996; 16:2325-2334.
76. Radi R, Beckman JS, Busch KM, Freeman BA. Peroxynitrite induced membrane lipid peroxidation: the cytotoxic potential of superoxide and nitric oxide. Arch Biochem Biophys 1991; 288:481-487.
77. Meister A, Anderson ME. Glutathione. Annu Rev Biochem 1983:711-760.
78. Zangerle L, Cuénod M, Winterhalter KH, Do KQ. Screening of thiol compounds: depolarization-induced release of glutathione and cysteine from rat brain slices. J Neurochem 1992; 59:181-189.
79. Lanius RA, Shaw C, Wagey R, Krieger C. Characterization, distribution and protein kinase C-mediated regulation of ^{35}S glutathione binding sites in mouse and human spinal cord. J Neurochem 1994; 34:155-160.
80. Ogita K, Enomoto R, Nakahara F, Ishitsubo N, Yoneda Y. A possible role of glutathione as an endogenous agonist at the N-Methyl-D-Aspartate recognition domain in rat brain. J Neurochem 1995; 64:1088-1096.
81. Ogita I, Yoneda Y. Temperature-dependent and independent apparent binding activities of ^{3}H glutathione in brain synaptic membranes. Brain Res 1988, 463: 37-46.
82. Guo N, McIntossh C, Shaw C. Glutathione: new candidate neuropeptide in the central nervous system. Neuroscience 1992; 51:835-842.
83. Guehmann S, Vorbrueggen F, Kalkbrenner F, Moelling K. Reduction of a conserved Cys is an essential for Myb DNA-binding. Nucleic Acids Res 1992; 20:2279-2286.
84. Mihm S, Galter D, Droge W. Modulation of transcription factor NFκ-B activity by intracellular glutathione levels and by variations of the extracellular cysteine supply. FASEB J 1995; 9:246-252.
85. Fernandes RS, Cotter TG. Apoptosis and necrosis: intracellular levels of glutathione influence mode of cell death. Biochem Pharmacol 1994; 48:675-681.
86. Han S-K, Mytilineou C, Cohen G. L-DOPA upregulates glutathione and protects mesencephalic cultures against oxidative stress. J Neurochem 1996; 66: 501-510.
87. Harlan JM, Levine JD, Callahan KS, Schwartz BR, Harker LA. Glutathione redox cycle protects culture endothelial cells against lysis by extracellularly generated hydrogen peroxide. J Clin Invest 1984; 73:706-712.
88. Kane DJ, Sarafian TA, Anton R et al. Bcl2 inhibition of neural death: decreased generation of reactive oxygen species. Science 1993:1274-1277.
89. Jain AJ, Martensson J, Stole E, Auld PAM, Meister A. Glutathione deficiency leads to mitochondrial damage to the brain. Proc Natl Acad Sci USA 1991; 88:1913-1917.
90. Philbert MA, Beiswanger CM, Waters DK et al. Cellular and regional distribution of reduced glutathione in the nervous system of the rat: histochemical localization by mercury orange and o-phtaldealdehyde-induced hisfluorescence.

Toxicol Appl Pharmacol 1991; 107: 215-227.
91. Pileblad E, Eriksson PS, Hansson E. The presence of glutathione in primary neuronal and astroglial cultures from rat cerebral cortex and brainstem. J Neural Transm 1991; 86:43-49.
92. Raps SA, Lai JCK, Hertz L, Cooper AJL. Glutathione is present in high concentrations in cultured astrocytes but not in cultured neurons. Brain Res 1989; 493: 398-401.
93. Sagara JI, Miura K, Bannai S. Cystine uptake and glutathione level in fetal brain cells in primary culture and suspension. J Neurochem 1993; 61:1667-1671.
94. Sagara JI, Miura K, Bannai S. Maintenance of neuronal glutathione by glial cells. J Neurochem 1993b; 64:1672-1676.
95. Walsh AC, Michaud SG, Malossi JA, Lawrence DA. Glutathione depletion in human T lymphocytes. Analysis of activation-associated gene expression and the stress response. Tox and Appl Pharmacol 1995; 133:249-261.
96. Bergelson S, Pinkus R, Daniel V. Intracellular glutathione levels regulate Fos/Jun induction and activation of glutathione S-transferase gene expression. Cancer Res 1994; 54:36-40.
97. Sato N, Iwata S, Nakamura K, Hori T, Mori K, Yodoi J. Thiol-mediated redox regulation of apoptosis. J Immunol 1995; 154:3194-3203.
98. Meister A. Glutathione metabolism and its selective modification. J Biol Chem 1988; 263:17205-17208.
99. Desagher S, Glowinski J, Premont J. Astrocytes protect neurons from hydrogen peroxide toxicity. J Neurosci 1996; 16(8): 2553-2562.
100. Murphy TH, Miyamoto M, Sastre A, Schnaar RL, Coyle JT. Glutamate toxicity in a neuronal cell line involves inhibition of cystine transport leading to oxidative stress. Neuron 1989; 2:1547-1558.
101. Bredesen DE. Neural apoptosis. Ann Neurol 1995; 38:839-851.
102. Slater AFG, Stefan C, Nobel I, van den Dobbelsteen DJ, Orrenius S. Intracellular redox changes during apoptosis. Death and Differentiation 1996; 3:57-62.
103. Yan CYI, Ferrari G, Greene LA. N-Acetylcysteine-promoted survival of PC12 cells is glutathione-independent but transcription-dependent. J Biol Chem 1995; 270(45):26827-26832.
104. Slivka AC, Mytilineou C, Cohen G. Histochemical evaluation of glutathione in brain. Brain Res 1987; 409:275-284.
105. Burdon RH, Alliangana D, Gill V. Endogenously generated active oxygen species and cellular glutathione levels in relation to BHK-21 cell proliferation. Free Rad Res Comm 1994; 21(3):121-133.
106. Chen Q, Ames BN. Senescence-like growth arrest induced by hydrogen peroxide in human diploid fibroblasts F65 cells. Proc Natl Acad Sci USA 1994; 91:4130-4134.
107. Fernandez A, Kiefer J, Fosdick L, McConkey DJ. Oxygen radical production and thiol depletion are required for Ca^{2+}-mediated endogenous endonuclease activation. J Immunol 1995; 155: 5133-5139.
108. Heales SJR, Bolanos JP, Clark JB. Glutathione depletion is accompanied by increased neuronal nitric oxide synthase activity. Neurochem Res 1996; 21(1): 35-39.
109. Heales SJR, Davies SEC, Bates TE, Clark JB. Depletion of brain glutathione is accompanied by impaired mitochondrial function and decreased N-Acetyl aspartate concentration. Neurochem Res 1995; 20:31-38.
110. Bolanos JP, Peuchen SP, Heales SJR, Land JM, Clark JB. Nitric-oxide-mediated inhibition of mitochondrial respiratory chain in cultured astrocytes. J Neurochem 1994; 63:910- 916
111. Bolanos JP, Heales SJR, Land JM, Clark JB. Effect of peroxynitrite on the mitochondrial respiratory chain: differential susceptibility of neurones and astrocytes in primary culture. J Neurochem 1995; 64:1965-1972.
112. Di Monte DA, Chan Piu Sandy MS. Glutathione in Parkinson's disease. A link between oxidative stress and mitochondrial damage. Ann Neurol 1992; 32: S111-S113.
113. Martensson J, Jain A, Frayer W, Meister A. Glutathione metabolism in the lung:

inhibition of its synthesis leads to lamellar body and mitochondrial defects. Proc Natl Acad Sci USA 1989; 86:5296-5300.
114. Garcia-Ruiz C, Colell A, Morales A et al. Role of oxidative stress generated from the mitochondrial electron transport chain and mitochondrial glutathione status in loss of mitochondrial function and activation of transcription factor nuclear Factor-κB: studies with isolated mitochondria and rat hepatocytes. Mol Pharmacol 1995; 48:825-834.
115. Zamzani N, Susin AS, Marchetti P et al. Mitochondrial control of nuclear apoptosis. J Exp Med 1996; 183: 1533-1544.
116. Bredt DS, Snyder SH. Nitric oxide, a novel neuronal messenger. Neuron 1992; 8:3-11.
117. Snyder SH. Nitric oxide. first in a new class of neurotransmitters? Science 1992; 257:494-496.
118. Garthwaite J. Glutamate, notric oxide and cell-cell signaling in the nervous system. Trends Neurosci 1991; 14:60-67.
119. Dawson VL, Kizushi VM, Huang PL et al. Resistance to neurotoxicity in cortical cultures from neuronal nitric oxide synthetase-deficient mice. J Neurosci 1996; 16:2479-2487.
120. Nowiski JP, Duval D, Poignet H, Scatton B. Nitric oxide mediates neuronal death after focal cerebral ischemia in the mouse. Eur J Pharmacol 1991; 204: 339-340.
121. Huang Z et al. Effects of cerebral ischemia in mice deficient in neuronal nitric oxide synthase. Science 1994; 265:1883-1885.
122. Dawson VL, Dawson TM, Bartley DA et al. Mechanisms of nitric oxide-mediated neurotoxicity in primary brain cultures. J Neurosci 1993; 13:2651-2661.
123. Hantraye P, Brouillet E, Ferrante R et al. Inhibition of neuronal nitric oxide synthase prevents MPTP-induced parkinsonism in baboons. Nature Med 1996; 2:1017-1021.
124. Schultz JB, Matthews RT, Muqit MMB et al. Inhibition of neuronal nitric oxide synthase by 7-nitroindazole protects against MPTP-induced neurotoxicity in mice. J Neurochem 1995; 64:936-939.
125. Predborski S et al. Role of nitric oxide in MPTP induced dopaminergic neurotoxicity. Proc Natl Acad Sci USA 1996; 93:4565-4571.
126. Schulz JB, Huang PL, Matthews RT et al. Striatal malonate lesions are attenuated in neuronal nitric oxide knock out mice. J Neurochem 1996; 67:430-433.
127. Strijbos PJLM, Leach M, Garthwaite J. Vicious cycle involving Na^+ channels, glutamate release, and NMDA receptors mediates delayed neurodegeneration through nitric oxide formation. J Neuroscience 1996; 16:5004-5013.
128. Stamler JS. Redox signaling. Nitrosylation and related target interactions of nitric oxide. Cell 1994; 78:931-936.
129. Zhang J, Dawson VL, Dawson TW, Snyder SH. Nitric oxide activation of poly(ADP-ribose) synthetase in neurotoxicity. Science 1994; 263:687-689.
130. Beckman JS, Beckman TW, Chen J et al. Apparent hydroxyl radical production by peroxynitrite: implications for endothelial injury from nitric oxide and superoxide. Proc Natl Acad Sci USA 1990; 87: 1620-1624.
131. Estevez AG, Radi R, Barbeito L et al. Peroxynitrite-induced cytotoxicity in PC12 cells: evidence for an apoptotic mechanism differentially modulated by neurotrophic factors. J Neurochem 1995; 65:1543-1550.
132. Mitrovic B, Ignarro LJ, Vinters HV et al. Nitric oxide induces necrotic but not apoptotic cell death in oligodendrocytes. Neuroscience 1995; 65:531-539.
133. Murphy ME, Sies H. Reversible convertion of nitroxyl anion to nitric oxide by superoxide dismutase. Proc Natl Acad Sci 1991; 88:10860-10864.
134. Barger SW, Fiscus RR, Ruth P et al. Role of cyclic GMP in the regulation of neuronal calcium and survival by secreted forms of beta-amyloid precursor. J Neurochem 1995; 64:2087-2096.
135. Peunova N, Enikolopov G. Nitric oxide triggers a switch to growth arrest during differentiation of neuronal cells. Nature 1995; 375:68-73.
136. Garg UC, Hassis A. Nitric oxide-generating vasodilatators inhibit mitogenesis and proliferation of BALB/c 3T3 fibro-

blasts by a cyclic GMP-independent mechanism. Biochem Biophys Res Commun 1990; 171:474-479.

137. Lepoivre M, Chenais B, Yapo A et al. Alterations of ribonucleotide reductase activity following induction of the nitrite-generating pathway in adenocarcinoma cells. J Biol Chem 1990; 265: 14143-14149.

138. Ferrari G, Greene LA. Proliferative inhibition by dominant-negative Ras rescues naive and neuronally-differentiated PC12 cells from apoptotic death. EMBO J 1995; 13:5922-5928.

139. Ferrari G, Yan CYI, Greene LA. N-acetylcysteine (D- and L-stereoisomers) prevents apoptotic death of neuronal cells. J Neurosci 1995; 15:857-2866.

140. Farinelli SE, Greene LA. The cell cycle blockers mimosine, ciclopirox and deferoxamine prevent the death of PC12 cells and post-mitotic sympathetic neurons following removal of trophic support. J Neurosci 1996; 16:1150-1162.

141. Kane DJ, Ord T, Anton R, Bredesen DE. Expression of bcl-2 inhibits necrotic neuronal death. J Neurosci Res 1995; 40: 269-275.

142. Hockenbery D, Nunez G, Milliman C et al. Bcl-2 is an inner mitochondrial membrane protein that blocks programmed cell death. Nature 1990; 348:334-36.

143. Monaghan P, Robertson D, Andrew T et al. Ultrastructural localization of Bcl-2 protein. J Histochem Cytochem 1992; 40:1819-25.

144. Jacobson MD, Burne JF, King MP et al. Bcl-2 blocks apoptosis in cells lacking mitochondrial DNA. Nature 1993; 361: 365-69.

145. Bredesen DE. Neuronal apoptosis: genetic and biochemical modulation. In: Tomei LD, Cope FO, eds. Apoptosis II: the molecular basis of apoptosis in disease. Plainview, NY: Cold Spring Harbor Laboratory Press 1994:397-421.

146. Steinman HM. The bcl-2 oncoprotein functions as a pro-oxidant. J Biol Chem W1995; 70:3487-3490.

147. Ellerby LM, Ellerby HM, Park SM et al. Shift of the cellular oxidation-reduction potential in neural cells expressing bcl-2. J Neurochem 1996; 67:1259-1267.

148. Myers KM, Fiskum G, Liu Y et al. Bcl-2 protects neural cells from cyanide/aglycemia-induced lipid oxidation, mitochondrial injury, and loss of viability. J Neurocham 1995; 65:2432-2440.

149. Shimizu S, Eguchi Y, Kosaka H et al. Prevention of hypoxia-induced cell death by Bcl-2 and Bcl-xL. Nature 1995; 374: 811-13.

150. Oltvai ZN, Milliman CL, Korsmeyer SJ. Bcl-2 heterodimerizes in vivo with a conserved homolog, Bax, that accelerates programmed cell death. Cell 1993; 74: 609-19.

151. Fernandez-Sarabia MJ, Bischoff JR. Bcl-2 associates with the ras-related protein R-ras p 23. Nature 1993; 366:274-75.

152. Wang H-G, Miyashita T, Takayama S et al. Apoptosis regulation by interaction of Bcl-2 protein and Raf-1 kinase. Oncogene 1994; 9:2751-56.

153. Shahin S, Bruce A, Sun N et al. p53 induction is associated with neuronal damage in the central nervous system. Proc Natl Acad Sci USA 1994; 91:7525-7529.

154. Miyashita T, Reed JC. Tumor suppressor p53 is a direct transcriptional activator of the human bax gene. Cell 1995; 80:293-99.

155. Hainaut P, Milner J. Redox modulation of p53 conformation and sequence specific DNA-binding in vitro. Cancer Res 1993; 53:4469-4473.

156. Muakkassah SF, Andresen JW, Shih JC et al. Decreased serotonine and spiperone binding consequent to lipid peroxidation in rat cortical membranes. Biochem Biophys Res Comm 1982; 194:1003-1010.

157. Van der Vielt A, Bast A. Effect of oxidative stress on receptors and signal transmission. Chem Biol Interactions 1992; 85:95-116.

158. Richter C and Frei B. Ca^{2+} release from mitochondria induced by pro-oxidants. Free Rad Biol Med 1988; 4:365-375.

159. Sandri E, Panfili E, Ernster L. Hydrogen peroxide production by monoamine oxidase in isolated rat-brain mitochondria: its effect on glutathione levels and Ca^{2+} efflux. Biochim Biophys Acta 1990; 1035: 300-305.

160. Jones DP, Thor H, Smith MT et al. Inhibition of ATP-dependent microsomal

$Ca2^+$ sequestration during oxidative stress and its prevention by glutathione. J Biol Chem 1983; 258:6390-6393.
161. Reeves JP, Bailey CA, Hale CC. Redox modification of sodium-calcium exchange activity in cardiac sarcolemmal vesicles. J Biol Chem 1986; 261: 4948-4955.
162. Kroemer G, Petit PX, Zamzani N et al. The biochemistry of apoptosis. FASEB 1995; 9:1277-1287.
163. Trump BF, Berezesky IK. Calcium-mediated cell injury and cell death. FASEB J 1995; 9:219-228.
164. Sundaresan M, Yu ZX, Ferrans VJ et al. Requirement for generation of H_2O_2 for platelet-derived growth factor signal transduction. Science 1995; 270:296-299.
165. Guyton KZ, Liu Y, Gorospe M et al. Activation of mitogen-activated protein kinase by H_2O_2. J Biol Chem 1996; 271: 4138-4142.
166. Monteiro HP, Stern A. Redox modulation of tyrosine phosphorylation-dependent signal transduction pathways. Free Rad Biol Med 1996; 21:323-333.
167. Sun Y, Oberley LW. Redox regulation of transcriptional activators. Free Rad Biol Med 1996; 21:335-348.
168. Lee SF, Huang YT, Wu Ws et al. Induction of c-*jun* protooncogene expression by hydrogen peroxide through hydroxyl radical generation and $p60^{SRC}$ tyrosine kinase activation. Free Rad Biol Med 1996; 21:437-448.
169. Freeman R, Estus SEM Jr. Analysis of cell cycle-related gene expression in post-mitotic neurons: selective induction of cyclin D1 during programmed cell death. Neuron 1994; 12:343-55.
170. Estus S, Zaks WJ, Freeman RS et al. Altered gene expression in neurons during programmed cell death: identification of c-*jun* as necessary for neuronal apoptosis. J Cell Biol 1994; 126:1717-27.
171. Anderson AJ, Pike CJ, Cotman CW. Differential induction of immediate early gene proteins in cultured neurons by β-amyloid (Aβ): association of c-Jun with Aβ-induced apoptosis. J Neurochem 1995; 65:1487-1498.
172. Anderson AJ, SU JH, Cotman W. DNA damage and apoptosis in Alzheimer's disease. Colocalization with c-jun immunoreactivity, relation to brain area and effect of post-mortem delay. J Neurosci 1996; 16:1710-1719.
173. Anderson AJ, Cummings BJ, Cotman CW. Increased immunoreactivity for Jun- and Fos-related proteins in Alzheimer's diseases: association with pathology. Exp Neurol 1994; 125:286-295.
174. Virgo L, de Belleroche J. Induction of the immediate early gene c-*jun* in human spinal cord in amyotrophic lateral sclerosis with concomitant loss of NMDA receptor NR-1 and glycine transporter mRNA. Brain Res 1995; 676:196-204.
175. Colotta F, Polentarutti N, Sironi M et al. expression and involvement of c-fos and c-*jun* protooncogenes in programmed cell death induced by growth factor deprivation in lymphoid cell lines. J Biol Chem 1992; 267:18278-18283.
176. Manome Y, Datta R, Fine HA. Early response gene induction following DNA damage in astrocytoma cell lines. Biochem Pharmacol 1993; 45:1677-1684.
177. Ham J, Babij C, Whitfield J et al. A c-*jun* dominant negative mutant protects sympathetic neurons against programmed cell death. Neuron 1995; 14: 927-939.
178. Bergelson S, Pinkus R, Daniel V. Intracellular glutathione levels regulate Fos/Jun induction and activation of glutathione S-transferase gene expression. Cancer Res 1994; 54:36-40.
179. Dragunow M, Young D, Hughes P, Mc Gibbon G et al. Is c-*jun* involved in nerve cell death following status epilepticus and hypoxic-ischemic brain injury? Molec Brain Res 1993; 18:347-352.
180. Schlingensiepen KH, Wollnik F, Kunst M et al. The role of *jun* transcription factor expression and phosphorylation in neuronal differentiation, neuronal cell death, and plastic adaptation in vivo. Cell Mol Neurobiol 1994; 14:487-505.
181. Gajate C, Alonso MT, Schimmang T et al. c-Fos is not essential for apoptosis. Biochem Biophys Res Commun 1996; 218:267-272.
182. Roffler-Tarlov S, Gibson Brown JJ, Tarlov E et al. Programmed cell death in the absence of c-Fos and c-Jun. Development 1996; 122:1-9.

183. Kong S-K, Yim MB, Stadtman ER et al. Peroxynitrite disables the tyrosine phosphorylation regulatory mechanism: lymphocyte-specific tyrosine kinase fails to phosphorylae nitrated cdc2(6-20)NH2 peptide. Proc Natl Acad Sci USA 1996; 93:3377-3382.
184. Ischiropoulos H, Zhu L, Chen J et al. Peroxynitrite-mediated tyrosine nitration catalyzed by superoxide dismutase. Arch Biochem Biophys 1992; 298:431-437.
185. Berlett BS, Friguet B, Yim MB et al. Peroxynitrite-mediated nitration of tyrosine residues in Escherichia coli glutamine synthetase mimmics adenylation: relevance to signal transduction. Proc Natl Acad Sci USA 1996; 93:1776-1780.
186. Russo T, Zambrano N, Esposito F et al. A p53-independent pathway for activation of WAF1/CIP1 expression following oxidative stress. J Biol Chem 1995; 270: 9386-29391.
187. Stevenson M, Pollock SS, Colleman N et al. X-irradiation, phorbol esters, and H_2O_2 stimulate mitogen activated protein kinase activity in NIH 3T3 cells through formation of reactive oxygen intermediates. Cancer Res 1994; 54: 12-15.
188. Busciglio J, Lorenzo A, Yeh J et al. β-amyloid fibrils induce tau phosphorylation and loss of microtubule binding. Neuron 1995; 14:879-888.
189. Zhang C, Qiu HE, Krafft GA et al. Protein kinase C and F-actin are essential for stimulation of neuronal FAK tyrosine phosphorylation by G-proteins and amyloid beta protein. FEBS Lett 1996; 386: 185-188.
190. Baumann K, Mandelkow E. Abnormal Alzheimer-like phosphorylation of tau protein by cyclin-dependent kinases cdk2 and cdk5. FEBS Lett 1993; 336:417-424.
191. Loo DT, Copani A, Pike CJ et al. Apoptosis is induced by β-amyloid in cultured central nervous system neurons. Proc Natl Acad Sci USA 1993; 90: 7951-7955.
192. Forloni G, Angeriti N, Chiesa R et al. Neurotoxicity of a prion protein fragment. Nature 1993; 362; 543-546.
193. Rabidazeh S, Bitler CM, Butscher LL et al. Expression of the low-affinity nerve growth factor receptor enhances β-amyloid peptide toxicity. Proc Natl Acad Sci USA 1994; 91:10703-10706.
194. Cotman CW, Anderson AJ. A potential role for apoptosis in neurodegeneration and Alzheimer's disease. Mol Neurobiol 1995; 10:19-45.
195. Duguid JR, Bohmont CW, Liu N et al. Changes in brain gene expression shared by scrapie and Alzheimer disease. Proc Natl Acad Sci USA 1989; 86:7260-64.
196. Rosen DR, Siddique T, Patterson D et al. Mutations in CuZnSOD superoxide dismutase gene are associated with familial amyotrophic lateral sclerosis. Nature 1993; 362:59-62.
197. Deng HX, Hentati A, Tainer JA et al. Amyotrophic lateral sclerosis and structural defects in Cu,Zn superoxide dismutase. Science 1993; 261:1047-1051.
198. Gurney ME, Pu H, Chiu AY et al. Motor neuron degeneration in mice that express a human CuZn superoxide dismutase mutation. science 1994; 264: 1772-1775.
199. Dal Canto MC, Gurney ME. Development of central nervous sysytem pathology in a murine transgenic model of human amyotrophic lateral sclerosis. Am J Pathol 1994; 145:1271-1279.
200. Roy N, Mahadevan MS, McLean M et al. The gene for neuronal apoptosis inhibitory protein is partially deleted in individuals with spinal muscular atrophy. Cell 1995; 80:167-178.
201. Sen CK, Packer L. Antioxidant and redox regulation of gene transcription FASEB J 1996; 10:709-720.
202. Bauerle PA. The inducible transcription factor NF-κB: regulation by distinct protein subunits. Biochim Biophys Acta 1991; 1072:63-80.
203. Molitor JA, Ballard DW, Greene WC. κB-specific DNA binding proteins are differentially inhibited by enhancer mutations and biological oxidation. New Biol 1991; 3:987-996.
204. Toledano MB, Leonard WJ. Modulation of transcription factor NF-κB binding activity by oxidation-reduction in vitro. Proc Natl Acad Sci USA 1991; 88: 4328-4332.
205. Bauerle PA, Henkel T. Function and activation of NF-κB in the immune system. Annu Rev Immunol 1994; 12:141-179.

206. Thanos D, Maniatis T. NF-κB: a lesson in family values. Cell 1995; 80:529-532.
207. Yan SD, Chen X, Fu J et al. RAGE and amyloid-β peptide neurotoxicity in Alzheimer's disease. Nature 1996; 382: 685-691.
208. Legrand-Poels S, Bours V, Piret B et al. Transcription factor NF-κB is activated by photosensitization generating oxidative DNA damage. J Biol Chem 1995; 270:6925-6934.
209. Schreck R, Rieber P, Bauerle PA. Reactive oxygen intermediates as apparently widely used messengers in the activation of NF-κB transcription factor and HIV-1. EMBO J 1991; 10:2247-2258.
210. Grilli M, Pizzi M, Memo M, Spano PF. Neuroprotection by aspirin and sodium salicylate through blockade of NF-κB activation. Science 1996; 274:1383-1385.
211. Sen CK, Roy S, Packer L. Involvement of intracellular Ca^{2+} on oxidant-induced NF-κB activation. FEBS Lett 1996:58-62.
212. Suzuki YJ, Packer L. Inhibition of NF-κB activation by vitamine E derivatives. Biochem Biophys Res Commun 1993; 193:277-283.
213. Yan SD, Schmidt AM, Anderson GM et al. Enhanced cellular oxidant stress by the interaction of advanced glycation end products with their receptors/binding proteins. J Biol Chem 1994; 269: 9889-9897.
214. Schenk H, Klein M, Erdbrugger W et al. Distinct effects of thioredoxin and antioxidants on the activation of transcription factors NF-κB and AP-1. Proc Natl Acad Sci USA 1994; 91:1672-1676.
215. Henkel T, Alkalay I, Machleid T et al. Rapid proteolytic degradation of IkBa induced by stimulation of cells with phorbol ester, cytokines and lipopolysaccharide is a necessary step in the activation of NF-κB. Nature 1993; 365: 182-185.
216. Sun S-C, Ganchi PA, Ballard DW et al. NF-κB controls expression of inhibitor IκBα: evidence for an inducible autoregulatory pathway. Science 1993; 259: 1912-1915.
217. Suzuki YJ, Mizuno M, Trishler HJ, Packer L. Redoc regulation of NF-κB DNA binding activity by dihydrolipoate. Biochem Mol Biol Int 1995; 36:241-246.
218. Kumar S, Rabson AB, Gelinas C. The RXXRXRXXC motif conserved in all Rel/κB proteins is essential for the DNA binding activity and redox regulation of the v-Rel oncoprotein. Mol Cell Biol 1992; 12:3094-3106.
219. Toledano MB, Chosh D, Trinh F et al. N-terminal DNA-binding domains contribute to differential DNA binding specificities of NF-κB p50 and p65. Mol Cell Biol 1993; 852-850.
220. Mathews JR, Wakasugi N, Virelizier JL et al. Thioredoxin regulates the DNA binding activity of NF-κB by reduction of a disulfide bond involving cysteine 62. Nucleic Acids Res 1992; 20:3821-3830.
221. Mitomo K, Nakayama K, Fujimoto K et al. Two different cellular redox sysytems regulate the DNA-binding activity of the p50 subunit of NF-κB in vitro. Gene 1994; 13:852-860.
222. Xanthodukakis S, Mioao G, Wang F et al. Redox activation of Fos-Jun DNA binding activity is mediated by a DNA repair enzyme. EMBO 1992; 11: 3323-3335.
223. Kaltschmidt C, Kaltschmidt B, Neumann H et al. Constitutive NF-κB in neurons. Mol Cell Biol 1994; 14:3981-3992.
224. Rattner A, Korner M, Walker MD et al. NF-κB activates HIV promoter in neurons. EMBO J 1993; 12:4261-4267.
225. Wood JN. Regulation of NF-κB activity in rat dorsal root ganglia and PC12 cells by tumor necrosis factor and nerve growth factor. Neurosci Lett 1995; 192: 41-44.
226. Kaltschmidt C, Kaltschmidt B, Bauerle PA. Stimulation of ionotropic glutamate receptors activates transcription factor NF-κB in primary neurons. Proc Natl Acad Sci USA 1995; 92:9618-9622.
227. Terai K, Matsuo A, Mc Geer PL. Enhancement of immunoreactivity for NF-κB in the hippocampal formation and cerebral cortex of Alzheimer's disease. Brain Res 1996; 735:159-168.
228. Yan SD, Yan SF, Chen X et al. Non-enzymatically glycated tau in Alzheimer's disease induces neuronal oxidant stress resulting in cytokine gene expression and release of amyloid β-peptide. Nature Med 1995; 1:693-699.

229. El Khoury J, Hickman SE, Thomas CA et al. Scavenger receptor-mediated adhesion of microglia to β-amyloid fibrils. Nature 1996; 382:716-719.
230. Barger SW, Horster D, Furukawa K et al. Tumor necrosis factor α and β protect neurons against amyloid β-peptide toxicity: evidence for involvement of a κB-binding factor and attenuation of peroxide and Ca^{2+} accumulation. Proc Natl Acad Sci USA 1995; 92:9328-9332.
231. Koop E, Ghosh S. Inhibition of NF-κB by sodium salicylate and aspirin. Science 1994; 265:956-959.
232. Weber C, Erl W, Pietsch A, Weber PC. Aspirin inhibits nuclear factor-κB mobilization and monocyte adhesion in stimulated human endothelial cells. Circulation 1995; 91:1914-1917.
233. Hyman BT, Marzloff K, Wenninger JJ et al. Relative sparing of nitric oxide synthase-containing neurons in the hippocampal formation in Alzheimer's disease. Ann Neurol 1992; 32:818-820.
234. Kowall NW, Beal MF. Cortical somatostatin, neuropeptide Y, and NADPH diaphorase neurons: normal anatomy and alterations in Alzheimer's disease. Ann Neurol 1988; 23:105-114.
235. Peng HB, Libby P, Liao JK. Induction and stabilization of IκBα by nitric oxide mediates inhibition of NF-κB. J Biol Chem 1995; 270:14214-14219.
236. Angel P, Karin M. The role of Jun, Fos and the AP-1 complex in cell-proliferation and transformation. Biochim Biophys Acta 1991; 471072:129-157.
237. Crawford D, Zbinden I, Amstad P, Cerutti P. Oxidant stress induces the protooncogenes c-fos and c-myc in mouse epidermal cells. Oncogene 1988; 3:27-32.
238. Morgan JI, Curran T. Stimulus-transcription coupling in the nervous system: involvement of the inducible proto-oncogene fos and *jun*. Annu rev Neurosci 1991; 14:421-451.
239. Handel ML, Watts CKW, DeFazio A et al. Inhibition of AP-1 binding and transcription by gold and selenium involving conserved residues in Jun and Fos. Proc Natl Acad Sci USA 1995; 92: 4497-4501.
240. Datta R, Hallahan DE, Kharbanda SM et al. Involvement of reactive oxygen intermediates in the induction of c-*jun* gene transcription by ionizing radiation. Biochemistry 1992; 31:8300-8306.
241. Galter D, Mihm S, Dröge W. Distinct effects of glutathione disulphide on the nuclear transcription factor κB and the activator protein-1. Eur J Biochem 1994; 221:639-648.
242. Friling RS, Bergelson S, Daniel V. Two adjacent AP-1 like binding sites from the electrophile-responsive element of the murine glutathione S-transferase Ya subunit gene. Proc Natl Acad Sci 1992; 89:668-672.
243. Frame MC, Wilkie NM, Darling AJ et al. Regulation of AP-1/ DNA complex formation in vitro. Oncogene 1991; 6: 205-209.
244. Meyer M, Schreck R, Bauerle PA. H2O2 and antioxidants have opposite effects on activation of NF-κB and AP-1 in intact cells: AP-1 as a secondary-responsive factor. EMBO J 1993:2005-2015.
245. Schmidt HHHW, Walter U. NO at work Cell 1994; 78:919-925.
246. Stamler JS. Redox signaling: nitrosylation and related target interactions of nitric oxide. Cell 1994; 78:931-936.
247. Haby C, Lisovoski F, Aunis D, Zwiller J. Stimulation of the cyclic GMP pathway by NO induces expression of the immediate early genes c-fos and junB in PC12 cells. J Neurochem 1994; 62:496-501.
248. Pilz RB, Suhasini M, Idriss S et al. Nitric oxide and cGMP analogs activate transcription from AP-1-responsive promoters in mammalian cells. FASEB J 1995; 9:552-558.
249. Dugan LL, Sensi SL, Canzoniero LMT et al. Mitochondrial production of reactive oxygen species in cortical neurons following exposure to N-methyl-D-Aspartate. J Neurosci 1995; 15:6377-6388.
250. Mills EM, Gunasekar PG, Pavlakovic G et al. Cyanide-induced apoptosis and oxidative stress in differentiated PC12 cells. J Neurochem 1996; 67:1039-1046.
251. Mills JC, Nelson D, Brecinska M et al. Metabolic and energic changes during apoptosis in neuronal cells. J Neurochem 1995; 65:1721-1730.

252. Myers KM, Fiskum G, Liu Y et al. Bcl-2 protects neural cells from cyanide/aglycemia-induced lipid peroxidation, mitochondrial injury, and loss of viability. J Neurochem 1995; 65:2432-2440.
253. Beal MF, Hyman BT, Koroshetz W. Do defects in mitochondrial energy metabolism underlie the pathology of neurodegenerative diseases? Trends Neurosci 1993; 16:125-131.
254. Zhang Y, Marcillat O, Giulivi C et al. The oxidative inactivation of mitochondrial electron transport chain components and ATPase. J Biol Chem 1990; 265:16330-16336.
255. Beal MF. Mitochondria, free radicals, and neurodegeneration. Current Opinion In Neurobiol 1996; 6:661-666.
256. Beal MF. Mitochondrial dysfunction and oxidative damage in neurodegenerative diseases. RG Landes Company. Springer Verlag.
257. Bondokas VP, Jordan J, Lee CC et al. Superoxide production in rat hippocampal neurons: selective imaging with hydroethidine. J Neurosci 1996; 16: 1324-1336.
258. Reynolds IJ, Hastings TG. Glutamate induces the production of reactive oxygen species in cultured forebrain neurons following NMDA receptor activation. J Neurosci 1995; 15:3318-3327.
259. Harrington J, Park YB, Babcock DF et al. Dominant role of mitochondria in clearance of large Ca^{2+} loads from rat adrenal chromaffin cells. Neuron 1996; 16: 219-228.
260. Patel M, Day BJ, Crapo DJ et al. Requirement for superoxide in excitotoxic cell death. Neuron 1996; 16:345-355.
261. Richter C. Prooxidant and mitochondrial Ca^{2+}: their relationship to apoptosis and oncogenesis. FEBS Lett 1993; 325: 104-107.
262. Hantraye P, Brouillet E, Ferrante R et al. Inhibition of neuronal nitric oxide synthase prevents MPTP-induced parkinsonism in baboons. Nature Med 1996; 2:1017-1021.
263. Wolvetang EJ, Johnson KL, Krauer K et al. Mitochondrial respiratory chain inhibitors induce apoptosis. FEBS Lett 1994; 339:40-44.
264. Hevner RF, Wong-Riley MTT. Mitochondrial and nuclear gene expression for cytochrome oxydase subunits are disproportinately regulated by functional activity in neurons. J Neurosci 1993; 13:1805-1819.
265. Liu S, Wong-Riley MTT. Nuclear-encoded mitochondrial precursor protein: intramitochondrial delivery to dendrites and axon terminals of neurons and regulation by neuronal activity. J Neurosci 1994; 14:5338-5351.
266. Vayssière JL, Petit PX, Risler Y et al. Commitment to apoptosis is associated with changes in mitochondrial biogenesis and activity in SV40 conditional cell lines. Proc Natl Acad Sci USA 1994; 91:11752-11756.
267. Liu X, Kim CM, Yang J et al. Induction of apoptotic program in cell-free extracts: requirement for ATP and cytochrome c. Cell 1996; 86:147-157.
268. Mayer A, Neupert W, lill R. Translocation of apocytochrome c across the outer membrane of mitochondria. J Biol chem 1995; 270:12390-12397.
269. Cossarizza A, Kalashnikova G, Grassilli E et al. Mitochondrial modifications during rat thymocyte apoptosis: a study at the single cell level. Exp Cell Res 1994; 214:323-330.
270. Cossarizza A, Franscheski C, Monti D et al. Protective effect of N-acetylcysteine in Tumor Necrosis Factor α-induced apoptosis in U937 cells: the role of mitochondria. Exp Cell Res 1995; 220:232-240.
271. Richter C, Schweizer M, Cossariza A et al. Control of apoptosis by the intracellular ATP level. FEBS Lett 1996; 376: 107-110.
272. Polla BS, Kentengwa S, Francois D et al. Mitochondria as targets for the protective effects of heat shock against oxidative injury. Proc Natl Acad Sci USA 1996; 93:6458-6463.
273. Zamzani NP, Marchetti M, Castedo M et al. Reduction in mitochondrial potential constitues an early irreversible step of programmed lymphocyte death in vivo. J Exp Med 1995; 181:1661-1672.
274. Zamzani NP, marchetti M, Castedo M et al. Sequential reduction of mitochondrial transmembrane potential and generation

of reactive oxygen species in early programmed cell death. J Exp Med 1995; 182:367-377.

275. Deckwert TL, Johnson EM. Temporal analysis of events associated with programmed cell death (apoptosis) of sympathetic neurons deprived of nerve growth factor. J Cell Biol 1993; 123: 1207-1222.
276. Petit PX, Susin SA, Zamzani N et al. Mitochondria and programmed cell death: back to the future. FEBS lett 1996; 396:7-13.
277. Sandstrom PA, Mannie MD, Buttke TM. Inhibition of activation-induced death in a T cell hybridoma by thio antioxidants-oxidative stress as a mediator of apoptosis. J Leukocyte Biol 1994; 55:221-226.
278. Hennet T, Richter C, Peterhans E. Tumor necrosis factor-alpha induces superoxide anion production in mitochondria of L929 cells. Biochem J 1993; 289:587-592.
279. Smets LA, Van der Berg D, Acton B et al. Bcl-2 expression and mitochondrial activity in leukemic cells with different sensitivity to glucocorticoid-induced apoptosis. Blood 1994; 5:1613-1619.
280. Marchetti P, Susin SA, Decaudin D et al. Apoptosis-associated derangement of mitochondrial function in cells lacking mitochondrial DNA. Cancer Res 1996; 56:2033-2038.
281. Skulachev VP. Why are mitochondria involved in apoptosis. Permeability transition pores and apoptosis as selective mechanisms to eliminate superoxide-producing mitochondria and cell. FEBS Lett 1996; 397:7-10.
282. Allen RG, Balin AK. Oxidative influence on development and differentiation: an overview of a free radical theory of development. Free Rad Biol Med 1989; 6:631-661.
283. Sohal RS, Allen RG, Nations C. Oxygen free radicals play a role in cellular differentiation: an hypothesis. Free Rad Biol Med 1986; 2:175-181.
284. Peunova N, Enikopolov G. Nitric oxide triggers a switch to growth arrest during differentiation of neuronal cells. Nature, 1995; 375:68-73.
285. Burdon RH. Superoxide and hydrogen peroxide in relation to mammalian cell proliferation. Free Rad Biol Med 1995; 18:775-794.
286. Shaw JP, Chou I-M. Elevation of intracellular glutathione content associated with mitogenic stimulation of quiescent fibroblasts. J Cell physiol 1986; 129: 193-198.
287. Clopton DA, Saltman P. Low-level oxidative stress causes cell-cycle specific arrest in cultured cells. Biochem Biophys Res Comm 1995; 210:189-196.
288. Griffith OW, Meister A. Origin and turnover of mitochondrial glutathione. Proc Natl Acad Sci USA 1985; 82:4668-4672.

CHAPTER 4

Oxidative Stress, Cerebral Aging and Longevity

INTRODUCTION

The aging process may be common to all living cells since the phenomena of aging and death are universal. It may also be the major determinant of the manifestations of aging even under optimal living conditions. Aging is under genetic control to some extent since the manifestations of aging and life span differ between species and individual members of a species. Aging is also subject to environmental influences. Progress in the qualitative and quantitative genetic analysis of longevity has led to a revolution in aging research. This research has revealed that extented longevity is frequently associated with enhanced metabolic capacity and response to stress. Moreover, it suggests that there are multiple mechanisms of aging. Because of its complexicity, the aging process takes us into the realm of integrative biology.

Many theories have been advanced to account for the aging process. For example, it has been attributed to molecular crosslinking, changes in immunologic function, damage by free radical reactions, and to senescence genes (for review see ref. 1). No single theory is generally accepted and this remarkable process remains a mistery. It is doubtful that a single theory will explain all the mechanisms of aging.

To elucidate the mechanisms of aging, any causal hypothesis should explain the following three conditions: (1) why organisms undergo progressive and irreversible physiological decline in the latter part of life; (2) why the life expectancy or rate of aging varies within and among species; and (3) why experimental regimens such as caloric restriction delay the onset of a variety of age-associated physiological and pathological changes and extend the average and maximum life-span of animals.

THE "FREE-RADICAL" THEORY OF AGING

A hypothesis ascribing one cause for aging postulates that the senescence-associated loss of functional capacity is due to the accumulation of molecular oxidative damages.[1–6] A chronic state of oxidative stress would exist in cells because of an imbalance between pro-oxidants and antioxidants. This hypothesis is based on the fact that oxygen is potentially a toxic substance, and its use by aerobes, although necessary for their immediate survival, also may be hazardous to their long-term existence. The phenomenon of oxygen toxicity, sometimes referred to as "the oxygen paradox," is inherent to the atomic structure of oxygen. Molecular oxygen is a biradical that upon single electron additions sequentially generates the partially reduced molecules $O_2^{\circ-}$, H_2O_2 and $^{\circ}OH$, which by further reactions can generate an array of

The Role of Oxidative Stress in Neuronal Death, by Irène Ceballos-Picot.

additional reactive oxygen metabolites and cause extensive oxidative damage to biological macromolecules (see chapter 2). This damage manifests as the peroxidation of membrane PUFA chains,[7] modifications of DNA (including base alterations, single-strand breaks, sister chromatin exchanges, and DNA-protein crosslinks),[8] and carbonyl modifiations[9,10] and loss of sulfhydryls in proteins among other changes.[11,12] Carbonyl modifications of proteins, occur in certain amino acid residues present near transition metal-binding sites and have been shown to cause enzymatic inactivation and enhance the likelihood of proteolysis.[13] Adverse effects of protein oxidation include memory impairment.[11] Oxidative changes in DNA and chromosomal material may shorten the life span, at least in part, by contributing to increases in improper gene expression with age. Increase in oxidative stress can result in sustained elevations of cellular Ca^{2+} levels with subsequent disruption of the cytoskeleton and activation of calcium dependent catabolic enzymes, including phospholipases, kinases, proteases, and endonucleases. Glycation is a common post-translational modification resulting from the nonenzymatic reaction between sugars, e.g., glucose, and primary amino groups on proteins. These products undergo a slow peroxidation to form so-called advanced glycosylation end products (AGE). Age-related increases in oxidative stress and protein glycation either individually or more probably in a synergistic manner could account for age-related brain cytoskeletal abnormalities.[14]

Support for this hypothesis includes the following observations:

(1) The level of molecular oxidative damage increases during aging.[1–6]

(2) Overexpression of antioxidative enzymes retards the age-related accrual of oxidative damage and extends the maximum life-span of transgenic *Drosophila melanogaster*.[15]

(3) Variations in longevity among different species inversely correlate with the rates of mitochondrial generation of the superoxide anion radical and hydrogen peroxide.[16]

(4) Restriction of caloric intake lowers steady-state levels of oxidative stress and damage, retards age-associated changes, and extends the maximum life-span in mammals.[17]

Oxidative Stress and Damage Increases During Aging; Possible Causes

The etiology of an age-associated increase in the amount of oxidative stress during aging can be linked to three different factors: an increase in the rate of generation of ROS, a decline in antioxidative defenses, and a decline in the efficiency of repair or removal of damaged molecules. Rates of $O_2^{\circ-}$ and H_2O_2 generation by mitochondria, which are the main source of production, increase with age in several different organs of mammals.[16–18] One mechanism contributing to this increase in ROS generated by mitochondria, is that it inflicts damage on the inner mitochondrial membrane, which by positive feedback induces a further increase in ROS generation. Another mechanism contributing to the increased production of $O_2^{\circ-}$ and H_2O_2 may be the age-related imbalance among different oxidoreductases; for example, the relative decrease in cytochrome oxidase activity that occurs during brain aging[2,19,21] would tend to increase the autoxidizability of the upstream respiratory components, thereby elevating the rate of $O_2^{\circ-}$ and H_2O_2 generation.[21] Age-associated abnormalities have also been observed an in the components of the electron transport chain in the brain.[22-25] Increase in Mt DNA deletions with age in the brain, observed in several species,[26-33] is also induced by ROS, resulting in mitochondrial functional abnormalities.[34-36] Simultaneous increase of mitochondrial DNA deletions and lipid peroxidation in human aging have been reported.[37] In the human brain particular mtDNA deletions or mutations appear in the areas particularly vulnerable during aging, in particular the substantia nigra.[31-33] Information about age-associated changes in the efficiency of processes that repair or remove oxidatively damaged molecules is quite fragmentary. A decrease in the activity of neutral and alkaline proteases, which preferentially degrade oxidized proteins, has been found only in certain tissues during aging.[38] Age-associated changes in the activities of individual antioxidant enzymes or low molecular weight antioxidant has been detected. A precise report for the brain is given below.

Overall, it seems that oxidative stress and damage increases during aging and particularly in the brain because of an increased rate of ROS generation and an increased susceptibility of this tissue to oxidative damages.

Experimental Models to Test the Oxidative Stress Hypothesis

Although the amount of accumulated oxidative damage during the normal adult life-span is extensive and rather substantial, no claims of causality can be made on this basis without the demonstration that an experimentally induced attenuation in oxidative stress does indeed retard the rate of aging and prolong the life-span. A recent approach to reduce oxidative stress and damage involved an experimental elevation in enzymatic antioxidant defenses so that the possibility of ROS generation and particularly $^{\circ}OH$ was minimized.[15] The effects of simultaneous overexpression of CuZnSOD and catalase, which act in tandem to remove $O_2^{\circ-}$ and H_2O_2, respectively, were determined in transgenic *Drosophila melanogaster*. As compared with diploid controls, transgenic flies carrying three copies of each of these genes exhibited: (1) up to one-third extension of the average and maximum life span; (2) a retardation in the age-related accumulation of oxidative damage to DNA and protein; (3) less DNA damage in response to X-ray exposure of live flies; (4) an attenuation in the age-associated increase in the rate of mitochondrial H_2O_2 generation; (5) an increase in the speed of walking; and (6) a 30% increase in the metabolic potential (total amount of oxygen consumed during adult life per unit body weight).

Together these results indicate, for the flies, that an experimentally induced decrease in oxidative stress or damage retards age-associated deteriorative changes at the organelle and organismic levels and extends the chronological as well as the metabolic life of the flies. These data are consistent with the notion that oxidative stress is one of the determinants of longevity. However, some caution is needed in interpretation.The excess enzymatic activities in the transgenic flies may do more than simply protect them from oxidative damage and could also intervene where ROS could act at the cellular level like in gene transcription and/or in cascade transduction pathways.

Relation Between Intra- and Inter-Species Variations in Life-Span and Oxidative Stress

Although information on the relation between oxidative stress and physiological age within a mammalian cohort is meager, an inverse relationship has been demonstratd in insects (For review see ref. 39). As compared with the age-matched, short lived flies, those with the longer life expectancy exhibited a lower rate of mitochondrial $O_2^{\circ-}$ and H_2O_2 generation, a lower level of protein oxidative damage, a lesser amount of DNA oxidative damage, higher activities of SOD and catalase, and an elevation of GSH. Variations in maximum life-span (MLS) among different species are often associated with differences in the metabolic rates (rate of oxygen consumption), metabolic potential and the level of oxidative stress. An inverse relationship was reported between metabolic rate and MLS in various phylogenetic groups.[1] Altogether, results of comparative studies indicate that metabolic potential and MLS are inversely related to oxidative stress.

Caloric Restriction, Attenuation of Oxidative Stress and Experimental Extension of Maximum Life-Span

Caloric restriction is now being increasingly used as a model regimen for understanding aging, primarily because it causes an unambiguous, robust and reproducible extension of MLS and delays many, although not all, age-associated biochemical, physiological and behavioral changes. Presently, no definitive information exists on whether or not caloric restriction similarly retards age-related deleterious changes and extends longevity in humans.

A decrease in caloric intake by laboratory rats and mice, performed without malnutrition, extends the MLS.[1] Decreasing the caloric intake of rats by 40% while maintaining essential nutrients increased average life span by 40% and maximum life span by 47%. This study demonstrated that caloric restriction slows production of aging changes by the inborn aging process.

If the hypothesis that oxidative stress is a major causal factor in the aging process is indeed valid, then one can predict that caloric restriction would lower the steady-state level of oxidative stress and thus decrease the rate of production of random free-radical damage throughout the cells and tissues, and in particular in the mitochondria where over 90% of the oxygen utilized by a mammal is consumed. Experimental results support these predictions. Rodents subjected to caloric restriction show attenuation of age-associated increases in rates of mitochondrial $O_2^{\circ-}$ and H_2O_2 generation, slowed oxidative damage, delayed loss of membrane fluidity, and decrease in the in vitro susceptibility of tissues to acute oxidative stress.[1,17] The highest degree of oxidative damage, as well as its attenuation by caloric restriction, occurs in tissues such as brain, heart and skeletal muscle, which are composed primarily of long-lived, post-mitotic cells. These tissues are also the targets of several age-related degenerative disorders in which oxidative stress has been implicated. The effects of caloric restriction on oxidative stress and damage in the brain appear to be rapidly inducible and reversible.

Thus, several lines of evidence suggest that the accrual of oxidative damage is a common feature of senescence, transcending phylogenetic boundaries. Characteristic variations observed in the rate of mitochondrial $O_2^{\circ-}$ and H_2O_2 generation among different species and tissues suggests that the rates of ROS production are genetically determined. Similarly, species-specific variations in the enzymatic antioxidative defenses together with the MLS extension by the overexpression of CuZnSOD and catalase in transgenic Drosophila implicate specific genes in the control of oxidative stress and aging. The extention of MLS by experimental regimens such as caloric restriction in mammals, which decrease the rates of ROS generation, point toward the involvement of environmental-genetic interactions in the governance of longevity. Although the present discussion has focused on the role of the damaging effects of ROS in aging, it is now extensively documented that ROS also modulate gene expression and cell signaling processes (see chapter 2), which further broadens their possible role in the aging process.

OXIDATIVE STRESS AND GENES IN AGING AND LONGEVITY

The life maintenance reserve is a genetically determined functional potential that allows the organism to survive, at least to reproductive maturity. Exhaustion of this reserve is a stochastic phenomenon with a probability that increases with age. There are two physiologic antecedents of life maintenance reserve: metabolic capacity and stress responses. The expansion of metabolic capacity results in the generation of oxidants, which demands the elaboration of the oxidative stress response; thus the response to oxidative stress is a primary stress response. The mitochondrial genome is especially vulnerable to oxidative damage and suffers deletions during brain aging.[26-33] The life maintenance reserve is modulated by epigenetic factors; the environment during development is of fundamental importance. The postdevelopmental environment exerts its effects by promoting aging through damage, stress and disease. As the somatic cell progresses through its life span, an epigenetic change in the regulatory states of chromatin set up during development becomes more probable. Deficits in the silencing apparatus lead to small changes in silencing.[39] Beyong a certain threshold, this becomes a self-accelerating process, resulting in gene dysregulation. Stress further accentuates this process; indeed, there is a reciprocal relation between environmental insult and dysregulation, one accentuating the other. Stress, particularly oxidative stress, has a profound effect on telomere metabolism and cellular senescence.[39] The resulting telomere changes are likely to affect the transcriptional status of subtelomeric genes, and they may also induce checkpoint controls.

The achievements of genetics in the analysis of aging in mammals are less impressive than in other organisms. Recently the APOE locus has also been associated with human longevity in a study of French centenarians.[40] The $ApoE_4$ isoform of apolipoprotein E is correlated with an increased susceptibility to both coronary heart disease and Alzheimer's disease. It is tempting to associate the increased frequency of the $ApoE_2$ isoform in centenarians with its antioxidant activity.[41] The association of ApoE with longevity raises important issues. When we uncover human longevity genes, are we actually defining the genes that affect susceptibility to disease with an age-dependent onset? Thus the factors that affect aging are genetic, epigenetic, and environmental, and the limiting factor for longevity are metabolic capacity, efficiency of stress responses.

OXIDATIVE PROCESSES AND ANTIOXIDANT DEFENSE MECHANISMS IN THE AGING BRAIN

Oxidative stress may play an important role in the pathogenesis of brain injuries associated with aging. Protein oxidation products are increased in both aged rodent and human brain, and the activity of enzymes vulnerable to oxidation, such as glutamine synthetase, is decreased.[9–11] In the human brain there is an age-related alteration from oligodendrocytes to astrocytes in the cellular distribution of ferritin, an iron-binding protein[42] and there is an overall reduction in the content of human brain membrane lipids with increasing age.[7] A clear indication of age-related alterations in brain lipid metabolism and possible increased lipid peroxidation is the accumulation of intraneuronal lipofuscin. The accumulation of this "aging pigment" has an essentially linear correlation with the age of the organism. Although the exact cause of the deposition is not known, the accumulation of lipofuscin in the nervous system and other tissues is exacerbated in animals fed a diet deficient in vitamin E, and markedly reduced when animals are treated with vitamin E supplements.[43,44] These reports provide evidence that oxidative stress occurs in the aging brain and that when the endogenous antioxidant mechanisms are able to minimize free radical damage, the aged brain remains relatively nonpathologic.

Any factor that weakens the brain antioxidant defense systems would render the brain more susceptible to extensive oxidative damage. The activity of antioxidant enzymes in normal dividing cells has been reported to increase or to decrease, depending on the tissue and age of the animals studied.[45,46] Furthermore, aging may affect each individual antioxidant enzyme in a particular way.[46-50]

The individual contribution of the antioxidants in neuronal protection during aging remains unknown. An approach to this question was to establish a comprehensive profile of the brain antioxidant defense potential during aging in control mice.[46] We have first determined the age-related profiles of CuZnSOD, MnSOD and glutathione-related enzyme (GSHPx and GSSG-Rd) activities from 2 months and continuing to senescence (28 months) in the brain of control mice. To elucidate the physiological significance of the two SODs in the aging process, we developed sensitive assays for CuZnSOD (cyanide sensitive) and MnSOD (cyanide-insensitive) activities in the mouse brain. The results are reported in Table 4.3.

We found a strong positive correlation between age and CuZnSOD activity and age and GSHPx activity. Another positive correlation was found between CuZnSOD and GSSG-Rd activities when the 3 groups were considered altogether. Morever, when this correlation was studied for each age group, this positive correlation was only found at 28 months. No modification with age in

Table 4.3. Age-related changes in antioxidant enzymes activities in the brains of mice

	Age (months) 2	12	28
CuZnSOD	43.2 ± 5.6 (n = 13)	50.3 ± 5.1a*** (n = 12)	50.3 ± 5.4a** (n = 15)
MnSOD	35.6 ± 6.6 (n = 13)	35.8 ± 2.6 (n = 12)	32.4 ± 2.7 (n = 15)
GSHPX	10.4 ± 1.9 (n = 13)	12.6 ± 1.8a** (n = 12)	14.6 ± 1.4a***,b (n = 15)
GSSG-Rd	41.0 ± 6.5 (n = 13)	40.2 ± 7.0 (n = 12)	41.7 ± 7.1 (n = 15)

Results are mean±SD; n: number of mice; CuZnSOD and MnSOD: U/mg of protein; GSHPx and GSSG-Rd: nmole of oxidized/min/mg of NADPH protein. NS, not significant; [a,b] Significant differences among age groups (Student t- test or Wilcoxon test); [a] significantly different from 2 months, [b] significantly different from 12 months; *P < 0.005; **P < 0.001; ***P < 0.001.

MnSOD and GSSG-Rd activities was observed. Thus, it is clear that the cellular antioxidant system is increased rather than weakened in senescent mouse brain.

Our data showing increased CuZnSOD activity with age are compatible with the other results in rats.[51,52] However, these findings contradict other authors who report no age-related increase in rat[45,53,54] or in mice[55a] or even a decrease in rats[55a–62] or mice.[60] It was demonstrated that the rate of mitochondrial $O_2^{\circ-}$ generation is correlated with aging.[63–65] Thus, the observed increase in CuZnSOD acitivity in the central nervous system of mice during aging may constitute a brain protection against superoxide anion elevation. Moreover, cytochrome oxidase activity increases in the brains of aging rats;[50] this suggests a close relationship between CuZnSOD activity and the rate of oxygen reduction by brain mitochondria. We found no changes in MnSOD activity with age. A possible explanation for the different CuZn and MnSOD activity profiles during aging is that the two enzymes are differently regulated in the brain. Few data are available for MnSOD activity in brain: an increase of MnSOD activity in the brain of old rats and mice has been reported[45,66] although another group showed a decrease.[67]

The pattern of changes in glutathione-related enzyme activities in the mouse brain are of particular interest: GSHPx activity showed a significant positive correlation with increasing age, whereas GSSG-Rd activity did not vary with age. Vitoria et al[68] also found increasing GSHPx activity with age in the soluble fraction of mitochondria. Other studies were unable to detect any variation in GSHPx activity in the aging rat brain[51] or mouse brain;[69] a decrease has even been reported.[64] For GSSG-Rd, a small but significant increase with age in rat brain has been reported[64,70] and confined to some brain areas such as the hippocampus and the cerebellum.[61] A decrease has also been reported.[55b,56] In brain, the GSHPx/GSSG-Rd catalyzed GSH-GSSG cycle plays a key role in metabolizing H_2O_2 and lipid peroxides. The observed increase of GSHPx activity with age suggests an increase in H_2O_2 formation in the cytoplasm and/or mitochondria, an adaptive phenomenon,[71] and could underlie the age-related increase in the rate of pro-oxidant generation, such as that of H_2O_2 as a candidate for being a biomarker of aging.[72] Both CuZnSOD and GSHPx activities increase with age. However, no direct correlation has been shown between the two activities. The strong positive correlation we have observed between CuZnSOD and GSSG-Rd in the aged group has no evident biological explanation, but it was also observed in human erythrocytes of subjects aged 65 years and over.[48]

The tripeptide GSH, the most abundant non-protein thiol in the cell, has been shown to participate in several protective mechanisms including the enzymatic reduction of intracellular peroxides via GSHPx, maintenance of reduced disulfide bounds in cellular proteins via GSH S-transhydrogenase, facilitation of the nonenzymatic reduction of free radicals, and in the direct conjugation of toxicants via GSH S-transferases. In this manner, GSH plays an important role in the prevention of oxidative damage that is induced either by exogenous toxicants or by endogenous products of oxidative metabolism, such as lipid peroxides and the products of monoamine degradation.[73] While the role of GSH in oxidation-reduction and conjugation reactions as well as in amino acid transport has been extensively studied in other tissues, little is known regarding the brain. Measurement of GSH and its oxidized form, GSSG, and the GSSG/GSH ratio are sensitive indices of redox status[74,75] of the cell and reflect the equilibrium between the generation of pro-oxidant challenge and the antioxidants to counteract such a challenge. Thus, their determination complements other antioxidants for the study of oxidative stress during aging.

The GSH level was higher in liver (6.50 μmol/g wt) than in the brain (1.18), lung (1.07) and cerebellum (0.70). The GSSG level was higher in lung (0.340 μmol/g wt) and liver (0.152) than in brain (0.022) and cerebellum (0.022). The role of increased levels of GSH within liver as compared to other tissues analyzed remains to be determined. It is reasonable, however, to predict that a larger intracellular pool of GSH reflects a greater potential to supply GSH to those reactions involved in detoxification and cellular protection from xenobiotic and oxidative damage. To determine whether or not brain aging is associated with variations of GSH and GSSG levels, we examined these three paramaters in young (3 months) and old (24 months) mice. No variation in GSH level was observed according to age in the different tissues analyzed. However, aging was associated with a significant increase in GSSG levels in brain, cerebellum and lung, but not in liver (Fig. 4.1).

Previous reports have shown a decrease in GSH levels in several peripheral organs with age[76-79] and in some brain regions of aged rats, such as cerebellum, cortex and olfactory bulb[80] and hippocampus.[80,81] GSH depletion decreases protein synthesis and alters DNA synthesis because GSH provides the reducing equivalents for the glutaredoxin system supplying electrons to ribonucleotide reductase. The GSH depletion affects many GSH-dependent enzymes, making the cells more susceptible to any further challenges. In contrast but in agreement with our results, GSH levels in the aged

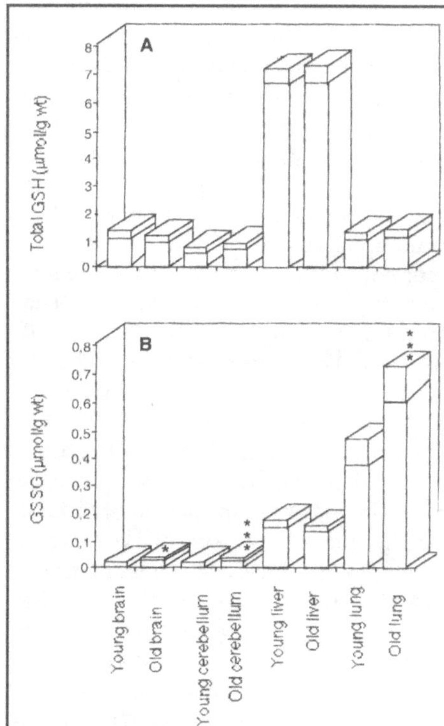

Fig. 4.1. Total (GSH) and oxidized (GSSG) glutathione levels in different tissues of mice and variations with aging. A: total GSH levels; B: GSSG levels. For each tissue, variation with age was compared using t-test. * $p < 0.05$; *** $p < 0.001$.

animals were not found to be significantly different from those in the neonatal rats in either the olfactory epithelium or hippocampus.[82] Farooqui et al[83] found low levels of GSH in whole brain homogenates at 1 month, which increase 6-fold at 2.5 months and return to the 1 month levels in 3-year-old rats. Benzi et al[55b] reported similar changes in GSH levels in rat forebrain, although these increases were less pronounced and followed a different time course. The use of homogenates provides a general idea of overall concentration, but such techniques cannot resolve the relative contribution of distinct cell populations and subtypes to the total GSH content within the tissue. Since the brain exhibits important regional and cellular heterogeneity, it will be necessary to quantify GSH in different regions/areas of the young and aged mice brains, which will allow GSH distribution to be examined at the cellular level. Differences in regional and cellular distribution of glutathione and its associated enzymes in heterogeneous tissue such as the nervous system are likely critical determinants of differential sensitivity to oxidative stress (see chapter). Reduced glutathione in the rat brain appears primarily localized in the neuropil and white matter tracts; the neuronal somata do not appear to contain appreciable amounts of GSH.[84] Similarly, Slivka et al[85] suggested that GSH may be primarily localized to glia and/or axons and nerve terminals. Raps et al[86] have shown high concentrations of GSH in cultured astrocytes but not in cultured neurons. Interestingly, GSH staining in the cerebellum became localized to the Purkinje cell layer in only the aged animals.[82] Enzymes associated with GSH functions might be expected to colocalize with the tripeptide. Although data are limited, the distribution of these proteins appears to follow the general pattern of GSH distribution. GSHPx and GSH-S-transferases have both been observed in glial cells in rat brain[87-88] and human brain (chapter 2). These data suggest that glia may play a dominant role in the detoxification of oxidants and/or excretion of xenobiotics and that neurons may play a more passive role. However, CuZnSOD protein and mRNA was preferentially expressed in neurons in mice[89] and humans, in areas particularly vulnerable during aging, i.e., the substantia nigra[90] and hippocampus.[91,92] The heterogeneous distribution of GSH and enzymes involved in the neutralization of oxygen free radicals and/or excretion of xenobiotics in the nervous sytem may form a basis for selective cellular and/or regional expression of neuronal vulnerability.

Concentrations of GSH are partly regulated by GSHPx and GSSG-Rd. GSSG-Rd reduces GSSG to GSH. The increased activity of GSHPx together with the unchanged GSSG-Rd activity in the brain during aging could explain the age-associated increase in GSSG concentration reported here. This finding is in agreement with those of previous authors on the brain.[93] This change in glutathione redox status denotes an imbalance between proxidant and antioxidant in favor of oxidation, the so-called oxidative stress. GSH plays a salient role in the regulation of intracellular redox status, and thus the higher level of GSSG in the aging brain may reflect an increased susceptibility to oxidative injury.

Although GSH is widely distributed in the brain, its biological function in the central nervous system is not fully understood. However, its large concentration[94] and rapid synthesis suggests an important role in this organ. Concentrations of GSH in tissues of the mouse and rat decline with age for some authors[69,80,95] as well as in peripheral tissues in humans.[78,79] Moreover, a decrease in total glutathione has been reported in the brains of parkinsonian patients as compared to control subjects (see chapter 8) and this decrease appeared to be selective for the substantia nigra. These changes have been interpreted as the result of increased oxygen radical production in the nigrostriatal pathway of parkinsonian patients, and have led to the speculation that Parkinson's disease may be attributable to nigral glutathione deficiency.[96]

It is clear that a number of links exist between oxidative stress and mitochondrial damage and that glutathione may represent one such link:

(1) Although most glutathione is localized in the cytosolic fraction, approximately 10% of the total glutathione is compartmentalized within mitochondria and depletion of GSH in the brain of newborn rats by administration of buthionine sulfoximine (L-BSO: an inhibitor of GSH synthesis) caused mitochondrial damage assessed by both electron microscopy and measurements of the activity of citrate synthase, a mitochondrial matrix marker enzyme.[97]

(2) the levels of a specific MtDNA deletion increases with age, particularly in three regions of the brains: caudate, putamen and substantia nigra.[30,31] These brain areas are characterized by a high dopamine metabolism. The breakdown of dopamine by mitochondrial MAO produces H_2O_2 which can lead to free radical formation. Like the levels of MtDNA, mitochondrial oxygen radical production appears to increase with age.[98] It was suggested that MtDNA deletions may be the "tip of the iceberg" of the spectrum of somatic mutations produced by oxidative damage.

(3) Excessive production of H_2O_2 within mitochondria may lead to depletion of mitochondrial GSH, oxidation of protein thiols, and impairment of mitochondrial function, providing an example of the relationship between the glutathione status, oxidative stress and mitochondrial damage.

(4) If the loss of GSH may cause mitochondrial damage, it is also concevable that impairment of mitochondrial function can lead to a decrease in cytosolic GSH. GSH synthesis requires ATP, and thus a deficiency of energy supplied by mitochondria is likely to affect the cellular turnover of glutathione.

The effects of GSH depletion on mitochondrial function and neuronal death is developed in chapter 8.

MITOCHONDRIA, OXIDATIVE STRESS AND AGING

Oxidative Stress and Mitochondrial Function in the Aging Brain

There is a substantial evidence that mitochondrial function declines with age. This evidence comes from direct biochemical measurements of mitochondrial function, studies of damage to mitochondrial DNA, and position emission tomography studies of glucose utilization during normal aging. Mitochondria are the main sources of ROS, and this chapter focuses first on whether and how these organelles contribute to age-related changes.

The mitochondrial electron transfer chain (ETC) conveys a flux of electrons from the donors (NADH and succinate, produced in the tricarboxylic acid cycle) to molecular oxygen, causing the release of considerable amounts of energy available for ATP synthesis, ion translocation, protein importation and so on. A series of molecular complexes (consisting of various equipotential subunits located in the mitochondrial inner membrane) (see Fig. 1.2) support the electron transfer that arises with the vectorial translocation of protons and gives rise to an electrochemical potential, thus providing the transduction of oxidative energy for ATP synthesis. Within complexes I, III and IV, the energy transduced from electron transfer is conserved by means of coupled vectorial proton translocation, which generates a membrane electrochemical potential of protons used in complex V to convert ADP to ATP. The final stage of electron transfer in complex IV (cytochrome aa3 to oxygen) is irreversible, so that the equilibrium in the system is shifted towards ATP synthesis. The functioning of ETC is, in any case, conditioned by the integrity of the phospholipid structures in the mitochondrial inner membrane.

Cytochrome aa_3 in complex IV retains all the partially reduced oxygen intermediates bound to its active site until the O_2 itself is completely reduced to water. This suggests that, although oxygen undergoes tetravalent reduction, partially reduced O_2 intermediates do not exist in a free form because they remain tightly bound to complex IV until water is formed. However, through an auto-oxidation process affecting their reduced forms, other elements in the mitochondrial ETC (ubiquinones and cytochrome b family) could transfer the electrons directly to oxygen, but do not retain the partially reduced oxygen intermediates in their active sites until the O_2 itself is completely reduced to water. Because oxygen accepts only one electron at a time, the superoxide radical is released. Isolated mitochondria generate 0.6 to 1.0 nmol of H_2O_2 min^{-1} mg of $protein^{-1}$, which is estimated to account for 2 to 4% of oxygen uptake under physiological conditions.[99]

During aging, increasing quantities of ROS alter the chemical and physical state of the inner mitochondrial membrane where ubiquinone redox cycling operates. A significant increase in the generation of superoxide,[21,100] hydrogen peroxide,[21] and hydroxyl radicals[101] is present in the brain mitochondria of aged rats as compared to young ones. This age-related increase in the formation of superoxide radicals and its dismutation product is based on the concept that aging modifies the electron flux conditions in the component of ETC, thus making it easier for electrons to escape the normal flow sequence. In parallel with the lower concentration of cyt aa_3 in complex IV, the activity of cytochrome oxydase is decreased, particularly in synaptic mitochondria from frontal and parieto-temporal cortex, cerebellum, hippocampus of aged rats.[102–103] The stoichiometric ratio of the components of ubiquinone population to cytochrome aa_3 exhibits an age-related increase. This, in turn, could result from the increase in ROS release inducing changes in the physico-chemical membrane properties and favoring the auto-oxidation of ubiquinones in mitochondria. Thus, the loss of an efficient control impairs the energy conservation and free ROS may be formed during aging. In addition ROS produced from other metabolic pathways such as NO could also modulate mitochondrial respiration.[104–106] This could be a possible factor of mitochondrial alteration during aging.

A detailed analysis of the mitochondrial inner membrane protein composition of cerebellar mitochondria revealed that the 75 kDa protein which corresponds to NADH dehydrogenase, a 16 kDa protein (subunit IV or cyt aa_3) or a 14 kDa protein (cyt c) decreases with aging.[107] It would be interesting to know whether these changes in mitochondrial constituents in the aged brain alter respiration and oxidative phosphorylation.The respiratory control ratio (RCR) determined in the presence of NADH-generating substrates (glutamate + malate) but not of $FADH_2$-generating substrate (succinate) is significantly slower in mitochondria from the hippocampus and striatum respectively from old compared to young rats.[108] The effects of aging on mitochondrial ETC in cerebral cortex of rhesus monkeys across their full life span were studied.[23] There were significant decreases in complex I and IV activities, but complex II-III, complex V, and citrate synthase were unaffected.

Age-Related MtDNA Mutation Accumulation in the Brain

Human MtDNA is a naked circular double-stranded DNA, which is continually exposed to the matrix that contains high levels of ROS and free radicals. High oxidative stress and lack of proofreading during mtDNA replication and efficient DNA repair mechanisms in the mitochondria have rendered mtDNA extremely vulnerable to oxidative damage. More than one dozen large-scale deletions in mtDNA have been identified in various tissues of old humans. The 4,977-bp and 7,436-bp deletions are the most prevalent and abundant ones.[26–32] The onset age of various mtDNA deletions varies greatly with tissues of each individual and type of deletion, and the frequency of occurrence and the proportion of the 4,977-bp and 7,436-bp deleted MtDNAs are significantly increased with the age of human. The MtDNA deletions are not detectable in any tissues from young healthy subjects from normal individuals, which indicates that the deletions are generated and accumulated only in postmitotic cells upon aging. Moreover, these MtDNA deletions occur more frequently and abundantly in tissues with high energy demand (e.g., brain) as compared to those with low energy demand. In brain tissue there are age-dependent increases in the 4,977-bp deletion, which are greatest in the striatum and substantia nigra, perhaps due to free radicals generated by MAO.[31,32,37] On the other hand, the amount of lipid peroxides and the activity of MnSOD in the mitochondria exhibit an age-dependent increase[36] with a positive correlation between the proportion of 4,977-bp deleted mtDNA and lipid peroxide content in the mitochondria of human tissues during aging. Mitochondrial DNA, which encodes 13 protein-coding genes specifying subunits of the mitochondrial respiratory enzyme complex, is particularly susceptible to oxidative damage due to its close proximity to the respiratory chain, limited repair mechanisms, few noncoding sequences, and absence of histones.

Pathogenic mutated MtDNAs are usually "heteroplasmic," coexisting intracellularly with wild-type MtDNAs. Several mitochondrial point mutations also increase with normal aging.[109] Some of these nucleotide substitutions were missense and nonsense mutations, which potentially have deleterious effects on mitochondrial function. The frequency of nucleotide substitutions in the striatum of three patients with Parkinson's disease was also significantly higher than in control tissues[109] together with an increased protein modification by 4-hydroxy-2-nonenal, a lipid peroxidation by-product. Large-scale deletions and tandem duplications of mtDNA have recently been found in various tissues of aged individuals. By use of PCR techniques with back to back primers, it was demonstrated that small tandem duplications occur in the D-loop of MtDNA in an age-dependent manner in human tissues.[110] The aging-associated 4,977-bp and 7,436-bp mtDNA deletions and some of the small tandem duplications could occur alone or in different combinations in human tissues in the aging process. The incidence and abundance of some of the tandem duplications as well as the large-scale deletions could be early molecular events of the human aging process.

Direct evidence for age-dependent increases in oxidative damage to MtDNA comes from measurements of 8-hydroxy-2-deoxyguanosine, which is an oxidized form of deoxyguanosine that occurs following attack by a variety of free radicals. Marked age-dependent increases in this marker of oxidized MtDNA that correlate with mitochondrial deletions occur in mitochondrial DNA from human brain tissue, and were 15-fold increased compared with nuclear DNA in patients older than age 70.[33] Therefore, mitochondrial mutations may increase oxidative stress and cerebral aging.[111]

CONCLUSIONS

Taking these results together, the enhanced generation of ROS and lipid peroxides in the mitochondria during the aging process occur simultaneously with large-scale deletions and the other types of mutations in mtDNA. It is therefore possible that a decline in oxidative phosphorylation with normal aging might be additive, with subclinical defects to produce delayed onset of neurodegeneration. These studies suggest that a vicious cycle contributes to the progression of degenerative process. In this cycle, first a primary mitochondrial mutation(s) or deletion(s) induces a mitochondrial respiratory defect, which increases the leakage of ROS from the respiratory chain. Then the ROS would trigger accumulation of secondary mtDNA mutations in post-mitotic cells, and damage mitochondrial membranes, leading to further aggravation of mitochondrial respiratory defects and increased production of ROS and lipid peroxides from mitochondria, resulting in degeneration of cellular components.

REFERENCES

1. Sohal RS, Weindruch R. Oxidative stress, caloric restriction and aging. Science 1996; 273:59-63.
2. Harman D. Aging a theory based on free radicals and radiation chemistry. J Gerontol 1956; 11:298-300.
3. Harman D. Free radical theory of aging: Beneficial effect of antioxidants on the life span of male NZB mice: role of free radicals in the deterioration of the immune system with age and in the pathogenesis of systemic lupus erythematosus. Age 1980; 3:64-73.
4. Harman D. The aging process. Proc Natl Acad Sci USA 1981; 78:7124-7128.

5. Harman D. Free radical theory of aging: the "free radical diseases." Age 1984; 7:111-131.
6. Harman D. Free radical theory of aging: role of free radicals in the origin and evolution of life, aging and diseases processes. In: Liss AR, ed. Free radicals, aging, and degenerative diseases. 3-49.
7. Svennerholm L, Bostrom K, Helander CG et al. Membrane lipids in the aging human brain. J Neurochem 1991; 56: 2051-2059.
8. Richter C, Park JW, Ames BN. Normal oxidative damage to mitochondrial DNA and nuclear DNA is extensive. Proc Natl Acad Sci USA 1988; 85:6465-6467.
9. Smith CD, Carney JM, Starke-Reed PE et al. Excess brain protein oxidation and enzyme dysfunction in normal aging and in Alzheimer disease. Proc Natl Acad Sci USA 1991; 88:10540-10543.
10. Fucci L, Oliver CN, Coon MJ, Stadtman ER. Inactivation of key metabolic enzymes by mixed function oxidation reactions: possible implication in protein turnover and ageing. Proc Natl Acad Sci USA 1983; 80:1521-1525.
11. Carney JM, Starke-Reed PE, Oliver CN et al. Reversal of age-related increase in brain protein oxidation, decrease in enzyme activity, and loss in temporal and spatial memory by chronic administration of the spin trapping compound N-tert-butyl-a-phenylnitrone. Proc Natl Acad Sci USA 1991; 88:3633-3636.
12. Mann DMA, Yates PO, Stamp JE. Relationship between lipofuscin pigment and aging in the human nervous system. J Neurol Sci 1978; 37:83-93.
13. Youngman LD, Kim Park J-Y, Ames BN. Protein oxidation associated with aging is reduced by dietary restriction of proteins or calories. Proc Natl Acad Sci USA 1992; 89:9112-9116.
14. Radical Ageing in Alzheimer's disease. Trends Neurosci 1995; 18:172-176.
15. Orr WC, Sohal RS. Extention of life-span by overexpression of superoxide dismutase and catalase in Drosophila melano*gaster*. Science 1994; 263:1128-1130.
16. Sohal RS, Dubey A. Mitochondrial oxidative damage, hydrogen peroxide release, and aging. Free Rad Biol Med 1994; 16:621-626.
17. Sohal RS, Ku H-H, Agarwal S et al. Oxidative damage, mitochondrial oxidant generation and antioxidant defenses during aging and in response to food restriction in the mouse. Mech Ageing Dev 1994; 74:121-133.
18. Sohal RS, Brunk UT. Mitochondrial production of pro-oxidant and cellular senescence. Mut Res 1992; 275:295-304.
19. Curti D, Giangare MC, Redolfi ME et al. Age-related modifications of cytochrome c oxidase activity in discrete brain regions. Mech Ageing and Dev 1990; 55: 171-180.
20. Martinez M, Ferrandiz ML, de Juan E et al. Age-related changes in glutathione and lipid peroxide content in mouse synaptic mitochondria: relationship with cytochrome c oxidase decline. Neurosci Lett 1994; 170:121-124.
21. Sohal RS. Aging, cytochrome oxidase activity, and hydrogen peroxide release by mitochondria. Free Rad Biol Med 1993; 14:583-588.
22. Bowling AC, Mutisya EM, Walker LC et al. Age-dependent impairment of mitochondrial function in primate brain. J Neurochem 1993; 60:1964-1967.
23. Di-Monte DA, Sandy MS, Delanney et al. Age-dependent changes in rat brain mitochondrial energy production in striatum and cerebellum of the monkey brain. Neurodegeneration 1993; 2:93-99.
24. Gadaletta MN, Petruzella V, Renis M et al. Reduced transcription of mitochondrial DNA in the senescent rat. Eur J Biochem 1990; 187:501-506.
25. Fernandez-Silva P, Petruzella V, Fracasso F et al. Reduced synthesis of MtRNA in isolated mitochondria of senescent rat brain. Biochem Biophys Res Comm 1991; 176:645-653.
26. Cortopassi GA, Arnheim N. Detection of a specific mitochondrial DNA deletion in tissues of older humans. Nucleic Acids Res 1990; 18:6927-6933.
27. Linnane AW, Baumer A, Maxwell RJ et al. Mitochondrial gene mutation: the ageing process and degenerative diseases. Biochem Intl 1990; 22:1067-1076.
28. Simonetti S, Chen X, DiMauro S et al. Accumulation of deletions in human mitochondrial DNA during normal hu-

man aging: analysis by quantitative PCR. Biochim Biophys Acta 1992; 1180: 113-122.
29. Brossas J-Y, Barreau E, Courtois Y et al. Multiple deletions in mitochondrial DNA are present in senescent mouse brain. Biochem Biophys Res Comm 1994; 202: 654-659.
30. Cortopassi GA, Shibata D, Soong N-W et al. A pattern of accumulation of somatic deletion of mitochondrial DNA in aging human tissues. Proc Natl Acad Sci USA 1992; 89:7370-7374.
31. Lee H-C, Pang C-Y, Hsu H-S et al. Differential accumulation of 4,977 bp deletion in mitochondrial DNA of various tissues in human ageing. Biochim Biophys Acta 1994; 1226:37-43.
32. Corral-Debrinski M, Horton T, Lott MT et al. Mitochondrial DNA deletions in human brain: regional variability and increase with advanced age. Nature Genet 1992; 2:324-329.
33. Mecocci P, McGarvey U, Kaufman AE et al. Oxidative damage to mitochondrial DNA shows marked dependent increases in human brain. Ann Neurol 1993; 34: 609-616.
34. Bolanos JP, Peuchen S, Heales SJR, Land JM, Clark JB. Nitric oxide-mediated inhibition of the mitochondrial respiratory chain in cultures astrocytes. J Neurochem 1994; 63:910-916.
35. Bolanos JP, Simon JR, Heales SJR, Land JM, Clark JB. Effect of peroxynitrite on the mitochondrial respiratory chain: differential susceptibility of neurones and astrocytes in primary culture. J Neurochem 1995; 64:1965-1972.
36. Brown GC. Nitric oxide regulates mitochondrial respiration and cell functions by inhibiting cytochrome oxidase. FEBS Lett 1995; 369:136-139.
37. Wei YH, Kao SH, Lee HC. Simultaneous increase of mitochondrial DNA deletions and lipid peroxidation in human aging. Ann NY Acad Sci 1996; 786:24-43.
38. Verity MA. Oxidative damage and repair in the developing nervous system. Neurotoxicology 1994; 15:81-92.
39. Jazwinski SM. Longevity, genes and aging. Science 1996; 273:54-59.
40. Schächter F, Faure-Delanef L, Guenot F et al. Genetic association with human longevity at the ApoE and ACE loci. Nature Genet 1994; 6:29-32.
41. Miyata M, Smith JD. Apolipoprotein E allele-specific antioxidant activity and effects on cytotoxicity by oxidative insults and β-amyloid peptides. Nature Genet 1996; 14:55-61.
42. Connor JR, Menzies SL, Martin SM et al. Cellular distribution of transferrin, ferritin, and iron in normal and aged human brains. J Neurosci Res 1990; 27:595-611.
43. Dowson JH. Neuronal lipofuscin accumulation in ageing and Alzheimer dementia: a pathogenic mechanism? Brit J Psychiat 1982; 140:142-148.
44. Dowson JH, Fattoretti P, Cairns M et al. The effects of ageing and a vitamin E-deficient diet on the lipopigment content of rat hippocampal and Purkinje neurones. Arch Gerontol Geriatr 1992; 14:239-251.
45. Cao Danh H, Strolin Benedetti M, Dostert P. Differential changes in superoxide dismutase activity in brain and liver of old rats and mice. J Neurochem 1983; 40:1003-1007.
46. Ceballos-Picot I, Nicole A, Sinet PM. Age-related changes in antioxidant enzymes and lipid peroxidation in brains of control and transgenic mice overexpressing copper-zinc superoxide dismutase. Mut Res 1992; 275:281-293.
47. Ceballos-Picot I, Trivier JM, Nicole A et al. Age-correlated modifications of copper-zinc superoxide dismutase and glutathione-related enzyme activities in human erythrocytes. Clin Chem 1992; 38:66-70.
48. Berr C, Nicole A, Godin J, Ceballos-Picot I et al. Relationships between plasma selenium, erythrocyte selenium, oxygen metabolizing enzyme concentrations and age in elderly community residents: a pilot epidemiologic study. Am J Geriatr Soc 1993; 41:143-148.
49. Ceballos-Picot I, Merad-Boudia M, Nicole A et al. Peripheral antioxidant enzyme activities and selenium in elderly subjects and in dementia of Alzheimer's type-place of the extracellular glutathione

peroxidase. Free Rad Biol Med 1996; 20:579-587.
50. Vanella A, Villa RF, Gorini A et al. Dismutase and cytochrome oxydase activities in light and heavy synaptic mitochondria from rat cerebral cortex during aging. J Neurosci Res 1989; 22: 351-355.
51. Cand F, Verdetti J. Superoxide dismutase, glutathione peroxidase, catalase and lipid peroxidation in the major organs of the aging rats. Free Rad Biol Med 1989; 7:59-63.
52. Devasagayam TPA. Decreased peroxydative potential in rat brain microsomal fractions during aging. Neurosci Lett 1989; 103:92-96.
53. Barja de Quiroga G, Perez-Campo, Lopez-Torrez M. Anti-oxidant defenses and peroxidation in liver an brain of aged rats. Biochem J 1990; 272:247-250.
54. Kellogg EW, Fridovich I. Superoxide dismutase in the rat and mouse as a function of age and longevity. J Gerontol 1976; 31:405-408.
55a. Reiss U, Gershon D. Comparison of cytoplasmic superoxide dismutase in liver, heart and brain of aging rats and mice. Biochim Biophys Res Commun 1976; 73:255-262.
55b. Benzi G, Pastoris O, Marzatico F et al. Age-related effect induced by oxidative stress on the cerebral glutathione system. Neurochem Res 1989; 14:473-481.
56. Benzi G, Marzatico F, Pastoris O et al. Relationship between aging, drug treatment and the cerebral enzymatic antioxidant system. Exp Gerontol 1989; 24: 137-148.
57. Geremia E, Baratta D, Zafarana S et al. Antioxidant enzymatic systems in neuronal and glial cell-enriched fractions of rat brain during aging. Neurochem Res 1990; 15:719-772.
58. Gupta A, Hasan M, Chander R et al. Age-related elevation in lipid peroxidation products: diminution of superoxide dismutase activity in the central nervous system of rats. Gerontology 1991; 37:305-309.
59. Lebel CP, Bondy SC. Persistent protein damage despite reduced oxygen radical formation in the aging rat brain. Int J Dev Neurosci 1991; 9:136-146.
60. Massie HR, Aiello VR, Iodice AA. Changes with age in copper and superoxide dismutase levels in brains of C57BL/6J mice. Mech Ageing Dev 1979; 10:93-99.
61. Mizuno Y, Ohta K. Regional distribution of thiobarbituric acid-reactive products, activities of enzymes regulating the metabolism of oxygen free radicals and some of the related enzymes in adult and aged rat brain. J Neurochem 1986; 46: 1344-1352.
62. Semsei I, Govinda R, Richardson A. Expression of superoxide dismutase and catalase in rat brain as a function of age. Mech Ageing Dev 1991; 58:13-19.
63. Sawada M, Carlson JC. Changes in superoxide radical and lipid peroxide formation in the brain, heart and liver during the life time of the rat. Mech Ageing Dev 1987; 41:125-137.
64. Sohal RS, Arnold LA, Sohal B. Age-related changes in antioxidant enzymes and pro-oxidant generation in tissues of the rat with special reference to parameters in two insect species. Free Radical Biol Med 1990; 10:495-500.
65. Sohal RS, Allen RG. Oxidative stress as a causal factor in differentiation and aging: a unifying hypothesis. Experiment Gerontol 1990; 25:499-522.
66. Scarpa M, Rigo A, Viglino P et al. Age-dependence of the level of the enzymes involved in the protection against active oxygen species in the rat brain. Proc Soc Ex Med 1987; 185:129-133.
67. Ledig M, Fried R, Ziessel M et al. Regional distribution of superoxide dismutase in rat brain during postnatal development. Dev Brain Res 1982; 4: 333-337.
68. Vitoria J, Machado A, Satrustegui J. Age-dependent variations in peroxide-utilizing enzymes from rat brain mitochondria and cytoplasm. J Neurochem 1984; 42: 351-356.
69. Hazelton GA, Lang CA. Glutathione peroxidase and reductase activities in the aging mouse. Mech Ageing Dev 1985; 29:71-78.
70. Hothersall JS, El-Hassan A, McLean P et al. Age-related changes in enzymes of rat brain. Redox system linked to NADPH

and glutathione. Enzyme 1981; 26: 271-276.
71. Ceballos I, Delabar JM, Nicole A et al. Expression of transfected human CuZnSOD in mouse L cells and NS2OY neuroblastoma cells induces enhancement of glutathione peroxidase activity. Bio chim Biophys Acta 1988; 949:58-64.
72. Sohal RS. Hydrogen peroxide production by mitochondria may be a biomarker of aging. Mech Ageing Dev 1991; 60: 189-198.
73. Meister A, Anderson ME. Glutathione. Ann Rev Biochem1983; 52:711-760.
74. Griffith O. Determination of glutathione and glutathione disulfide using glutathione reductase and 2-vinylpyridine. Anal Biochem 1980; 106:207-212.
75. Anderson ME. Determination of glutathione and glutathione disulfide in biological samples. Methods Enzymol 1985; 113:548-555.
76. Abraham EC, Taylor JF, Lang CA. Influence of mouse age and erythrocyte age on glutathione metabolism. J Biochem 1978; 174:819-822.
77. Hazelton GA, Lang CA. Glutathione content of tissues in the aging mouse. J Biochem 1980; 188:25-30.
78. Naryshkin S, Miller L, Lindeman R et al. Blood glutathione: a biochemical index of human aging. Fed Proc 1981; 40:3179.
79. Schneider D, Naryshkin S, Lang CA. Blood glutathione, a biochemical index of aging women. Fed Proc 1982; 41:7671.
80. Ravindranath V, Shivakumar BR, Anandatheerthavarada HK. Low glutathione levels in brain regions of aged rats. Neurosci Lett 1989; 101:187-190.
81. Chen TS, Richie JP, Lang CA. The effect of aging on glutathione and cysteine levels in different regions of the mouse brain. Soc Biol Med 1989; 190:399-402.
82. Kudo H, Kokunai T, Kondoh T et al. Quantitative analysis of glutathione in rat central nervous system: comparison of GSH in infant brain with that in adult brain. Brain Res 1990; 511:326-328.
83. Farooqui MYH, Day WW, Zamorano DM. Glutathione and lipid peroxidation in the aging rat. Comp Biochem Physiol 1987; 88B:177-180.
84. Philbert MA, Beiswanger CM, Waters DK et al. Cellular and regional distribution of reduced glutathione in the nervous system of the rat: histochemical localization by mercury orange and O-phtaldialdehyde-induced histofluorescence. Toxicol and Applied Pharmacol 1991; 107:215-227.
85. Slivska A, Mytillineou C, Cohen G. Histochemical evaluation of glutathione in brain. Brain Res 1987; 209:275-284.
86. Raps SP, Lai JCK, Hertz L et al. Glutathione is present in high concentration in cultured astrocytes but not in cultured neurons. Brain Res 1989; 493:498-501,
87. Cammer W, Tansey F, Abramovitz M et al. Differential localization of glutathione-S-transferase Yp and Yb subunits in oligodendrocytes of rat brain. J Neurochem 1989; 52:876-883.
88. Ushikima K, Miyazaki H, Morioka T. Immunohistochemical localization of glutathione peroxidase in the rat brain. Rescucitation 1986; 13:97-105.
89. Ceballos-Picot I, Nicole A, Briand P et al. Neuronal-specific expression of human copper-zinc superoxide dismutase gene in transgenic mice: animal model of gene dosage effect in Down's syndrome. Brain Res 1991; 552:198-214.
90. Ceballos I, Lafon M, Javoy-Agid F et al. Superoxide dismutase and Parkinson's disease. The Lancet 1990; 335:1035-1036.
91. Ceballos I, Javoy-Agid F, Hirsch EC et al. Localization of copper-zinc superoxide dismutase mRNA in human hippocampus by in situ hybridization. Neurosci Lett 1989; 105:41-46.
92. Delacourte A, Defossez A, Ceballos I et al. Preferential expression of copper-zinc superoxide dismutase in the vulnerable cortical neurons in Alzheimer's disease. Neurosci Lett 1988; 92:247-253.
93. Vina J, Sastre J, Anton V et al. Effect of aging on glutathione metabolism, protection by antioxidants. In: I Emerit, Chance B, eds. Free radicals and Aging. Birhaüser Verlag, 1992:89-98.
94. Griffith O, Meister A. Glutathione:interorgan translocation, turnover and metabolism. Proc Natl Acad Sci USA 1979; 76:5605-5610.

95. Teramoto S, Fukuchi Y, Uejima Y et al. Age-related changes in GSH content of eyes in mice—a comparison of senescence-accelerated mouse (SAM) and C57BL/J mice. Comp Biochem Physiol 1992; 102A:693-696.
96. Di Monte DA, Chan P, Sandy MS. Glutathione in Parkinson's disease: a link between oxidative stress and mitochondrial damage. Ann Neurol 1992; 32:S111-S115.
97. Jain A, Martensson J, Stole E et al. Glutathione deficiency leads to mitochondrial damage in brain. Proc Natl Acad Sci USA 1991; 88:1913-1917.
98. Benzi G, Curti D, Pastoris O et al. Sequential damage in mitochondrial complexes by peroxidative stress. Neurochem Res 1991; 29:527-532.
99. Boveris A, Chance B. The mitochondrial generation of hydrogen peroxide. Biochem J 1970; 134:707-716.
100. Sawada M, SesterU, Carlson JC. Superoxide dismutase formation and associated biochemical alterations in the plasma membrane of brain, heart and liver during the lifetime of the rat. J Cell Biochem 1992; 48:296-304.
101. Zhang Y, Marcillat O, Giulivi C et al. The oxidative inactivation of mitochondrial electron transport chain components and ATPase. J Biol Chem 1990; 265:16330-16336.
102. Curti D, Benzi G. Cytochrome c oxidase activity in discrete brain regions. Mech Ageing Dev 1990; 55:171-180.
103. Benzi G, Pastoris, Marzatico et al. The mitochondrial electron transfer alteration as a factor involved in the brain aging. Neurobiol Aging 1992; 13:361-368.
104. Benzi G, Moretti A. Age- and peroxidative stress-related modifications of the cerebral enzymatic activities linked to mitochondria and the glutathione system. Free Rad Biol Med 1995; 1:77-101.
105. Cleeter MWJ, Cooper JM et al. Reversible inhibition of cytochrome c oxidase, the terminal enzyme of the respiratory chain, by nitric oxide. Implications for neurodegenerative diseases. FEBS Lett 1994; 345:50-54.
106. Radi R, Rodriguez M, Castro L et al. Inhibition of mitochondrial electron transport by peroxynitrite. Arch Biochem Biophys 1994; 308:89-95.
107. Turpeenoja L, Villa RF, Magri G et al. Changes of mitochondrial membrane proteins in rat cerebellum during aging. Neurochem Res 1988; 13:859-865.
108. Harmon HJ, Nank S, Floyd RA. Age-dependent changes in rat brain mitochondria of synaptic and nonsynaptic origins. Mech Ageing Dev 1987; 38:167-177.
109. Tanaka M, Kovalenko SA, Gong JS et al. Accumulation of deletions and point mutations in mitochondrial genome in degenerative diseases. Ann NY Acad Sci 1996; 786:102-111.
110. Wei YH, Pang CY, You Bjn, Lee HC. Tandem duplications and large-scale deletions of mitochondrial DNA are early molecular events of human aging process. Ann NY Acad Sci 1996; 786:82-101.
111. Bandy B, Davison AJ. Mitochondrial mutations may increase oxidative stress: implications for carcinogenesis and aging? Free Rad Biol Med 1990; 8:523-539.

CHAPTER 5

Oxidative Stress in Alzheimer's Disease

INTRODUCTION AND NEUROPATHOLOGY

It has been nearly 90 years since Alois Alzheimer first described the dementing illness associated with his name.[1] AD is the most common cause of progressive cognitive decline in the aged population and the most prevalent neurodegenerative disease affecting more than 15 million people wordwile. AD shares the triad hallmark features of β-amyloid plaques (senile plaques; SP), neurofibrillary tangles (NFT) with extensive neuronal loss, particularly in the hippocampus and cerebral cortex, and dementia, which appear to arise sporadically and are typically late in onset (after the age of 65). Familial ADs (FAD) exist and frequently display a much earlier onset of the disease (age 40-50 years).

While most of the behavioral, and many of the neuropathological features of AD have been known since the time of Alzheimer, the greatest progress in understanding the causes of the disorder has come only in the past ten years with the application of genetic and molecular biological analysis. Many hypotheses exists nd one of the main lines of research has been concerned with the pathological hallmarks, the SP and NFT, and progress has been made in their characterisation. However their importance in the neuropathogenesis of AD is debated. Is the SP, the NFT, or both important? Are these changes the causes of the disease? One of the main constituents of SP, amyloid peptide (Aβ), has received most interest. Researchers are divided into those who believe that amyloid deposition is the crucial factor for dementia and those who do not believe it. Other lines of research are concerned with the function of neurons and their synapses and the distribution of neuronal damage. It seems that the large pyramidal neurons are those most affected. However, why these cells are the most affected is not clear.

Attention has focused on the SP (also known as dense-core or neuritic plaque) as a key feature of the neuropathology of AD. The density of SP found in brain post mortem shows a significant correlation with the severity of the clinical dementia. The SP is a complex lesion of the cortical neuropil. A central deposit of extracellular amyloid fibrils (the core) is surrounded by dystrophic neurites (both dendrites and axon terminals) and by activated microglia and reactive astrocytes.[2,3] The major protein component of the amyloid core is the 39-43 amino acid residue called Aβ.[4] During AD, Aβ seems to aggregate after reaching a critical concentration, resulting in the formation of SP and vascular deposits in brain microvessels.

The Role of Oxidative Stress in Neuronal Death, by Irène Ceballos-Picot.

Aβ is derived from the larger membrane spanning glycoprotein called amyloid precursor protein (APP)[5] (Fig. 5.1).

The APP gene maps to the long arm of chromosome 21[6] (see Fig. 6.1), and four alternatively spliced pre-mRNAs encode isoforms of 695 (the predominant brain species), 714, 751 and 770 amino acids.[7] These isoforms differ in the presence or absence of the Kunitz-type protease inhibitor and 'OX-2' antigen domains, but are all potentially amyloidogenic. APP is cleaved by at least three different proteolytic activities. The α-secretase cleavage occurs within the Aβ region, thus preventing Aβ formation and resulting in the secretion of the extracellular domain of APP. APP is normally cleaved within the Aβ domain to generate a 90-100 kDa secreted form.[8] An alternative processing pathway in which APP is cleaved at different sites in the extracellular and transmembrane domains generates soluble Aβ, the putative source of plaque amyloid in AD.[9,10] Aβ is constitutively generated by two putative endoproteolytic activities: the aminoterminal cleavage is mediated by β-secretase, the carboxy terminal cleavage by γ-secretase.

NFT are intracellular bundles of straight or paired helical filaments (PHF) that form in the cytoplasm of degenerating neurons. The stability of these lesions in brain may be critical in the progression of the disease,[11] and it has been documented that NFT per-

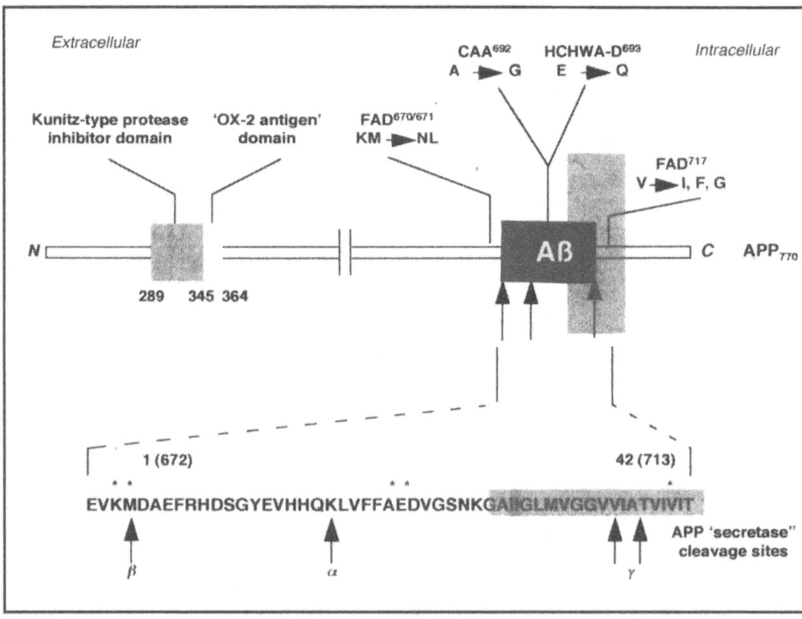

Fig. 5.1. Schematic representation of the human β-amyloid precursor protein (APP_{770}). Shorter isoforms of APP are produced by alternative splicing of pre-mRNAs to exclude the OX-2 antigen domain (APP_{751}), the Kunitz-type protease-inhibitor domain (APP_{714}) or both (APP_{695}). 'α-Secretase'-mediated cleavage of a proportion of cellular APP molecules at Lys687-Leu688 (Aβ Lys16-Leu17) precludes the formation of Aβ. Intracellular processing of other APP molecules by endopeptidases designated 'β-' and 'γ-secretase', however, yields intact Aβ, consisting of 28 amino acids derived from the extracellular domain and 12-14 amino acids from the transmembrane domain. A number of point mutations in the APP molecule have been identified in the vicinity of the Aβ sequence (indicated by asterisks). These mutations cosegregate with family pedigrees predisposed to early-onset familial AD (FAD), cerebral amyloid angiopathy (CAA) and hereditary cerebral hemorrhage with amyloidosis of the Dutch-type (HCHWA-D), and are characterized by alterations in the formation and deposition of Aβ. Reprinted with permission from Iversen et al. Biochem J 1995; 311:1-16. © Portland Press.[187]

sist, even after the death of the neuron that contained them.[12] Considerable evidence suggests that the major constituent of PHF is hyperphosphorylated tau that self-associates via its microtubule-binding domain.[13-15]

A central issue in the pathogenesis of AD is the relationship between amyloid deposition and NFT formation. It was recently demonstrated that fibrillar Aβ markedly induces the phosphorylation of tau in human cortical neurons,[16,17] resulting in the loss of microtubule binding capacity and somatodendritic accumulation, properties also exhibited by tau in the AD brain. Moreover, Aβ caused an increase in the activity of tau protein kinase that correlates with the appearance of NFT in rat hippocampal culture.[18,19] Amyloid fibril formation may therefore be a cause of abnormal tau phosphorylation.

The obligate coincidence of NFT and SP in Alzheimer's disease led to speculation that there is a direct biochemical relationship between the lesions and that this relationship is mediated by a common pathogenic mechanisms involving oxidative stress. Oxidative post-translational modification is probably one of the major mediators that, by disrupting cellular homeostatic balance, both promotes abnormal interactions and makes them resistant to proteolytic removal.

Several risk factors have been identified that may initiate or accelerate AD pathology, including genetic factors. Particularly in presenile AD (early onset; age <65 years), the evidence for the involvement of genetic defects in AD pathophysiology is compelling. Several large families have been identified in which presenile AD segregates as a fully penetrant autosomal dominant trait. Linkage analysis studies in presenile AD families have identified three genes, on chromosomes 1, 14 and 21, that, when mutated, cause presenile AD.[20] FAD is otherwise clinically and neuropathologically indistinguishable from cases with later onset.

In sporadic cases with later onset an interesting new area emerges for a few years. The possibility of an important association between ROS and β-amyloidosis has been strengthened by recent studies. Several lines of experimental evidence suggest a free radical theory of AD pathology, wherein oxidative stress may originate in part from ROS generated by Aβ or glycated cytoskeletal proteins or from metabolic dysfunction.[21-26] A radical-mediated aggregation of Aβ monomers could also produce toxic forms of Aβ.[27]

This chapter describes recent advances in the molecular genetics and cell biology of AD and discusses the role of oxidative stress as a potential common pathogenic mechanism.

GENETIC HETEROGENEITY

APP Mutations

The first presenile AD gene was identified in 1991.[28] It maps to chromosome 21 and codes for the APP. Mutations in this gene account for, at most, 5% of the families. Several missense mutations has been described in the *APP* gene. In early-onset FAD, V717I/F/G mutations have been identified in the transmembrane domain in APP 695, four amino acids downstream from Aβ (see Fig. 5.1) and account for most, if not all, of the evidence showing a linkage of AD to chromosome 21.[28,29] Recently Games et al[30] have reported that overexpression of V717F-APP mimics the neuropathology of AD in transgenic mice. Expression of these mutants causes transfected cells to undergo apoptosis which is sensitive to Bcl-2.[31]

A double mutation found in Swedish families (K670N and M671L) is located before the Aβ region of APP.[32,33] Of the four autosomal dominant mutations of the *APP* gene that have been identified as causes of FAD, none occur within the Aβ region of the protein (Fig. 5.1). Expression of APP containing the Swedish mutation in transfected cells increases the production of Aβ about 5-fold resulting in abnormal intracellular accumulation of Aβ[34,35] but also causes altered secretion of APP.[36] Transgenic mice overexpressing this mutated APP isoform showed a 5-fold increase in Aβ(1-40) and 14-fold increase in Aβ(1-42) accompanied by the appearance of behavioral deficits.[37]

These results demonstrate a causal link between altered APP processing and the development of AD.

Presenilins

The two remaining genes, the presenilin (*PS*) genes, were identified recently. Presenilin-1 (*PS-1*), on chromosome 14, harbors an estimated 70% of the disease-causing mutations, making it the major gene for presenile FAD (estimated 10% of all AD cases). PS-1 is predicted to be an integral membrane protein with at least 7 transmembrane domains. Several missense mutations have been identified in the coding region in families of Caucasian, Japanese, Ashkenazi-Jewish and Hispanic origin (for review see ref. 38). A third gene for presenile AD termed *PS-2* maps to chromosome 1 in the Volga-German AD families, a group of families in which AD is the result of a founder effect.[39] Mutations in *PS-2* genes are minor genetic causes of presenile dementia.

More than 30 different mutations in *PS-1* and 2 in *PS-2* have been identified to date. Most of the mutations are localized within, or proximal to, a predicted transmembrane boundary, raising the possibility that they disrupt membrane anchorage or oligomeric protein interactions. The biological functions of the presenilins are unknown. PS-1 and PS-2 show significant sequence homology to integral membrane proteins from *C. elegans*, SEL-12.[40] SEL-12 was identified as a protein that facilitates signaling through the Notch/LIN-12 family of receptors that dictate cell fate during development.[41] The high degree of homology (50-60% identity) between PS-1, PS-2, and SEL-12 suggests that the presenilins may also be involved in this signaling pathway, either by regulating signal transduction or by affecting the processing of ligands or receptors. A recent study suggest that PS-1 and PS-2 may be localized to the endoplasmic reticulum and Golgi complex in transfected cells.[42] If this localization is confirmed in vivo, then it would raise the possibility that presenilins affect the transport or processing of proteins in the secretory pathways, potentially affecting the processing of APP and the production of Aβ. One hint about the pathological cosequence of presenilin mutations comes from the observation that fibroblasts of patients with mutations in *PS-1* secrete more Aβ.[43] This could provide evidence for a role of presenilins in the amyloidogenic cascade leading to AD neuropathology.

At least one other locus must exist for presenile AD since in some families mutations in *APP* and the *PS* genes have been excluded.[44]

RISK FACTORS

An age-dependent and complex heterogeneous interaction of genetic and environmental influences could provoke and/or accelerate Alzheimer neuropathogenesis.

Apolipoprotein E

While these findings have firmly established a series of genetic causes in a number of early onset familial forms of the disease, understanding the causes of the far more prevalent late-onset form of the disorder has proven more elusive. Recent studies have suggested that allelic variation in the gene for apolipoprotein E (*ApoE*) on chromosome 19 is associated with an increased risk and earlier age of onset in both sporadic and familial late-onset AD.[45,46]

There are three major alleles of ApoE on chromosome 19, ε2, ε3 and ε4. ApoE-ε3 is the most prevalent allele. A disproportionate increase in the frequency of the ApoE-ε4 allele in patients with late-onset was first described by Strittmatter et al[47] and has been confirmed by many subsequent studies. Gene dose was shown to have an effect on risk of developing AD, age of onset, accumulation of senile plaques in the brain, and reduction of choline acetyl transferase (ChAT) activity in the hippocampus of AD subjects.[48,49] The association between ApoE-ε4 and AD, however, appears to vary across ethnic groups. The presence of ApoE-ε4 has been shown to modulate the age of onset in AD families with the APP mutation but not in families with chromosome 14-linked early-onset. The presence of an ApoE-ε2 allele was found to have the opposite effect, decreasing the risk of AD and delaying disease onset.[45] By contrast, the ε4 allele frequency is decreased, and the ε2 frequency is increased in centenarians compared to younger control populations.[50] The fact that these studies report a consistent presence of subjects that are homozygous for ε4 without AD, and subjects with the disease carrying no ε4 allele, leads to the conclusion that the ε4 allele itself is neither necessary nor sufficient to cause late-onset AD.

The mechanism by which ε4 is associated with AD is not known. ApoE is a component of the very low density lipoprotein (VLDL) and high density lipoprotein (HDL) complexes that mediate the cellular uptake and metabolism of cholesterol. In the central nervous system, ApoE plays a key role in mobilization and redistribution of cholesterol and phospholipid during membrane remodeling associated with synaptic plasticity. ApoE binds to soluble Aβ in vitro and promotes amyloid fibril formation in an isoform specific manner, with ApoE-ε4 promoting Aβ fibrillogenesis more potently than ApoE-ε3.[51,52] These in vitro results are consistent with immunochemical and biochemical studies which demonstrate that ApoE is associated with amyloid plaques.[53] It has been proposed that ApoE may function as a pathological chaperone, leading to conformation shifts in Aβ peptides that causes fibrils formation.[54] An alternative pathogenic mechanism for the role of ApoE in AD has been proposed based on the observation that ApoE binds to the microtubule-associated protein tau in vitro.[55] Individual with an ApoE-ε4 genotype would be at increased risk for AD because of decreased levels of the ApoE-ε3 allele that inhibits tau phosphorylation.

Thus, isoform specific differences in ApoE production at an incipient stage of Aβ deposition may increase the rate of Aβ fibril formation, potentially accelerating the rate of onset of AD.

An observation that plasma lipoproteins from ApoE deficient mice are more susceptible to in vitro oxidation than the lipoprotein from wild type mice[56] opens the idea that ApoE may play a protective role against oxidative reactions. Recently, Miyata et al[57] demonstrated that ApoE has allele-specific effects in protecting a rat neuronal cell line from oxidative cell death, with ε2 the most effective, ε3 moderately effective, and ε4 being the least effective. In the same allele order, ApoE at levels found in the cerebrospinal fluid also protected cells from Aβ peptides, while higher doses of ApoE led to increased cytotoxicity, particularly for ε4. The allele-specificity of these effects correlates with the allele specificity of ApoE's antioxidant activity in three in vitro assays.[57] ApoE is also a metal binding protein, which could be involved in the mechanism of ApoE's antioxidant activity.

In conclusion, ApoE-ε4 with less antioxidant activity than ApoE-ε2 or ApoE-ε3, may be the least effective in protecting neurons from the oxidative damage caused by aggregated amyloid peptides or other oxidative or age-related insults, and in this way play a role in AD pathogenesis. Thus, the decreased antioxidant activity of ApoE-ε4 could contribute to its association with AD, and decreased longevity.

Mitochondrial DNA Mutations

It was recently reported an A→G mutation of the mitochondrial tRNAGln gene at position 4336 occurs at increased frequency in a cohort of late-onset cases of AD.[58] The frequency data, combined with those of Shoffner et al,[59] include 17 4336G occurences in 367 patients with AD (4.6%) but only 1 occurrence in 322 ethnic and age-matched controls (0.33%), still consistent with an increased risk factor for AD of 15.6.

One hypothesis to explain the neutotoxicity of the 4336G mutation is that this mutation may lead to defects in complex I and increased free radical damage to neurons. Mutations in tRNAs have been observed in multiple mitochondrial genetic diseases, but no general model for why these mutations are deleterious has been proven. Since most of the mitochondrial genome encodes complex I subunits, defects in mitochondrial protein synthesis are more likely to impair complex I (NADH dehydrogenase) than any other mitochondrial gene product.

This raises the possibility that these and additional, as yet unidentified, mtDNA point mutations might be involved as causes or risk factors in late-onset AD.

Age-Related Oxidative Stress

Although it is likely that multiple molecular pathways can lead to AD, a central issue is whether all causes of the disease lead to a final common mechanism of neuronal death. Any explanation of AD must take into account that age is the primary risk factor for the disease. Since normal aged individuals usually acquire the histological manifestations of AD in the form of a limited number of plaques as well as some decrease in cognitive ability, it is possible that the mechanisms responsible for AD simply exacerbate those caused by aging.

One well documented change in the CNS associated with aging is oxidative damage caused by free radicals (see chapter 4). On the basis of these observations, age-associated perturbation of ROS production and/or detoxification may be also involved and/or may be the common mediator in the neurodegeneration in AD. Specifically Aβ-induced cytotoxicity expressed on a background of already compromised cells may be sufficient to precipitate the clinical and pathological manifestations of the disease.

Environment

Of the potential environmental toxins implicated in the etiology of AD aluminium has been the prime suspect agent[60] although increased brain levels of mercury have also been detected in AD brains.[61]The identification of aluminosilicate deposits within the core of SP in the brain[62] is of special significance to the etiopathogenesis of AD. Zinc play an important role in the CNS,[63] aluminium uptake into the brain being enhanced in rats fed a zinc-deficient diet.[64] Regional accumulation of aluminum, zinc or other polyvalent cations may accelerate Aβ aggregation and senile plaques formation through electrostatic crosslinking of Aβ monomers independent of redox chemistry. The mammalian brain is heterogeneous with respect to regional iron and copper content and possibly metal-chelation state.[65,66] The rates of spontaneous lipoperoxidation of brain regional homogenates correlate well with regional iron concentration.[66] It is interesting that iron-catalyzed oxidation accelerates Aβ aggregation in vitro,[67] iron levels are somewhat increased in the AD,[68] and ferritin-iron complexes have been immunolocalized to senile plaques.[69] Aβ aggregation is greatly facilitated in vitro by low concentrations of either aluminum or zinc.[70] Using model aluminosilicate particulate zeolites of different morphological, dimensional and chemical composition, it has also been demonstrated that murine microglial cells are also capable of being stimulated to generate ROS.[71,72] Aluminum reportedly exacerbates iron-catalyzed lipoperoxidation, possibly by indirect physical effects associated with crosslinking of phospholipids.[73]

Hence, polyvalent cation accumulation in certain regions of the AD brain may promulgate Aβ toxicity via redox-dependent or redox-independent pathways.

This oxidative insult may arise from known or postulated AD risk factors cited above. Release of Aβ engenders further oxidative stress, either directly through peptide radicalization processes or indirectly through disruption of metabolic pathways or via senile plaque recruitment of active microglia. The consequence of this hypothetical cycle is an exponential increase in amyloid burden and related tissue damage, particularly protein oxidation, as a function of time. The dynamics of the cycle would be further modulated by chaperone proteins such as ApoE-ε3/ε4, which have been suggested to affect Aβ clearance and toxicity. Moreover, the observed regional differences in senile plaque density and protein oxidation may reflect regional variation in metal content and/or in anti-oxidant capacity influenced by age.

The "Oxidative Stress" Hypothesis in Alzheimer's Disease

Other than the described differences in the pattern of inheritence and age of onset, early onset familial AD is indistinguishable from the late onset AD in phenotype, both clinically and neuropathologically. This suggests that the different genetic causes described intersect at some intermediate point before the appearance of the shared phenotypic alterations of the disease. Such a point of intersection would represent the earliest point in the neurodegenerative cascade shared by all forms of AD, independent of their individual genetic or environmental causes. The characterization of such a bottleneck would almost certainly provide a conceptual "high ground" from which the various forms of the disease could be viewed as being unified, and from which new prospects for therapeutic intervention might be considered.

A substantial body of evidence indicates that oxidative stress may play the central role in AD.

Evidence of Oxidative Stress in the Alzheimer's Brains

One unifying hypothesis suggests that oxidative processes, individually and synergistically, play contributory or acceleratory roles in accounting for many of the salient features of Alzheimer disease (i.e., NFT, SP, regionally and neuronal selective neurodegeneration) in AD.

Evidence of increased oxidative stress in the AD brain has come from studies showing increased lipid peroxidation,[74,75] increased carbonyl modifications of proteins,[76] and increased oxidation of mitochondrial DNA.[77,78] Oxidative stress leads to the production of reactive lipid peroxidation-derived carbonyl compounds, such as MDA and 4-HNE, which have the capacity for inducing a spectrum of crosslink and non-crosslink protein modifications through a time-dependent oxidative "cementing." Indeed, malondialdehyde and related carbonyl modification epitopes were previously immunolocalized to NFTs.[26,79] Oxidative stress has been implicated in NFT formation by the observation that oxidation of tau at residue Cys-322 is required for PHF formation in vitro[80] and by the demonstration that oxidized proteins are present in neurofibrillary tangles in vivo.[79]

There is other evidence that oxidative stress may contribute to the formation of SP and NFT. Inhibition of energy metabolism increases the amyloidogenic pathway of APP processing,[81] and oxidation of Aβ in vitro increases its aggregation.[67] Some of the Aβ in plaques shows advanced glycosylation modifications, a posttranslational modification that is promoted by oxidation and results in increased Aβ aggregation.[79,82] There is now considerable evidence that neuronal damage by Aβ is mediated by free radical damage,[25,83] and that this toxicity might be mediated by the spontaneous generation of peptidyl radicals and consequent oxidative stress.[84,85]

The brain regional variation of oxidation-sensitive biomarkers such as the decreased glutamine synthetase (GS) and creatine kinase (CK) corresponds to established histopathological features of AD (SP and NFT densities) and is paralleled by an increase in immunoreactive microglia.[86] Both GS and CK are crucial metabolic enzymes, whose impairment is likely to engender a cellular energy deficit. CK catalyzes reversible transfer of phosphoryl groups between phosphocreatine pools and ADP, which buffers ATP concentration and enables a constant supply of high-energy phosphoester bonds. GS consumes ATP while incorporating nitrogen into glutamate to form glutamine. Dysfunction of GS would therefore affect nitrogen use and glutamate/glutamine ratios, with concomitant threat of glutamate excitotoxicity or excitotoxic potentiation of amyloid cytotoxicity as reported by Mattson et al.[87] Energy-starved cells may be more susceptible to oxidative insult or less capable of repairing oxidative damage; hence, damage to GS and CK may be both a consequence of, and contributing factor to, oxidative lesions.

The brain responds to oxidative assaults in part through the activation of NF-κB[88] and by the induction of specific antioxidant enzymes. For example, in AD, two such enzymes, SOD and heme oxygenase-1 (HO-1), are associated with intraneuronal NFT.[89-93] Moreover, specific induction of HO-1 (an inducible isoform), but not HO-2 (constitutively produced), was demonstrated to parallel regional susceptibility and pathology in AD.[94] Significantly, in addition to the antioxidant bilirubin, HO generates carbon monoxide and free iron, both of which could underlie neurotoxicity and further free radical damage in AD.[95]

Indirect support in favor of the oxidative stress hypothesis is provided by the constant occurrence of AD-like neuropathology in Down's syndrome (see chapter 6). The gene for CuZnSOD is encoded by chromosome 21, so there is a 50% increase in CuZnSOD activity and protein in Down's syndrome tissues, including brain. This appears to be associated with a concomitant increase in H_2O_2 production and with increased peroxidative damage, which is consistent with a role for such changes in the production of AD neuropathology.

Several reports indicate increased activity of SOD in the brains of Alzheimer's patients. There is also an increased staining of microglial cells surrounded SP with MnSOD antibodies and increased staining of degenerating neurons containing NFT with antibodies to CuZnSOD[89,90] (see chapter 2). Thus, these data suggest that oxidative stress may be involved in the pathogenesis of these lesions and in the mechanism of neuronal death.

In conclusion, these data indicate that AD brain regions containing abundant SP and NFT are microenvironments of elevated oxidative stress.These studies document a pattern of oxidative modifications in the AD brain that suggests a correspondence among amyloid deposition, NFT formation, microglial accumulation, and tissue oxidation. In addition, variation in the measured endpoints within the AD brain is partially reflected by similar but smaller trends in control aged brain, in particular for protein oxidation.[96-99] This observation supports the hypothesis that Alzheimer's pathology is a quantitative exacerbation of oxidative processes that occur to some extent in aged brains.

Source of Oxidative Stress in Alzheimer's Brain

Numerous reports implicate free ROS as causative agents of Aβ toxicity. Glycated cytoskeletal proteins have also been suggested as molecular sources of oxidizing equivalents in the AD brain.

Senile plaque-associated NFTs are composed largely of posttranslationally modified tau protein, which has been shown to be glycated and, like Aβ, is capable of generating oxyradicals via the metal-catalyzed Maillard reaction.[26,76,79,80,82] NFT and SP are sites of specific oxidative damage that are initiated upon covalent modification by reactive sugar and lipid-derived carbonyl products. Carbonyl moieties arising from direct oxidation of protein side-chains may also contribute to the overall generation of pathological intra- and intermolecular protein crosslinks. Either type of modification may underlie the proteolytic resistance of NFT and SP in human brain, and either may explain why associated elements are integral to the structures and why they persist, even following major structural reorganization. Further redox chemistry undergone by the initial oxidative stress-derived modifications of NFT and SP can lead to further intramolecular protein damage as well as serving as sites of elevated production of ROS responsible for inducing oxidative responses in adjacent neurons, as evidenced by induction of HO-1. Moreover, glycated tau induces lipoperoxidation when coincubated with cultured neuroblastoma cells.[26]

One major source of oxidative stress in AD brain is the Aβ peptide,[25,83,85,100] which under some conditions can acquire a free radical state in vitro.[27,101] This "radicalization" of the Aβ peptide has been hypothetized to mediate Aβ toxicity. Aβ is also capable of producing elevation of intracellular ROS such as H_2O_2 and NO (see below).

Oxyradical generation triggered by microglial activation is another potentially significant source of ROS to consider in conjunction with Alzheimer's brain histopathology. Activated microglia that express complement receptors and major histocompatibility antigens[102] are closely associated with amyloid plaques[103,104] and it has become increasingly evident that microglial cells play an important role in plaque pathology.[105-108] Activated microglia proliferate, exhibit phagocytic activity removing infectious agents or remnants of dying brain tissue, and produce a number of secreted factors that may be involved in the pathogenesis of CNS disease, such as ROS,[108] proteases and cytotoxic agents that induce death of neurons.[109] These cells may either directly contribute to the pathology of AD or they may exarcerbate an already preexisting problem. This may occur through several possible mechanisms: first microglia may directly contribute to amyloidosis by participating in the formation of Aβ fibrils; second, microglia may normally be engaged in phagocytosis of amyloid fibers or aggregates,[110] but in AD brains, microglia may fail to clear Aβ protein and thus cause a buildup of the protein that aggregates into large plaques; and third, Aβ may activate microglia and trigger the production of proinflammatory and potentially cytotoxic mediators.[111,112] In cell culture, Aβ acts synergistically with interferon-γ to activate microglia, resulting in the production of neurotoxic nitrogen free radicals.[112]

ANTIOXIDANT DEFENSE MECHANISMS IN ALZHEIMER'S DISEASE

Brain Regional Variation

Brain regional variation in competency of free radical defense mechanisms, or concentration of antioxidant compounds, could also explain some of the regional and cellular variations in AD brain protein oxidation or lipoperoxidation reported. Activities of the antioxidant enzymes SOD,[113] CAT, GSHPx, and GSSR-Rd are often reportedly elevated in specific AD brain. Markesbery et al[114] reported increased GSHPx, CAT and GSSR-Rd activities in the AD parahippocampal gyrus relative to control brain tissue. Balazs and Leon[115] also documented increased SOD and CAT activity in AD cortex relative to control cortex. A conflicting study by Chen et al[116] reported that SOD activity is decreased in several AD brain regions and that CAT and GSHPx activity were similar between corresponding regions of AD and control brain. A significant decreased CAT activity in parietotemporal cortex, basal ganglia, and amygdala was reported in AD compared with controls.[117] Glucose-6-phosphate dehydrogenase (G6PD), the rate-limiting enzyme in the hexose monophosphate shunt whose function is the generation of NADPH a major source of cytoplasmic reducing power, is elevated in the frontal cortex of Alzheimer's patients,[118] a finding consistent with increased oxidative activity.

It is also possible that the brain area which has increased cell loss shows few signs of oxidative response because the affected cells are gone. Moreover, the selection of any one brain region may not reveal an oxidative challenge throughout the brain. However, the general implication of the literature is that antioxidant enzyme activity is elevated in certain AD brain regions, consistent with an attempt by the AD brain to compensate for elevated oxidative stress.

Heterogeneous Cellular Expression

Most important are the studies of antioxidant expression at the cellular levels in AD brain. A preferential localization of the CuZnSOD protein and mRNA[89,90,91,93,119] has been demonstrated in hippocampal neurons known to be the most vulnerable in AD and to overlap the distribution of APP mRNA.[120] Thus, high levels of CuZnSOD in neurons which are preferentially vulnerable in AD suggest that oxidative stress may be involved in the pathogenesis of these lesions and in the mechanism of neuronal death.

In AD hippocampi MnSOD is highly expressed in microglial cells which infiltrate SP (see chapter 2).This change in the localization of MnSOD in the brain of AD was the main difference between controls and AD tissue and is possibly due to metabolic activation of reactive glia. Thus, this enzyme could have a role during plaque formation but we don't know if these changes in the cellular expression are causally related or a primary mechanism in the early stage of senile plaque formation.

The differential localization of CuZnSOD and MnSOD in cerebral sites of degeneration suggests that cellular responses to oxidative stress is antioxidant enzyme-specific and cell-type specific and that these two forms of SOD may have different functions in antioxidant mechanisms. Alternatively high expression of these enzymes could exacerbate oxidative stress and consequently SP and NFT formation.

Table 5.4. Peripheral CuZnSOD, GSSG-Rd, GSHPx, TBARS and selenium from AD patients and controls.

	n	Controls	n	AD patients	p value
Age (years)	34		40		
mean ± SD[a]		79.1 ± 8.3		84.2 ± 7.5	0.007
range[b]		[60 - 93]		[66 - 96]	
E-CuZnSOD (UI/mg Hb)	34		40		
mean ± SD[a]		1.37 ± 0.20		1.37 ± 0.18	0.92
range[b]		[1.12 - 2.20]		[1.08 - 2.18]	
adjusted mean[c]		1,36		1,38	0.71
E-GSSG-Rd	34		40		
mean ± SD[a]		4.31 ± 0.85		4.36 ± 0.66	0.78
range[b]		[2.43 - 5.98]		[3.11 - 5.77]	
adjusted mean[c]		4,31		4,35	0.82
E-GSHPx[d]	34		40		
mean ± SD[a]		7.86±2.01		7.76 ± 2.62	0.86
range[b]		[3.5 - 13.1]		[4.12 - 17.7]	
adjusted mean[c]		7,75		7,86	0.85
P-GSHPx[e]	25		29		
mean ± SD[a]		56.9 ± 20.5		65.2 ± 17.6	0.08
range[b]		[30.4 - 109.4]		[41.5 - 98.4]	
adjusted mean[c]		56,0		66,1	0.04
P-TBARS	24		29		
mean ± SD[a]		3.22 ± 0.37		3.16 ± 0.35	0.54
range[b]		[2.49 - 4.23]		[2.46 - 3.83]	
adjusted mean[c]		3,22		3,16	0.55
P-Se (ng/ml)	34		40		
mean ± SD[a]		49.7 ± 9.6		54.1 ±12.9	0.09
range[b]		[30.9 - 71.9]		[34.8 - 92.7]	
adjusted mean[c]		48,4		55,2	0.01
E-Se (ng/mg Hb)	34		40		
mean ± SD[a]		0.291 ± 0.038		0.286 ± 0.048	0.60
range[b]		[0,222 - 0,432]		[0.193 - 0.398]	
adjusted mean[c]		0,287		0,289	0.87

a. Means with standard deviation (Student's -tailed t-test);
b. Range [minimum-maximum];
c. Means adjusted for age (analysis of variance);
d. E-GSSG-Rd and E-GSHPx, the unit is μM NADPH /min/g Hb
e. For P-GSHPx, the unit is μM NADPH /min/l
These significant differences mostly appeared in AD subjects under 80 years (Fig. 5.2)

Peripheral Enzymatic Antioxidants

The finding of a specific peripheral marker would be of the greatest interest in the field of primary cognitive disorders, since a large part of the current research on AD has involved the postmortem characterization of pathological and biochemical alterations in the brain. Thus, we search for a peripheral marker of oxidative stress in AD.[121] Since synergistic interactions occur between the various antioxidant enzymes, a complete understanding of the antioxidant system requires measurement of the combined enzymes implicated in free radical detoxification. Therefore, we measured the activities of CuZnSOD, GSSG-Rd in red blood cells, Se-GSHPx and selenium (Se) concentration in red blood cells and plasma and TBA-reactive materials (TBARS) in plasma from 40 AD patients, and we compared these activities with those of 34 healthy aged controls with normal cognitive function.

No crude significant difference in erythrocyte CuZnSOD (E-CuZnSOD), seleno-dependent GSHPx (E-GSHPx), GSSG-Rd (E-GSSG-Rd) activities and Se concentration was found between AD cases and control subjects. When controlling for age, a significant increase in plasmatic P-GSHPx activity and P-Se was observed in AD patients as compared to controls. Mean value of P-GSHPx adjusted for age was 66.1 UI/l in AD and 56.0 UI/l in controls. The mean value of P-Se adjusted for age was increased in AD patients (55.2 ng/ml) compared to controls (48.4 ng/ml). All other comparisons were performed controlling for age but no additional significant difference appeared (Table 5.4).

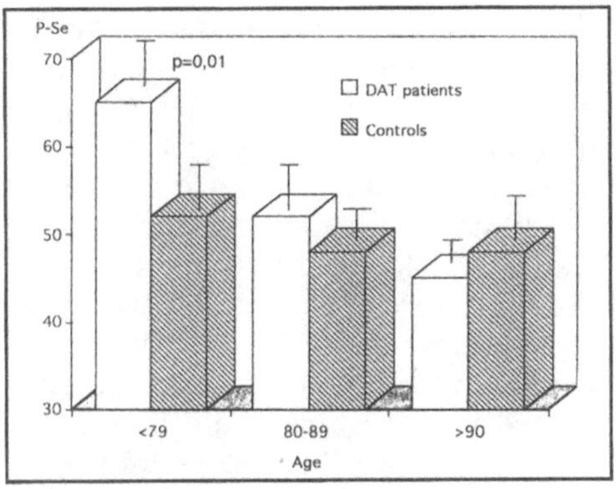

Fig. 5.2. Mean level of plasma-selenium (ng/ml) according to age in three classes (<79 years, 80-89 years, >90 years).

In the AD group, P-GSHPx activity and P-Se were negatively correlated with age ($r = -0.58$; $p < 0.001$ and $r = -0.63$; $p < 0.001$ respectively). Moreover, E-GSHPx activity and E-Se were also negatively correlated with age ($r = -0.40$; $p < 0.01$ and $r = -0.46$; $p < 0.01$ respectively) (Table 5.5b) No significant correlation with age was observed in the controls (Table 5.5a). Some correlations were only observed in the AD group such as P-GSHPx and E-GSHPx; ($r = +0.68$; $p < 0.001$); P-GSHPx and E-Se ($r = +0.79$; $p < 0.001$). Correlations between P-GSHPx and P-Se, E-GSHPx and P-Se and P-Se with E-Se are greater in the AD group ($r = +0.84$; $p<0.001$; $r = +0.76$; $p < 0.001$ and $r = 0.75$; $p < 0.001$) than in the control group ($r = 0.54$, $p < 0.01$; $r = 0.43$, $p < 0.01$ and $r = +0.34$, $p < 0.05$ respectively).

Activity of E-CuZnSOD is similar in the AD group as compared to controls of the same age, in agreement with reports.[122-124] However, they do not agree with other contradictory results of increased[125-128] or decreased[129] SOD activity in AD patients. Our data clearly showed no differences in erythrocytes for GSHPx and GSSG-Rd activities and for Se between controls and AD group. This observed normality of erythrocyte enzymes is in accord with two reports in AD patients.[122,129] However, Anneren et al[130] and Jeandel et al[123] reported, respectively, an increase or a decrease in patients with AD.

The contradictory results in other studies in dementia may have several reasons. One might be the selection of the control group. The risk that a randomly chosen control group of aged people contains individuals with an early process of dementia is high because the frequency of dementia increases with age. In this study the control group was composed of aged subjects who did not have

Table 5.5a. Correlations between biological parameters and age in the control group. (Pearson's correlation coefficient)

	Cu Zn SOD	GSSG-Rd	E- GSHPx	P-GSHPx	TBARS	P-Se	E-Se
GSSG-Rd	+ 0.21						
E-GSHPx	- 0.13	- 0.19					
P-GSHPx	- 0.29	- 0.25	- 0.02				
TBARS	+ 0.31	+ 0.36	+ 0.13	- 0.36			
P-Se	- 0.18	+ 0.01	+ 0.43**	+ 0.54**	+ 0.17		
E-Se	+ 0.19	- 0.10	+ 0.75***	- 0.11	+ 0.25	+ 0.34*	
Age	- 0.13	+ 0.02	+ 0.22	- 0.29	+ 0.02	+ 0.07	+ 0.04

p value : *$p < 0.05$; **$p < 0.01$; ***$p < 0.001$

Table 5.5b. Correlations between biological parameters and age in the AD patient group. (Pearson's correlation coefficient)

	Cu Zn SOD	GSSG-Rd	E-GSHPx	P-GSHPx	TBARS	P-Se	E-Se
GSSG-Rd	+ 0.37*						
E-GSHPx	- 0.06	+ 0.10					
P-GSHPx	- 0.02	- 0.02	+ 0.68***				
TBARS	- 0.03	+ 0.01	+ 0.17	+ 0.11			
P-Se	- 0.03	+ 0.03	+ 0.76***	+ 0.84***	+ 0.03		
E-Se	- 0.03	+ 0.12	+ 0.77***	+ 0.79***	+ 0.13	+ 0.75***	
Age	- 0.08	+ 0.01	- 0.40 **	- 0.58***	- 0.03	- 0.63***	- 0.46**

any symptoms of dementia and who lived in the same institution as the AD group (same environmental and nutritional factors).The difficult choice of controls in case-control studies accounted for discrepancies observed between previous reports and also with our results.

The plasma Se concentration was higher in AD patients than in control patients. As a trace mineral, Se is an integral part of the enzyme GSHPx. Plasma and erythrocyte selenium level are demonstrated to show great regional variations depending on its availability in the diet. Moreover, plasma levels of selenium and other trace elements such as aluminium, cadmium and mercury were also found to be increased in the plasma of AD patients by Basun et al.[131] A possible role of metals and trace elements as environmental determinants has been suggested.[67-73] The imbalance in metals or trace elements may reflect a role for a selected element as a pathogenic cofactor in AD. It may also represent an epiphenomenon or a biological marker of the disease, like a blood-brain barrier disturbance leading to metals entering the brain. Other support for membrane derangement are reported in AD which could be due to increased lipoperoxidation.[132-133] Nearly all of the selenium in tissues or cells is associated with protein. Selenoproteins with known enzymatic activity are redox enzymes and contain selenocysteine in their active sites. Thus, effective function of these enzymes depends on their selenium.

P-GSHPx activity was slightly elevated in our AD group. There are two distinct forms of Se-dependent GSHPx in mammals, a cellular form and an extracellular or plasma form. The two enzymes are structurally, functionally, and antigenically distincts. Virtually all peroxidase activity in plasma is accounted for by the extracellular form of Se-dependent GSHPx.[134] In regard to the observed major correlation between P-GSHPx and P-Se or E-Se in this study, its activity in plasma is a convenient index of selenium nutritional status. The fact that its reducing substrate, glutathione, is present at very low concentrations in extracellular fluids had led to suggestions that this enzyme might have a function other than the erythrocyte peroxidase activity. Little is known about the enzymatic antioxidant defense mechanisms in extracellular fluids, although extracellular SOD and GSHPx have been described and it was described that the kidney supplies a major portion of plasma GSHPx protein.[134]

The observed increase of P-GSHPx in DAT patients suggests an increase in H_2O_2 formation since the relationship between the critical level of oxidative stress and the GSHPx activity is now well known.[135-137] P-GSHPx increase could reflect the importance of oxygen radical formation in the extracellular space and underline one important aspect of the ROS metabolism: the compartmentalization. In brain tissues of AD patients, there are many activated microglia or infiltrating macrophages, especially around the amyloid in the center of neuritic plaques. Because the respiratory burst of activated phagocytic leukocytes produces superoxide, it is natural to assume that activated microglial cells produce superoxide. Amyloid is supposed to aggregate in the extracellular space. If this is so, the extracellular GSHPx is more important than cytoplasmic GSHPx to detoxify the hydrogen peroxide demonstrated recently to be responsible to the cytotoxic action of Aβ.[25,83,84,100] However, the slight increase observed in this population is mostly observed in the youngest AD cases and interpretation must be made with great caution.

There is no significant difference concerning the peroxidation marker TBARS between the two groups and this marker is not associated with age. The same result was reported by Jeandel et al.[123] On the other hand, TBARS does not seem to be correlated with other protective systems against free radicals.

Our analysis of the relationship between selenium and GSHPx activity both in plasma and erythrocyte suggests that high selenium values are associated with increased GSHPx activity in both groups. A correlation between selenium (in plasma and erythrocytes) and GSHPx activity has been reported by some authors[138] but not by others.[123,139,140] A threshold for selenium levels in humans has been proposed.[138,141] Below this level of Se, a correlation exists showing that GSHPx activity is dependent on selenium status; above this level the relationship exhibits a plateau, indicating that the selenium requirement for the function of the enzyme is satisfied. In a previous study,[138] we found a clear-cut division in the plasma around 77 ng/ml. In the present study, we are largely under this value for both groups and we observed positive correlations between GSHPx and Se in erythrocytes and plasma. Most European people have normal levels in the range of 50-110 ng/ml, but significant differences occur between countries. Here we demonstrated that differences also occur in Se concentrations between different regions of France since a higher concentrations in P-Se and E-Se (76.9 ± 15.7 ng/ml and 0.48 ± 0.14 ng/mg Hb respectively) were found in our previous study on elderly subjects residing in southwestern France.

Most of the studies have shown a significant decrease in selenium with age.[138,140] This is not the case for this control population but a net decrease with age was found in the AD group for P-Se and E-Se. It is tempting to speculate that nutritional factors could occur for this difference. Correlations observed between parameters are mostly consistent in the AD group, although they do not appear in the controls. Negative correlations with age were also observed for GSHPx and Se, while no changes with age were detected in healthy controls for all the parameters studied.

In conclusion, the free radical scavenging systems tested here are not altered in erythrocytes of AD patients, while plasma GSHPx and Se are slightly increased. These findings suggest that in AD a change in free radical scavenging processes, in relation with H_2O_2 and lipoperoxides detoxification, occurs in the extracellular space such as the plasma. This significant relationship between AD and P-GSHPx has not been previously described and could serve as a peripheral marker of oxidative stress in AD if this result is confirmed in larger studies. Moreover, some modifications in the relationships between antioxidant parameters, and the observation of an age-dependent decrease of glutathione-peroxidase activities and their cofactor, appeared only in the AD group. Taken together these data suggest that AD is associated with an oxidative stress due to an imbalance between reactive oxygen species and the peripheral antioxidant opposing forces.

AMYLOID PRECURSOR PROTEIN IN THE PATHOGENESIS OF ALZHEIMER'S DISEASE

Physiological Role of APP

While the brain is found to express high levels of APP, the physiological role of this protein in the central nervous system is largely unknown. APP consists of a single transmembrane structure with a glycosylated extracellular domain and a short cytoplasmic domain which is well conserved in evolution (Fig. 5.1). APP is involved in signaling of cell adhesion, neurite outgrowth, and synaptic contact.The secreted form of APP can function as an autocrine factor to stimulate cell proliferation (for review see ref. 142).

The biological functions of APP in vivo have recently been examined in homozygous APP-knock out mice. APP-knock out mice proceed through gestation normally, suggesting that APP is not absolutely required during development. As adults, they exhibit a slight decrease in weight, decreased locomotor activity, and forelimb motor abnormalities. The only detectable abnormality in the brain is diffuse reactive gliosis without clear evidence of neuronal degeneration. Another mouse model has been developed in which a mutant form of APP lacking exon 2 is expressed. These animals exhibit cognitive impairment and agenesis of corpus callosum.[143]

An important question is whether loss of APP function contributes to the pathogenesis of AD. Secretion cerebrospinal fluid levels of APPS have been demonstrated to be significantly decreased in AD.[144] APPs can protect cultured neurons from a variety of stresses, including glucose deprivation, glutamate toxicity, and oxidation by a mechanism that may involve stabilization of intracellular calcium.[145] Furthermore overexpression of APP in transgenic mice can increase the number of cortical synapses and protect against the neurotoxicity of the HIV gp120 protein.[146] Thus decreased levels of APP in AD could potentially render neurons more vulnerable to neurotoxic insults. Moreover, the altered processing of mutated-APP gives rise to amyloid deposition[37,147] resulting in increased neuronal vulnerability.

APP and Copper Homeostasis

APP isoforms (containing the KPI domain) that form complexes with extracellular proteases are internalized by the apoE receptor LRP (low density lipoprotein receptor-related protein).[148] A Zn(II)-binding site resides within APP residues 181 to 200 and a Cu(II)-binding site resides within APP residues 135 to 155.[149,150] Zinc(II) and Cu(II) binding influence APP conformation, stability and homophylic interaction. In neurons APP is first delivered from the cell body to the axonal cell surface and then to the dendritic plasma membrane.[151] If APP undergoes trancytosis from axons to dendrites, APP could function as a trancytotic receptor in neurons and could facilitate the transcellular transport of cerebral Zn(II) and Cu(II). Copper is an important component of various redox enzymes. Free Cu is also a toxic ion, as exemplified by its ability to inactivate proteins through tyrosine nitration, and both deficiency and excess lead to disorders such as Wilson's disease, Menkès syndrome, and possibly FALS. The candidate genes for these diseases encode membrane Cu(II) and Cd(II) transporters and CuZnSOD respectively.

Because transmembrane APP binds Cu(II), APP could be involved in Cu homeostasis and in accumulation of APP in neurites, such as occurs in AD,[152] which may lead to dysruption of Cu compartimentalization and hence to Cu toxicity. Multhaup et al[153] have identified a Cu(II)- binding site in APP and ascertained that APP has an intrinsic redox potential specific for Cu(II). Structural change of APP is initiated by Cu(II)[150] and because Cu(II) is reduced to Cu(I) by APP, this conformational change may result from cystine formation involving Cys 144. By using matrix-assisted laser desorption mass spectrometry to directly analyze the possible participation of this cysteine in Cu(II) reduction, it was shown that Cu(II) specifically induced oxidation of the single cysteine residue contained within peptide APP 135-155.[153] Thus, Cu(II) binding led to oxidative modification of APP, giving rise to cystine and Cu(I) formation. Accordingly, in vitro APP has a function in electron transfer to Cu(II). The reduction of Cu(II) to Cu(I) by APP involves an electron-transfer reaction and could enhance the production of hydroxyl radicals, which could then attack nearby sites. The Cu(II)-reducing activity of APP may occur on or near the cell surface because the redox state of cytosol in viable cells is reducing in the absence of exogenous reductant.

APP-Cu(I) complexes on the surface of neurons may be particularly vulnerable to peroxides generated by SOD and by other various mechanisms generating ROS, including activated microglia. In Down's syndrome a gene dosage effect involving CuZnSOD and APP may be responsible for premature onset of AD.[22] Copper homeostasis could be adversely affected in an interplay of APP and SOD, as has been suggested for a similar mechanism of Cu toxicity with mutant SOD in FALS.[154] Copper reaches higher concentrations in cerebrospinal fluid of AD patients and has been discussed as a possible effector of Aβ aggregation in AD.[155]

Two major hypotheses have been proposed concerning the mechanism of amyloid deposition in AD. In one hypothesis, synaptic and axonal injury leading to the formation of dystrophic neurites may precede the deposition of Aβ.[152,156] The alternative view is that Aβ deposits constitute the first abnormality.[157] A mechanism of disease mediated by disruption of vesicular-mediated APP-Cu(I) complex trafficking is consistent with both hypotheses. In the presence of hydrogen peroxide, the key feature of such an APP-Cu(I) accumulation would be the gain of Cu-mediated toxicity. The final common pathway in AD may be oxygen radical-induced cellular damage and radical-mediated aggregation of Aβ monomers, which produce toxic forms of Aβ.

Alteration in APP Processing

Alterations in the processing of APP appear central to the pathogenesis of AD for several reasons:

(1) Glutamate and APP processing. Stimulation of glutamate metabotropic receptors (mGluR) can, by generating diacylglycerol and activating protein kinase C, accelerate the breakdown of the APP to form soluble, nonamyloidogenic derivatives.[158] These data show that selective mGluR agonists that initiate signal-transduction events can regulate APP processing in primary neurons and transfected cells. Recent findings by Copani et al[159] showed that activation of mGluR attenuates Aβ-induced neuronal apoptosis. As glutamatergic neurons in the cortex and hippocampus are damaged in AD, amyloid production in these regions may be enhanced by deficits in glutamaergic neurotransmission.

(2) Stress induces transcriptional activation of APP gene. Aβ also accumulates in the brains of older patients affected with

Down's syndrome, connected with a duplication of the APP gene. This suggests that overexpression of the APP gene could be a contributing factor in the aberrant processing of the precursor protein and make it important to study the factors that influence and induce its expression. Several mechanisms could control the regulation of the APP gene. Nerve growth factor, interleukin-1 and the transcription factor jun/AP1[160] are among cellular mediators that demonstrate an increase in APP mRNA levels in neuronal cells. There is a heat shock consensus sequence located at position -317 bp on the APP gene promoter, and stress in the form of heat shock, ethanol and probably oxidative stress transcriptionally activates the APP gene.[161] It was recently demonstrated that the nuclear factor that mediates this activation is heat shock factor-1 (HSF-1).[162] HSF-1 activation is mediated by numerous stresses. These include hypoxia, elevated $[Ca^{2+}]i$, decreased ATP and exposure to ROS. This could contribute to the stress-exacerbate neuron loss and neuropathology in AD.[163]

(3) APP overexpression and energy metabolism defect. Mutations in the APP gene, which have been identified as causes of AD, may lead to the disorder by impairing the normal function of APP, rather than by overproducing Aβ. A number of findings appear to be consistent with such hypothesis. The promoter of the APP gene contains both heat-shock and interleukin-responsive elements, suggesting a role in response to injury. Consequently, APP expression has been found to be upregulated in response to stress. Several components of amyloid plaques have been identified as stress-response proteins.[164] These findings suggest that APP may be recruited as part of a cellular stress response and may serve under these conditions to assist in re-establishing cell/tissue homeostasis. If such a role includes recruitment in response to oxidative stress, it would assign a role to APP in the homeostatic maintainence of energy metabolism. A persisting recruitment of APP, induced by the oxidative stress accompanying normal aging, would then provide an age-dependent pathology in individuals harboring an APP mutation from birth.

An alternative hypothesis is a defect in energy metabolism due to the APP gene product.[165] This hypothesis suggest that gene products of APP play a role in energy metabolism. The transfer of the APP gene using an adenovirus vector, leading to APP overproduction, causes structural mitochondrial abnormalities and decreased complex IV activity in muscle fiber cultures.[166] The degree of abnormalities correlated with the duration of APP overexpression. Decreased complex IV activity preceded mitochondrial structural abnormalities. These data suggest that excessive APP may be responsible for mitochondrial and complex IV abnormalities in muscle fibers and perhaps in AD and DS brains.

Alternatively, inhibitors of oxidative energy metabolism markedly increase the proteolytic cleavage of APP[81] and may predispose to increased amyloid production in the Alzheimer brain. Thus, this dramatic effect of inhibition of energy metabolism on the proteolytic processing of APP may have significance for the pathogenesis of AD. Recent findings suggest that there is a selective deficit in the azide-sensitive mitochondrial complex IV in the cerebral cortex of patients with AD.[167] Furthermore, deficits in oxidative energy metabolism have been shown to accompany normal aging and may be related to an age-dependent increase in MtDNA mutations.[168]

Impairment of oxidative energy metabolism as in AD[169] may therefore contribute to the pathogenesis of AD by altering the metabolism of APP, leading to decreased APP secretion and increased generation of a potentially amyloidogenic derivative.

On the basis of a hypothesis of overproduction of APP causing AD, one might expect that transfected cells or transgenic animal engineered to overexpress APP would readily model the pathology. To investigate the consequences of increased intracellular production of various regions of APP on neuronal physiology, PC12 cells were transfected with the APPC100 protein (C-terminal 100 amino acids of APP, containing the entire Aβ, transmembrane and intracellular domains of APP, a potentially amyloidogenic fragment).[170] Expression and accumulation of APPc100 protein results in an abnormal response to growth factor stimulation. Exposure to nerve growth factor (NGF) did not promote differentiation. Transfectants continued to divide and failed to elaborate extensive neurites. Unlike NGF treatment, basic fibroblast growth factor (bFGF) profoundly accelerated neurite outgrowth in transfected cells. Also a dramatic increase in a tyrosine phosphatase activity was reported. SP, NFT and neuronal loss was reported in the brain of transgenic mice overexpressing the C-terminal fragment of human APP.[171] Transgenic mice expressing the 751 amino acid isoform of human APP have been shown to develop early AD-like histopathology and exhibit age-dependent deficits in spatial learning.[172] La Ferla et al[173] have introduced a transgene into mice to direct expression of Aβ peptide to neurons. The transgene is expressed in brain regions which are severely affected in AD-resulting in extensive neuronal degeneration. Morphological and biochemical evidence indicates that the death of these cells occurs by apoptosis. Over 50% of the transgenic mice die by 12 months of age, half of the normal life span of the controls. These data show that APP overexpression and Aβ are neurotoxic in vivo and also suggest that apoptosis may be responsible for the accompaning neuron loss.

GLYCATED TAU IN THE NEUROPATHOGENESIS OF ALZHEIMER'S DISEASE

Glycated Tau Induces Oxidative Stress, NF-κB activation and Aβ Release

Paired helical filament (PHF)-tau is the principal component of NFT, a characteristic feature of the neurodegenerative pathology in AD. Post-translational modification of tau, especially phosphorylation, has been considered a major factor in aggregation and diminished microtubule interaction of PHF-tau. Recently, it has been recognized that PHF-Tau is also subject to nonenzymatic glycation, with formation of advanced glycation end-products (AGE).[174,175] Nonenzymatic glycation and oxidation of polypeptides, with the eventual formation of irreversible AGE, occur most readily when relatively long-lived proteins rich in lysine, such as tau, are exposed to high concentration of aldose/aldose phosphates,[176] as occurs intraneuronally. AGE-modified polypeptides are indeed associated with PHF-tau isolated from AD brains,[26] and Amadori adducts are also present in PHF-tau.[177]

As a consequence of glycation, PHF-tau from AD and recombinant AGE-tau generate ROS and induce oxidant stress in neuronal cells, thereby activating transcription via NF-κB.[178] In order to characterize free radical generation from PHF-tau, electron paramagnetic resonance (EPR) spin-trapping measurements were performed in the presence of the spin trap 5,5-dimethyl-1 pyrroline-N-oxide (DMPO). The spectra suggest that superoxide is generated by PHF-tau and reacts to form DMPO-OOH, which rapidly degrades to DMPO-OH. In the presence of SOD the DMPO-OH quartet signal was decreased by 50-60%. In addition, the residual DMPO-OH signal seen in the presence of SOD suggests that hydroxyl radical is also generated in these preparations.[178]

Although cellular oxidant stress generating high levels of oxygen free radicals can result in cytotoxicity, lower concentrations of ROI can affect signal transduction pathways. In particular, the transcription factor NF-κB is activated by ROS (see chapter 3). NF-κB is in a state of activation in cultured neuroblastoma cells (SK-N-SH cells) into which AGE-tau has been introduced. These data suggest that ROS generated by AGE-tau initiate mechanisms that result in nuclear translocation of a p50/p65 complex.

As a consequence of glycation, PHF-tau from AD and recombinant AGE-tau, formed after nonenzymatic glycation, could potentially impact on neuronal gene expression when introduced into cultured neuroblastoma cells, especially of cytokines, including interleukin-6 (IL-6) whose expression is observed in neurons in AD. Enhanced IL-6 gene expression in response to AGE-tau reflects transcriptional activation at the NF-κB site in the IL-6 promoter. Moreover there was a 3- to 4-fold increase in APP expression in the presence of AGE-tau, promoting proteolytic processes and leading to the release of Aβ.[178]

These data provide insight into how PHF-tau disturbs neuronal function, and add to a growing body of evidence implicating oxidant stress in the pathogenesis of AD, which is thereby related to other chronic disease states in which oxidant stress and the activation of the transcription factor NF-κB are believed to be the common pathogenetic mechanisms

Tau/APP Interaction and Oxidative Stress

The appearance of AGE-related modifications in the mature pathological lesions of AD brain suggests that AGE modifications might be an important primary, and therefore pathological, event in lesion formation.[174,175] AGE-related crosslinks may play a critical role in sequestering NFT from ubiquitin-mediated proteolysis and, following neuronal death, in protecting them from microglial removal.[12]

In AD, where tau phosphorylation[13-15] likely affects microtubule assembly,[179] the interaction of tau with APP might be of particular relevance. Although not directly affecting the interaction of tau and APP, increased phosphorylation of tau might serve as a "pathological switch" by decreasing the interaction of Tau with tubulin, thereby creating cytoskeletal disturbances characteristic of Alzheimer's disease[180,181] and at the same time increasing availability for interaction with APP. Importantly, and paralleling the known function of tau to promote fibrils with tubulin,[182] actin[183] and neurofilament,[184] tau/APP interaction promotes fibrillogenesis.[185] Therefore, the effect of phosphorylation would be to free a pool of tau from microtubule interaction, promoting APP binding and in turn inducing tau-APP polymer formation and promoting fibril formation. The interaction of Tau with APP (and senile plaques) is β-sheet conformation-dependent since tau binding is blocked by either stabilization of an α-helical conformation or incubation with Congo red, a dye that avidly binds to β-sheets.[185] It seems more than coincident that the one element common to all amyloids, i.e., β-sheet conformation, is necessary for the interaction of tau and APP. Another interesting aspect regarding the tau-APP interaction is that the tau-binding domain spans a mutation site associated with FAD, i.e., APP717, and one hypothesis is that mutations at this site alter the interaction with tau in a manner that induces disease development.[185]

As evidenced, there is an increasing body of evidence to suggest that protein-protein interactions and oxidative modifications are important in the pathogenesis of AD. The putative chronology of this pathogenesis is likely to be that a perturbation of normal physiological interactions leads to the establishment of novel pathological interactions. Such pathogenic interactions might be in equilibrium with normal interaction and with the possibility that an etiological factor either (1) stabilizes the pathological conjugates (i.e., by oxidative crosslinking) or (2) alters the equilibrium against nominal interactions and favors pathological ones (i.e., through phosphorylation). In conclusion, understanding of either (1) how normal homeostatic interactions can be maintained or (2) how pathological interactions could be destabilized or prevented, might provide novel strategies to lower the prevalence of AD.

AMYLOID PEPTIDE IN THE PATHOGENESIS OF ALZHEIMER'S DISEASE

Increased Aβ Production and Aggregation: Key Pathogenic Events

According to the "amyloid theory,"the factor that initiates and promotes AD is Aβ. The theory is supported by several main lines of evidence; progressive deposition of Aβ in the brain seems to precede the onset of dementia by many years. Several AD-linked missense mutations in the *APP* gene have been shown in cell culture or in a transgenic mouse model to increase the cellular production of Aβ. Also, the recently cloned *PS-1* gene that is responsible for the most common form of early-onset AD appears to lead to elevated cellular production of Aβ.[186] Aβ induced oxidative stress and disruption of cellular ion homeostasis are implicated in the neurodegenerative process. The peptide also activates microglia, the resident immune effector cells of the brain, which are mobilized in AD and may play a central role in inflammatory processes associated with Aβ. The increase in Aβ plaques in AD brains could be due to several factors, such as an increase in cellular production, an increase in the stability of Aβ aggregates or fibrils, or a decrease in the normal clearance of Aβ peptides or aggregates.

Aβ isolated from senile plaques is hydrophobic, has limited solubility even in strongly denaturing aqueous solvents and is resistant to crude collagenase and pepsin digestion. Aβ peptides form antiparallel β-pleated sheets that aggregate into 7.5-10 nm fibrils that, by X ray diffraction, have a cross β-sheet conformation. Synthetic peptides corresponding to Aβ can also spontaneously form amyloid-like fibrils in a β-sheet conformation (for review see ref. 187).

The general cytopathic effect of amyloid fibrils suggests that the structure of the fibrils, rather than the sequence of their constituent peptide, is responsible for toxicity. In primary neuronal cultures and neuronal cell lines, neurotoxicity is associated with synthetic fragments of Aβ that exhibit a significant degree of aggregation. Evaluation of the structure-activity requirements for age-dependent aggregation and neurotoxicity further revealed that peptides containing the hydrophobic Aβ(29-35) region slowly formed stable aggregates on standing in solution and were neurotoxic to neuronal cultures upon aging, but not in freshly prepared solution.[187] The most notable exception to this age-dependence toxicity is Aβ(25-35) which aggregates and displays maximal toxicity immediately upon dissolution. Aging of Aβ by incubation at 37°C for several days leads to the formation of stable oligomeric peptide aggregates which contain a significantly increased proportion of β-sheet structure. This time-dependent conformational transition from random coil to β-sheet structure appears to be an important contributory factor to enhanced toxicity.

Hence, aggregation of Aβ is a key pathogenic event in Aβ toxicity. It is important to understand the mechanisms by which (1) the neurotoxic aggregated species is produced and (2) it is neurotoxic.

Free Radical-Induced Aggregation of Amyloid Peptides

Among possible factors affecting aggregation are ROS. The synthetic Aβ(1-40) and the fragment Aβ(25-35) both react with the spin trap phenyl-tert-butyl nitrone (PBN) to generate a nitroxide product, indicating the presence of a peptidyl radical.[27,189] Aβ cleaved the C=N bond of PBN which is a radical addition fragmentation reaction.[101] In solution Aβ(1-40) and Aβ(25-35) both react with salycilate to form 2,3-dihydroxybenzoate, implying the presence of a ROS in the peptide incubate.[100,101] The kinetics of appearance of the signal and the aggregation kinetics of the two Aβ peptides were tightly correlated. The Aβ radicals promote covalent crosslinking of Aβ monomers to form amyloid fibrils.

Moreover the oxidation sensitive enzymes glutamine synthetase (GS) and creatine kinase (CK) are rapidly inactivated upon exposure to Aβ, suggesting that Aβ toxicity may be a consequence of its ability to promote oxidation reactions.[100,101] Peptidyl radical generation was shown to be independent of metals, but dependent on oxygen concentration, and exhibited a time-dependence similar to that over which neurotoxicity is observed to increase. To assess the nature of the oxygen dependence on Aβ-derived free radical reactivity, Hensley et al[101] reported that the C-terminal methionine residue of Aβ(25-35) was converted to a species that co-eluted with Met-sulfoxide in the time frame of free radical production, and that the free radical scavenger PBN completely blocked formation of this putative sulfoxide. EPR spin trapping studies with PBN were used with Aβ(25-35) to detect the three line-EPR spectrum of the resulting spin adduct. This three line spectrum is not affected by the presence of SOD, suggesting that superoxide radical anion is not involved in the resulting EPR spectrum. However, if catalase is added to the system, no EPR spectrum is observed, consistent with the notion that a peroxyl radical, though not necessarily hydrogen peroxyde, is formed from Aβ(25-35).

In conclusion, the structure of Aβ allows formation of free radicals on a certain amino acid(s) and that these radicals are involved in the covalent crosslinking of soluble Aβ to form toxic amyloid aggregates. Although pure synthetic Aβ can aggregate spontaneously in the presence of oxygen, the crosslinking process can be accelerated when Aβ is modified by AGE.[82] This suggest the possibility that glycation precedes and plays a role in Aβ aggregation. Thus, the specific events in the AD brain that promote "radicalization" and aggregation of Aβ may thus include an increased oxidizing environment resulting from age and AD-related changes in oxidants/antioxidant systems, AGE formation, a reduction in pH and "ischemia-like" conditions and local concentration in metals such as zinc, which was shown to promote rapid aggregation of human Aβ.[190] Importantly, the Aβ radicals can be propagated to cell proteins and membranes increasing the intracellular ROS production.

Amyloid "Ox-Tox" Transducers

Recent reports now identify mechanisms for interactions of Aβ with the cell surface that may explain how Aβ damages neurons and vascular endothelial cells, and activates microglia.

Papers showed that microglia can internalize Aβ microaggregates via a scavenger receptor (SR)[191,192] (Fig. 5.3a) and could degrade Aβ(1-42) added to the culture medium.[193] Thus, uptake of Aβ by microglia via a SR could play a role in the clearance of Aβ, which is produced continuously in the normal and AD brains. The SR receptor is not expressed on astrocytes, neurons or vessel-associated structures. In AD there is a strong expression of the SR on activated microglia in the vicinity of senile plaques.[194] Class A SR mediate adhesion of rodent microglia and human monocytes to Aβ fibril-coated surfaces leading to secretion of ROS and cell immobilization.[192] It was proposed that deposition of Aβ fibrils initiates a similar sequence of events in AD. Microglia adhere to fibrillar Aβ via their class A SR, perhaps in an attempt to clear them from the extracellular milieu. The high density of SR ligands present in these Aβ deposits causes immobilization of microglia that come into contact with them, and induces these microglia to secrete cytokines, ROS and nitrogen species that injure neighbouring neurons.[112]

Using a similar experimental approach, Yan et al[195] provide compelling evidence that Aβ interacts with a cell surface receptor called RAGE (receptor for advanced glycation end products) on neurons, vascular endothelial cells and microglia (Fig. 5.3b). Aβ bound to RAGE could trigger reactions generating cytotoxic oxidizing stimuli that could focus its effect on these target cells. The interaction with both vascular cells and neurons causes oxidant stress and brings about perturbation of cellular function. Incubation of synthetic Aβ(1-40) with endothelial cells from murine cerebral vasculature, or with PC12 cells, resulted in dose-dependent generation of thiobarbituric acid-reactive substances (TBARS) which was blocked by pretreatment of cultures with the antioxidant probucol or N-acetylcysteine, by trypsin or by blocking access to RAGE using either anti-RAGE IgG or excess soluble receptor or AGE albumin. Activation of NF-κB was also supressed by blocking RAGE. These data are consistent that Aβ-like AGE ligands interact with RAGE resulting in oxidant stress and cytotoxicity.

It was also shown that stimulation of such a receptor by other ligands which do not themselves generate ROS induces intracellular generation of oxidants in target cells, suggesting that signal transduction mechanisms may be recruited after ligand-RAGE interaction. Data suggest a possible feedback loop whereby binding of Aβ to RAGE induces oxidant stress which further upregulates RAGE, thereby augmenting further Aβ-RAGE interaction. RAGE expression would seem to be significantly upregulated in neurons around or participating in the neuritic plaques, as well as in the vasculature of the affected Alzheimer's disease brain.[195]

RAGE was also present in cultured rat brain microglia, as well as in the immortalized mouse microglial cell line, BV-2.[195] Aβ-RAGE interaction induces migration of microglia and mediates microglial activation. Activated microglia, which produce ROS and cytotoxic cytokines, may be another important source of neuronal damage in AD and may underlie a chronic inflammatory component in AD. Aβ stimulates microglial expression of tumor necrosis factor α (TNFα). Blockade of RAGE prevented the appearance of TNFα mRNA and diminished TNFα antigen to levels seen in untreated cells.

It will be important to establish how binding of Aβ to RAGE or SR triggers oxyradical production. A signal transduction pathway linked to RAGE or SR has not been identified, and these receptors are thought to function mainly as mediators of endocytosis or transcytosis of macromolecules.

Taken together, these studies suggest that RAGE and SR are key players in the neurodegenerative process in AD, and that these receptors are potential drug targets to reduce the devastating neuronal injury, inflammation and vascular damage that occurs in the brains of AD victims.

OXIDATIVE STRESS IN AMYLOID PEPTIDE TOXICITY

Hydrogen Peroxide Production and Antioxidant Protection

Aβ neurotoxicity is accompanied by an increase in intracellular ROS production inferred from the relative increase in dichlorofluorescein (DCF) fluorescence in primary cultured neurons and neuronal cells in culture.[25,196,197] Aβ causes overproduction of H_2O_2 or related peroxides that leads to cell death in PC12 cells, a clonal cell line B12, and CNS primary cultures.[25,196,197,198] The primary site at which Aβ induces ROS formation in cells is the plasma membranes. A potential source of Aβ-induced ROS in membranes is the inositol phospholipid-arachidonic acid system; indeed inhibitors of lipoxygenases protect cultured neurons against Aβ toxicity.[25,197]

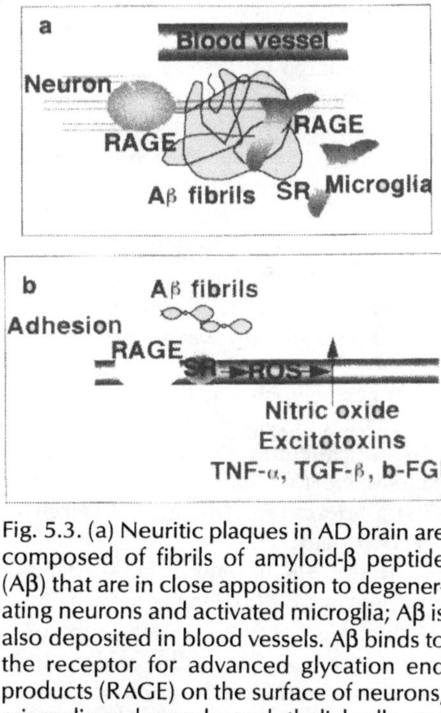

Fig. 5.3. (a) Neuritic plaques in AD brain are composed of fibrils of amyloid-β peptide (Aβ) that are in close apposition to degenerating neurons and activated microglia; Aβ is also deposited in blood vessels. Aβ binds to the receptor for advanced glycation end products (RAGE) on the surface of neurons, microglia and vascular endothelial cells, and to the scavenger receptor (SR) on microglia. (b) Aβ interaction with, and activation of, microglia. Binding of Aβ fibrils to RAGE or the SR on the surface of microglia promotes cell adhesion, and induces activation of the microglia resulting in microglial generation of nitric oxide, excitotoxins, cytokines such as tumor necrosis factor-α (TNF-α) and transforming growth factor-β (TGF-β), and the neurotrophic factor basic fibroblast growth factor (b-FGF) Reprinted with permission from Mattson and Rydell. Nature 1996; 382. © Nature.

The neurodegenerative pathways induced by Aβ in vitro have been reported to be attenuated by antioxidants, and the degree of lipid peroxidation, ROS generation and cytotoxicity could be prevented or attenuated by propylgallate, a free radical scavenger,[25] α-tocopherol,[83] norihydroguaiaretic acid, a lipoxygenase inhibitor[197] and antioxidants PBN[84] and EUK-8,[199] a synthetic catalytic superoxide and hydrogen peroxide scavenger. The H_2O_2-degrading enzyme catalase protects cells from Aβ toxicity.[25,196,200] The protection by catalase is dependent upon Aβ concentration, suggesting that the H_2O_2 generating system is activated by an Aβ dose-dependent manner. Clonal cell lines selected for their resistance to Aβ cytotoxicity also become resistant to the cytotoxic action of H_2O_2 and an increase in H_2O_2-degradating enzymes such as catalase, and GSHPx was reported in these Aβ-resistant cell lines.[200] It is likely that H_2O_2 induces these enzyme systems which protect the cells from itself. These cell line that are selected for their resistance to Aβ cytotoxicity also became resistant to exogenously applied H_2O_2.

Aβ has been shown to increase lipid peroxidation in rat primary cortical[25] and hippocampal cultures[197] and rat neuronal cell lines,[25] further implicating H_2O_2 or one of its more reactive metabolites in its mechanism of toxicity.

The source of Aβ-induced H_2O_2 production could be produced in the nervous system directly by various oxidases, superoxide dismutase, and xanthine oxidase and nonenzymatically from the autooxidation of catecholamines. Alternatively, since H_2O_2 is broken down to H_2O_2 and O_2 by catalase and GSHPx, it is possible that the 2-fold increase in H_2O_2 accumulation is due to the inhibition of its breakdown. Aβ-induced H_2O_2 production and Aβ toxicity are blocked by reagents that inhibit flavin oxidases, suggesting that Aβ activates a member of this class of enzymes.[25]

The role of H_2O_2 in the response of cells to Aβ may be 2-fold: (1) It may serve as a second message system to induce a protective stress response or genes involved in Aβ-cytotoxicity; and (2) at higher concentrations H_2O_2, secondary and tertiary peroxides, or hydroxyl radicals may lead directly to cell death via oxidative damage. Potential evidence for this second message role of H_2O_2 is the ability of Aβ to induce NF-κB activity.[25,195] NF-κB is a transcription factor that is activated by oxygen radicals and serves to activates genes whose products are involved in inflammatory defense mechanisms (see chapter 3). Evidence is accumulating that these inflammatory components play a significant role in exacerbating neuronal damage in AD.

These results not only support a role for ROS in Aβ neurotoxicity but also highlight the potential of synthetic or natural radical scavengers and antioxidants in the prevention or treatment of AD and expend the widening field of free radical research in AD.

Nitric Oxide and Inflammatory Processes

Microglial cells are ubiquitously spread in the CNS and the main immune effector cell population in the brain. In AD, neuritic plaques are associated with clusters of activated microglial cells.[106,107,201] These cells associated with amyloid plaques show intense immunostaining for IL-1, IL-6 and TNF-α, and the cytokine immunoreactivity appears to be related to the state of activation. The activated microglial cells can produce superoxide anions and NO[202] and mediate neurotoxicity.[203] The inducibility of MnSOD expression in the microglial cells in the vicinity of plaques (Fig. 2.8) is in favor of a high production of superoxide anions in these cells.

In AD, it has been suggested that activated microglial cells might be the link between Aβ deposition and neuronal degeneration.[204] An amyloid-induced microglial respiratory response is another potential source of oxidizing equivalents to consider in relation to plaque-associated protein alteration. New mechanisms for pathogenesis of AD were recently proposed in which microglial cells activated by Aβ produce a large amount of NO and may cause neuronal death through a mechanism mediated by NO.[112,205,206] The ability of NO synthase inhibitors and the NO scavenger hemoglobin to interfere with the cell death cascade are in accord with a key role for NO in β-amyloid cell death.

Aβ(1-40) and Aβ(1-42), which are deposited in SP in the AD brain, induced NO production in rat microglial cells.[205-207] Aβ(25-35), the minimal fragment generating neurotoxicity, induced NO production in mouse microglial cells but not in rat microglial cells.[112] It must be noted that there are species differences in NO production. Previous reports demonstrated that human microglial cells don't have the capacity to produce detectable levels of NO when stimulated with Aβ.[206]

Further investigations involving the effect of Aβ on the production of NO by human microglial cells are needed to establish a pathophysiological role of Aβ-induced NO in AD. A number of biological properties have been identified for Aβ, with activity being associated with different portions of the polypeptide. Thus, the direct neurotoxicity and NO production caused by Aβ are likely to be mediated by different peptide regions. Moreover, the differences between microglial cells and astrocytes in NO production[205] suggest that Aβ initiates distinct signals in a cell-specific manner.

Because activated microglia, which are the main inflammatory response cells in the brain, are enriched around amyloid plaques, it has been proposed that plaques are the sites of chronic inflammation. There have been reports that Aβ can activate microglia and trigger the production of proinflammatory and potentially cytotoxic mediators.[112] Aβ enhances microglial release of IL-1 and b-FGF[208,209] and potentiates cytokine secretion by interleukin-1β-activated astrocytoma cells.[210] A synergistic effect between Aβ and IFN-γ in triggering the production of reactive nitrogen intermediates and TNF-γ from microglia was reported. Furthermore, using co-culture experiments, activation of microglia with IFN-γ and Aβ leads to neuronal cell injury in vitro.[112] These findings suggest that Aβ and IFN-γ activate microglia to produce reactive nitrogen intermediates and TNF-γ, and this may have a role in the pathogenesis of neuronal degeneration observed in AD. The production of TNF-γ from microglia activated by Aβ could be the source of TNF-γ, and probably of other cytokines (IL-1α/β and IL-6) present in senile plaques, but the role of TNF-γ on neuronal behavior remains controversial.

Aβ also induces NF-κB[25,195] a transcription factor that can be actived posttranslationally by oxidative stress. NF-κB is also one of the most ubiquitously conserved components of TNF signaling. In addition, the genes for TNF-α and TNF-β are themselves actived by NF-κB. Therefore, it is possible that tissues damaged by an oxidative attack from Aβ respond with an increase in the levels of cytokines that offer protection against the insult. Barger et al[211] tested the hypothesis that TNFs affect neuronal vulnerability to Aβ. Aβ(25-35) and Aβ(1-40) induced neuronal degeneration in a concentration and time-dependent manner. Pretreatment of cultures for 24 hr with TNF-α or TNF-β resulted in significant attenuation of Aβ-induced neuronal degeneration. Accumulation of peroxides induced in neurons by Aβ was significantly attenuated in TNF-pretreated cultures, and TNFs protected neurons against iron toxicity, suggesting that TNFs induce antioxidant pathways.

These data suggest that TNFs protect hippocampal neurons against Aβ toxicity by suppressing accumulation of ROS and that κB-dependent transcription is sufficient to mediate these effects. A modulatory role for TNF in the neurodegenerative process in AD was thus proposed. The connection of NF-κB to the control of genes that increase cellular antioxidant capacity provides a reasonable explanation for the neuroprotective activity of the TNFs. TNF-α increases the expression of MnSOD in neural cells, probably through κB enhancer sites in its promoter. NF-κB and/or other κB-dependent transcription factors may act to coordinate antioxidant expression with the oxidative load of a cell. This type of feedback mechanism may even utilize peptide factors such as the TNFs in order to more diffusely propagate a signal more stable than the ROS themselves.

The Lipoperoxidation Product 4-Hydroxynonenal

Previous studies of neuroblastoma cells,[25] synaptosomes,[212] and cultured rat hippocampal neurons [213,214] suggest that Aβ can induce lipid peroxidation in cell membranes. Mark et al[215] directly showed that Aβ induces large increases in levels of free and protein-bound 4-hydroxynonenal (HNE), a major aldehyde product of lipid peroxidation, in cultured hippocampal cells. The concentrations of HNE generated in response to Aβ were within the range of concentrations at which HNE disrupts ion homeostasis and kills neurons. Using antibodies that recognize HNE-protein conjugates, the authors showed that HNE binds to multiple cell proteins in neurons exposed to Aβ, and that the induction of the anti-HNE immunoreactivity was prevented by the antioxidant propylgallate, demonstrating a requirement for oxyradicals. Glutathione, which was previously shown to play a key role in HNE metabolism[216] (HNE can be conjugated to GSH by glutathione S-transferase or by direct reaction with GSH), attenuated the neurotoxicities of both Aβ and HNE. HNE can mediate cytoskeletal alterations since a recent study of the effects of HNE on cultured postnatal neuroglial cells provided evidence that HNE can induce crosslinking of cytoskeletal proteins, including tau, the major component of PHF.[217]

HNE produced in response to Aβ appears to play an important role in the disruption of ion homeostasis and cytotoxicity induced by the peptide. Both Aβ and HNE impaired the function of the plasma membrane Na^+,K^+-ATPase, the main protein involved in maintenance of resting membrane potential in neurons. As with Aβ,[218] the elevation of $[Ca^{2+}]i$ in response to HNE involved Ca^{2+} influx through plasma membrane channels because it was blocked by removal of extracellular Ca^{2+}. Moreover, NMDA antagonists protected neurons against HNE toxicity, suggesting that HNE promoted glutamate release and/or NMDA receptor activation.

These findings are consistent with a key role for HNE in the mechanism of Aβ toxicity.

Ionic Homeostasis Perturbation: Relationships with Free Radical Formation and Excitotoxicity

Studies in some laboratories have shown that Aβ can disrupt cellular ion homeostasis. The peptide radicals can be propagated to the cell membrane (via SR or RAGE) where they induce lipid peroxidation and damage enzymes involved in regulation of ion homeostasis and signal transduction.

The ability of synthetic Aβ to form weakly selective cationic 'channels' in planar lipid bilayers has been suggested to be responsible for some of its neurotoxic effects.[219]

Calcium influx is mechanistically involved in Aβ toxicity since (1) addition of Aβ to primary neurons results in a slow increase in $[Ca^{2+}]i$, an effect that has been proposed to mediate neurotoxicity[220] and (2) removal of extracellular calcium[220] and agents that block voltage-dependent calcium channels[82] can protect neurons against Aβ toxicity. Neurons exposed to Aβ exhibit heightened calcium responses to excitatory amino acids (EAA) and membrane depolarization,[218,221] an action which would make neurons more vulnerable to reduced energy availability and excitotoxicity. Actions of the neurotoxic fragment Aβ(25-35) on calcium homeostasis and viability of PC12 cells are blocked by antioxidants but not by calcium channel antagonists in PC12 cells.[222] These results indicate that both the disruption of calcium homeostasis and the reduction of cell viability produced by Aβ in PC12 cells are mediated by free radical-based processes. Patch clamp analyses of cultured neurons have shown that synthetic Aβ peptide causes membranes to become « leaky » to several ions including Na^+ and Ca^{2+}.[223] The elevation of $[Ca^{2+}]i$ and increased sensitivity to excitatory stimuli induced by Aβ appears to be mediated by free radicals because antioxidants stabilize $[Ca^{2+}]i$ in cultured neurons exposed to Aβ.[213]

The specific events whereby Aβ induces free radical accumulation and disrupts Ca^{2+} homeostasis are unknown, but impairment of plasma membrane ion-motive ATPases plays a role. Exposure of cultured rat hippocampal and cortical neurons to Aβ resulted in impairment of Na^+/K^+-ATPase which preceded and contributed to the elevation of $[Ca^{2+}]i$. Elevation of [Ca2+]i results from impairment of ion-motive ATPases, resulting in membrane depolarization and opening of voltage-depedent channels. Also contributing to elevation of $[Ca^{2+}]i$ is glutamate receptor activation reversal of Na^+/Ca^{2+} exchange. The reduced energy availability documented in the brains of AD patients would be expected to further compromise energy driven ion pumps and increase neuronal vulnerability to excitotoxicity.

The ROS impair ion-motive ATPase activities leading to disruption of calcium homeostasis and cell death.[224] The impairment of ion-motive ATPase activities was prevented by antioxidants, again suggesting that induction of free radicals is an early event in the Aβ toxicity pathway.

In addition to its direct neurotoxic effect, Aβ can potentiate the toxic effect of a variety of different neuronal insults including EAA, particularly glutamate. It was demonstrated that the Aβ-derived free radicals are able to inhibit glutamate uptake in cultured astrocytes,[225] suggesting a possible link between free radical injury and excitotoxicity in AD. At present, it is not known whether Aβ-derived free radicals inhibit the glutamate transporter directly or through indirect effects on ion-motive ATPases. An implication of inhibited astrocyte glutamate uptake in brain is increased extracellular levels of glutamate, which is excitotoxic to neurons. Then higher levels of extracellular glutamate might lead to cell death through well-established mechanisms involving excitotoxicity and NMDA receptors. Stimulation of such receptors is known to lead to large increases in $[Ca^{2+}]i$ and to the generation of oxygen-dependent free radicals.[226]

Such alterations in AD brains could conceivably expose neurons to greater risk as a result of excess glutamate and may be important in mechanisms of neuronal death in this disorder, including the Aβ-induced increase in neuronal $[Ca^{2+}]i$ and the reported synergy between Aβ and glutamate. Whether or not the dysruption of cell membrane integrity by the insertion of Aβ into the lipid bilayer and the consequent current leakage through these rogue channels could explain or account for these changes has yet to be determined. Nevertheless, sustained elevation of $[Ca^{2+}]i$ is known to have many deleterious consequences, such as the activation of cellular proteases and nucleases, the disruption of protein phosphorylation/dephosphorylation cycles, and increased formation of free radicals and ROS (see chapter 3). Many of the alterations in cytoskeletal components observed in AD may be explained by accumulation of free radicals and calcium in neurons. Studies in vitro and in vivo have shown that EAA and calcium influx can induce antigenic and ultrastructural changes in the cytoskeleton similar to those seen in neurofibrillary tangles in AD.[227-229]

Mitochondria and Redox Activity

Among the cytotoxic mechanisms of actions suggested, several lines of evidence suggest a role for oxidative abnormalities, specifically in mitochondrial enzymes, in the pathophysiology of AD.[169] Changes in cell morphology and inhibition of the bioreduction of the redox dye 3-(4,5-dimethylthiazol-2-yl)-2,5-dimethylthiazolium bromide (MTT) together represent the earliest changes detectable following exposure of rodent primary cortical cells and PC12 cells to Aβ.[25,230,231] Inhibition of MTT reduction, in particular, appears to be a specific indicator of the mechanism of Aβ toxicity.[230-232] Identification of the site of reduction of this dye and the redox system involved in cellular MTT reduction would therefore give valuable information on the nature of Aβ toxicity. Early studies on succinate-dependent MTT reduction by rat liver homogenates in the presence of respiratory-chain inhibitors led to the proposal that MTT reduction can take place at two sites on the mitochondrial electron transport chain, involving succinate dehydrogenase and cytochrome c.[233] Other studies have demonstrated in intact cells an intracellular, predominantly perinuclear, deposition of MTT formazan cristals.This characteristic pattern of deposition is disrupted following treatment with Aβ. The distribution of MTT-formazan within cells is suggestive of an organellar (possibly, but not necessarily, mitochondrial) site of reduction. Evidence that MTT reduction in intact cells is carried out primarily by redox enzymes the of mitochondrial inner membrane is however, mostly inferential. Recent work has shown that the majority of cellular MTT reduction is likely to be catalyzed by nicotinamide-nucleotide-dependent enzymes, and that MTT reduction by isolated mitochondria[234] and by intact PC12 and Hela cells[235] is not sensitive to respiratory chain inhibitors. These observations offer different interpretation on the sites of action of Aβ. If mitochondria are in fact the principal site of MTT reduction in intact cells, the inhibitory action of Aβ on mitochondrial MTT reduction is obviously different from that of a classical respiratory chain inhibitor. Alternatively, the majority of cellular MTT reduction may be catalyzed by NADPH and NADPH-dependent redox systems distinct from those of the mitochondrial inner membranes. Potential organelle sites of MTT reduction yet to be investigated in any detail include the microsomal redox system and the less well characterized redox system of the mitochondrial outer membrane and plasma membrane. In addition, cytoplasmic flavoprotein oxidases, such as those acting on xanthine, amines and aldehydes, could be candidates for enzymes involved in MTT reduction, as they contain complex prosthetic groups, components of which may be capable of directly reducing tetrazolium salts.

Interestingly, it is also known that superoxide is capable of reducing tetrazolium dyes directly. In HeLa cells, a Aβ-sensitive cell line,[232] 20-30% of MTT reduction has been proposed to be mediated by $O_2^{\circ-}$, generated by xanthine oxidase.[236] Alterations in $O_2^{\circ-}$ production may therefore provide a link between the observations on the effects of Aβ on cellular MTT reduction and on ROS formation.

Addition of the H_2O_2 scavenging enzyme catalase protects neurons in culture against Aβ-mediated toxicity;[25,196,200] however it does so by a mechanism that does not necessarily involve its ability to scavenge H_2O_2 since DCF fluorescence still exists.[196] ATP levels were significantly lower in Aβ-treated cells before a significant loss of viability.[196] This significant loss of intracellular ATP was also observed in differentiated cortical neurons, at which time the content of intracellular ADP had significantly increased, indicating that the reduction in the level of ATP was due to ATP hydrolysis and not to a neural decline in the adenine nucleotide pool. Thus the decreased ATP content following Aβ exposure is a relatively early event that precedes overt degeneration.[196] In this regard Shearman et al[230] found that an early, specific response of cells to Aβ was a potent inhibition of cellular redox activity.

A "Radical" Vascular Connection

Thomas et al[237] demonstrated that Aβ interacts with endothelial cells on blood vessels to produce an excess of superoxide radicals, with attendant alterations in endothelial structure and function. The superoxide radical can scavenge endothelium-derived relaxing factor (EDRF) which has properties that are virtually identical to NO[238] and can produce potent oxidizing agents, which can cause lipid peroxidation and other degenerative changes. Insight into the mechanisms by which the Aβ increases endothelial $O_2^{\circ-}$ production may be gleaned from the observation that exogenous SOD confers protection. This places the source of $O_2^{\circ-}$ in or at the plasma membrane, as the cell is not permeable to the superoxide anions. This possibility is supported by reports that Aβ activates the membrane-associated neuronal NADPH oxidase in neuronal cells.[25] SOD almost totally removed the enhancement of contraction by Aβ.

Endothelial damage is currently recognized as an early event in the development of vascular disease. By providing evidence for direct action of Aβ on blood vessels and probable mediation of the effect by oxygen radicals, this new finding points to the disruption of vascular function in AD. Deposits of Aβ are seen in the brain but also in cerebral blood vessels, and abnormalities in microvasculature have been shown to precede other neuropathological features of AD. Even though the action of Aβ delineated by Thomas et al is on peripheral blood vessels, such an effect on brain vasculature is likely. If brain or platelet-derived Aβ were to come into contact with the endothelium, the free radical production could induce local damage and reduce local blood flow, further increasing oxidative stress and upregulating neuronal Aβ production. Notwithstanding the known accumulation of insoluble Aβ aggregates in walls of cerebral blood vessels, the notion that age-related disorders of the cortical neuropile may arise from endothelial dysfunction is potentially highly important. Other factors that affect the local oxidative stress on cerebral vasculature would influence this process.

It remains to be resolved whether activation occurs through binding to receptors, or by virtue of the amphiphilic character of peptides that enables membrane-associated signaling pathways to be perturbed. The study of Thomas et al highlights a critical balance between NO and $O_2^{\circ-}$ in cells. Their constitutive production at the cell membrane probably reflects their involvement in redox signaling. A relative excess of NO over $O_2^{\circ-}$ would seem to be advantageous. This condition would facilitate regulation by S-nitrosylation of proteins, including the charybdotoxin-sensitive K^+ channel (direct activation by EDRF elicits relaxation), Glu-receptor coupled ion channels (inhibition may limit the neuronal toxicity of Aβ), p21ras and the NAD(P)H oxidoreductase for inhibitory feedback control.[239] As the production of $O_2^{\circ-}$ increases, oxidation raction and related signals would become more prevalent. A relative excess of $O_2^{\circ-}$ will lead to toxicity if cellular defenses are overwhelmed.[239,240]

The possibility that disorders associated with deposition of insoluble Aβ aggregates reflect an NO/ $O_2^{\circ-}$ imbalance has potential therapeutic implications. Treatments might include NO supplementation with $O_2^{\circ-}$-resistant NO donors, or the elimination of $O_2^{\circ-}$ with either NAD(P)H oxidase inhibitors or SOD mimetics. Scavengers of $ONOO^-$, the oxidizing product of NO/ $O_2^{\circ-}$, could be beneficial. Alternatively antioxidants (such as vitamin E), which both preserve endothelial function and attenuate the neurotoxicity of Aβ, could be tried.

Activation of Transcription Factors, IEGs Expression and Apoptosis

Aβ when associated with RAGE and SR could mediate an intracellular signaling pathway leading to cell death. This signaling pathway could involve induction of ROS as second messengers and could in turn activate some transcription factors such as NF-κB.[25,195] The transcription factor NF-κb is activated by various signals associated with brain injury, including TNF,[211] oxidative insults,[241] glutamate,[242] AGE[195] and Aβ.[25,195] C2-ceramide, a membrane permeant activator of NF-κB, protects cultured rat hippocampal neurons against death induced by Aβ.[243] Lipid peroxidation and accumulation of cellular peroxides induced by Aβ were significantly attenuated in neurons pretreated with C2-ceramide.

These results suggest a role of NF-κB acctivation in amyloid-induced neurodegeneration and protection and suggest novel approaches for reducing neuronal injury associated with Aβ toxicity.

Previous studies have shown that Aβ induces neurodegeneration via apoptosis[244,245] suggesting that Aβ may also initiate an apoptotic pathway of cell death in AD. Furthermore, there is evidence suggesting that apoptosis occurs in the AD brain.[246] A potential role for apoptosis in neurodegeneration and Alzheimer's disease is also suggested.[247] Apoptosis could play a protective role in disease by minimizing insult to surrounding cells and tissues attribuable to inflammation. Conversely, apoptosis may participate in the disease state, as a part of a cascade of pathological events

Apoptosis is dependent on the activation of a gene-directed program in several systems, and Aβ-mediated apoptosis is associated with the induction of genes that may regulate or play a role in cell death. Particularly, IEGs respond to cellular stimuli and participate in cellular signaling pathways. The protein products of some IEGs, e.g., c-Jun and c-Fos, are capable of forming dimers and acting as transcriptional regulatory proteins, and have been implicated in apoptosis. A direct relationship between *c-jun* expression and apoptosis has been identified in neuronal cells (see chapter 3).

Although it is difficult to establish a direct relationship between the expression of any single gene and apoptosis, the temporal association of an increase in the expression of a gene with a cellular signal of apoptosis is suggestive of a relationship between the expression of that gene and the process of cell death, particularly when viewed in the context of the overall decrease in protein synthesis that occurs during cell death.

The implications of such a relationship are further strengthened by the association between *jun* and *fos* expression and vulnerability to Aβ-mediated apoptosis. Increased immunoreactivity for the Jun protein has been reported in AD brains.[248,249] Neurons that initiate apoptosis in this paradigm also exhibit a selective and prolonged expression of c-jun, and neurons resistant to Aβ-induced apoptosis do not express c-jun in this model.[249] A double labeling experiment to examine whether c-Jun is associated with DNA strand breaks in AD brain described a strong colocalization between these c-Jun and apoptosis.[250] Moreover, exposure of a mouse hippocampal cell line to the Aβ(25-35) caused a rapid and sustained increase in nuclear c-fos immunoreactivity.[251]

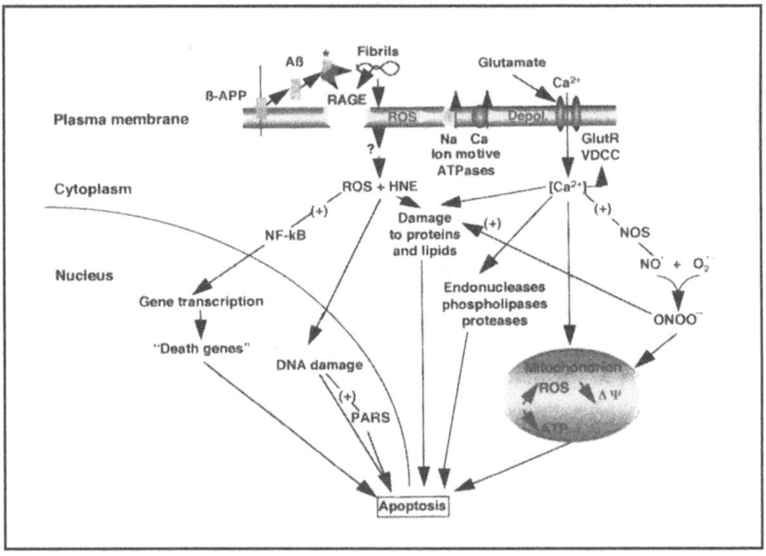

Fig. 5.4. Mechanisms of interaction of Aβ with neurons. Enzymatic processing of the β-amyloid precursor protein (APP) generates Aβ which is released from brain cells in a soluble form. Under oxidizing conditions Aβ may form free-radical moieties (*) and peptide fibrils. Interaction of Aβ with the cell surface of neurons induces production of reactive oxygen species (ROS) and 4-hydroxynonenal (HNE) in the cell membrane (lipid peroxidation); RAGE may mediate generation of ROS. The ROS and HNE impair the function of membrane proteins involved in maintenance of ion homeostasis resulting in membrane depolarization and calcium influx through glutamate receptor channels (GlutR) and voltage-dependent calcium channels (VDCC). ROS and calcium promote damage to cell proteins, DNA and lipids, to mitochondria leading to ATP depletion. ROS could activate gene transcription via the transcription factor NF-κB known to be activated by Aβ. Damaged DNA fragments activate poly-ADP-ribose synthetase (PARS) causing NAD depletion, the PARS substrate and subsequently ATP. All these events ultimately cause apoptosis.

However, as noted by Estus et al,[252] the diversity and functions regulated by c-Jun or other IEG products argues against the possibility that the induction of a single gene such as *c-jun* may be sufficient for cell death in and of itself. *c-jun* expression could be related to the activation of an apoptotic pathway of cell death,[253] perhaps by participating in an apoptosis-related gene cascade. Such a role for c-Jun in Aβ-induced apoptosis would be predicted to involve transcriptional regulation of downstream gene targets by c-Jun and/or c-Jun heterodimers.

The signal transduction pathway for *c-fos and c-jun* induction remains to be elucidated. Amyloid fibrils interacting with RAGE could result in aberrant activation of signal transduction pathways.[254] This mechanism will be consistent with observed increases in the phosphorylation of focal adhesion kinases[255] and the microtubule-associated protein tau.[17] Elevation of $[Ca^{2+}]i$ levels, activation of different protein kinases (e.g., protein kinase C, calcium/calmodulin dependent protein kinase) and oxidative stress lead to an induction of *c-fos* transcription.[256] On the other hand, it has been demonstrated that Aβ causes a rapid increase in intracellular hydrogen peroxide in neurons, which may be the signal for *c-fos* and *c-jun* activation, as has been shown in other cell systems.[257]

These data suggest a relationship between the expression of *c-jun* and *c-fos* and neuronal risk and/or cell death induced by Aβ. Aβ may participate in early events associated with DNA damage and/or apoptosis. Interestingly, immunoreactivity for the cell death inhibitor bcl-2[258] is elevated in AD neurons. These proteins could be components of a dynamic and extended competition between cell death processes and compensatory responses in AD brain.

CONCLUSION

Based on the data presented above, we can describe the following scenario for the mechanisms of interactions of Aβ with neurons and how they explain neurotoxicity (Fig. 5.4);

(1) An increased oxidizing environment resulting from age or the presence of heavy metals coupled with AD-related changes in the oxidant/antioxidant balance induces radicalization aggregation of Aβ and peptide fibrils.

(2) Interaction of Aβ with RAGE induces production of ROS and 4-HNE in the neuronal membrane (lipid peroxidation). The ROS and 4-HNE impair the function of membrane regulatory proteins involved in maintenance of ion homeostasis, including ion-motive ATPases, resulting in membrane depolarization and Ca^{2+} influx through glutamate receptor channels (GlutR) and voltage-dependent Ca^{2+} channels (VDCC). Sustained increases in levels of $[Ca^{2+}]i$ contribute to the cytotoxic action of Aβ. If RAGE mediates the oxidation-inducing affects of relatively low concentrations of Aβ, it may be particularly important in promoting excitotoxic cascades. Such an endangering action of Aβ-RAGE interaction could explain why some types of non-neuronal cells which do not express NMDA receptors (astrocytes and fibroblasts) are resistant to being killed by Aβ. An alternative possibility is that Aβ may act by forming a membrane pore, leading to increased ion influx.

(3) ROS and Ca^{2+} promote damage to cell proteins, DNA and lipids and to mitochondria leading to ATP depletion. Aβ can also impair mitochondrial redox activity increasing the formation of ROS.

(4) Another mechanism is that amyloid fibrils interacting with RAGE leads to aberrant activation of signal transduction pathways. This mechanism will be consistent with observed increases in the phosphorylation of focal adhesion kinases and the microtubule-associated protein tau. Neuronal apoptosis may be a consequence of persistent binding and activation of cell surface receptors by amyloid fibrils, analogous to the induction of apoptosis by interaction of the FAS ligand with its cell surface receptor.

(5) ROS could activate gene transcription via the transcription factor NF-κB known to be activated by Aβ. Damaged DNA fragments activate poly-ADP-ribose

synthetase (PARS) causing NAD depletion, the PARS substrate and subsequently ATP.

(6) β-amyloid fibrils stimulate microglia to produce neurotoxins, such as ROS and cytokines. Moreover SR mediate adhesion of microglia to β-amyloid fibrils leading to secretion of ROS and nitrogen species that injure neighbouring neurons.

(7) Factors, both genetic (ApoE, APP mutations) or environmental (heavy metals) may potentiate the toxicity of Aβ and may contribute to the formation, persistence, or toxicity of Aβ.

(8) Microvasculatur alterations consistent with reduced perfusion and reduced energy availability may increase neuronal vulnerability to the toxicities of Aβ.

All of these events ultimately cause apoptosis. It is now evident that ROS play a central role in the pathogenic cascade leading to AD. Thus, compounds that block interactions between class A SR or RAGE and/or inhibit ROS may be useful in preventing or in delaying the onset or attenuating the progression of Alzheimer's disease.

References

1. Alzheimer A. Uber eine eigenartige erkrantung der Hirnrinde. In: Schulte E, Snell eds. Allgemeine zeitschrift für psychiatrie und psychish-Gerichtliche Medizin. Georg Relmer, Berlin 1907; 64:146-148.
2. Kosik K. Alzheimer's disease sphinx: a riddle with plaques and tangles. J Cell Biol 1994; 127:1501-1504.
3. Goedert M. Tau protein and the neurofibrillary pathology of Alzheimer's disease. Trends Neurosci 1993; 16:460-465.
4. Glenner GG. Alzheimer's disease: its proteins and genes. Cell 1988; 52:307-8.
5. Haass C, Selkoe DJ. Cellular processing of β-amyloid precursor protein and the genesis of amyloid β-peptide. Cell 1993; 75:1039-42.
6. Blanquet V, Golgaber D, Turleau C et al. The amyloid protein precursor cDNA hybridizes in normal and Alzheimer individuals near the interface of 21q21 and 21q22.1. Ann Genet 1987; 30:68-69.
7. Dyrks T, Weideman A, Multhaup et al. Identification, transmembrane orientation and biogenesis of the amyloid A4 precursor of Alzheimer's disease. EMBO J 1988; 7:949-957.
8. Sisodia SS. β-amyloid precursor protein cleavage by a membrane-bound protease. Proc Natl Acad Sci USA 1990; 89: 6075-6079.
9. Busciglio J, Gabuzda DH, Matsudaira P et al. Generation of β-amyloid in the secretory pathway in neuronal and nonneuronal cells. Proc Natl Acad Sci USA 1993; 90:2092-2096.
10. Haass C, Schlossmacher MG, Hung AY et al. Amyloid β-peptide is produced by cultured cells during normal metabolism. Nature 1992; 359:322-325.
11. Perry G, Smith M. Senile plaques and neurofibrillary tangles: what role do they play in Alzheimer's disease? Clin Neurosci 1993; 1:199-203.
12. Cras P, Smith MA, Richey PL et al. Extracellular neurofibrillary tangles reflect neuronal loss and provide further evidence of extensive protein crosslinking in Alzheimer disease. Acta Neuropathol 1995; 89:291-295.
13. Grundke-Iqbal I, Iqbal K, Tung YC et al. Abnormal phosphorylation of the microtubule-associated protein tau in Alzheimer cytoskeletal pathology. Proc Natl Acad Sci USA 1986; 83:4913-4917.
14. Delacourte A, Flament S, Dibe EM et al. Pathological proteins tau 64 and 69 are specifically expressed in the somatodendritic domain of the degenerating cortical neurons during Alzheimer's disease. Acta Neuropathol 1990; 80:111-117.
15. Kosik KS, Orecchio LD, Binder L et al. Epitopes that span the tau molecule are shared with paired helical filaments. Neuron 1988; 1:817-825.
16. Greenberg SM, Koo EH, Selkoe DJ et al. Secreted β-amyloid precursor protein stimulates mitogen activated protein kinase and enhances tau phosphorylation. Proc Natl Acad Sci USA1994; 91: 7104-7108.
17. Busciglio J, Lorenzo A, Yeh J et al. β-amyloid fibrils induce tau phosphorylation and loss of microtubule binding. Neuron 1995; 14:879-888.
18. Takashima A, Nogushi K, Sato K et al. Tau protein kinase I is essential for amyloid b-protein-induced neurotoxicity. Proc Natl Acad Sci 1993; 90:7789-7793.

19. Hoschi M, Takashima A, Nogushi K et al. Regulation of mitochondrial pyruvate dehydrogenase activity by tau protein kinase I/glycogen synthase kinase 3β in brain. Proc Natl Acad Sci USA 1996; 93:2719-2723.
20. Tanzi RE, St George-Hyslop P, Gusella JF. Molecular genetics of Alzheimer's disease. J Biol Chem 1991; 266: 20579-20582.
21. Ceballos I, Agid F, Delacourte A et al. Parkinson's disease and Alzheimer's disease: neurodegenerative disorders due to brain antioxidant system defiencies? In: Emerit I, Packer L, Auclair C, eds. Antioxidant in Therapy and Preventive Medicine. Adv Exp Biol Med, Plenum Press 1990:493-498.
22. Sinet PM, Ceballos-Picot I. Role of free radicals in Alzheimer's disease and Down's syndrome. In: Packer L, Prilipko L, Christen Y, eds. Free radicals in the brain. Adv Exp Biol Med. Springer-Verlag:New York, 1992; 91-98.
23. Volicer L, Crino PB. Involvement of free radicals in dementia of the Alzheimer's type: a hypothesis. Neurobiol Aging 1990; 11:567-571.
24. Smith CD, Carney JM, Starke-Reed PE et al. Excess brain protein oxidation and enzyme dysfunction in normal aging and in Alzheimer's disease. Proc Natl Acad Sci USA 1991; 88:10540-10543.
25. Behl C, Davis JB, Lesley R et al. Hydrogen peroxide mediates amyloid β protein cytotoxicity. Cell 1994; 77:817-827.
26. Yan SD, Chen X, Schmidt A-M et al. Glycated tau protein in Alzheimer disease: a mechanism for induction of oxidant stress. Proc Natl Acad Sci USA 1994; 91:7787-7791.
27. Hensley K, Carney JM, Mattson MP et al. A model for beta-amyloid aggregation and neurotoxicity based on free radical generation by the peptide: relevance to Alzheimer disease. Proc Natl Acad Sci USA 1994; 91:3270-3274.
28. Goate A, Chartier-Harlin MC, Mullan M et al. Segregation of a missense mutation in the amyloid precursor protein gene with familial Alzheimer's disease. Nature 1991; 349:704-7066.
29. Hardy J. Framing β-amyloid. Nature Genet 1992; 1:233-234.
30. Games D, Adams D, Alessandrini R et al. Alzheimer-type neuropathology in transgenic mice overexpressing V717F β-amyloid precursor protein. Nature 1995; 373:523-527.
31. Yamatsuji T, Okamoto T, Takeda S et al. Expression of V642 APP mutant causes cellular apoptosis as Alzheimer trait-linked phenotype. EMBO J 1996; 15: 498-509.
32. Citron M, Vigo PC, Teplow DB et al. Excessive production of amyloid β-protein by peripheral cells of symptomatic patients carrying the Swedish familial Alzheimer's disease mutation. Proc Natl Acad Sci USA 1994; 91:11993-97.
33. Cai X, Golde TE, Younkin SG. Release of excess amyloid β protein from a mutant amyloid β protein precursor. Science 1993; 259:514-16.
34. Haass C, Selkoe D. Cellular processing of β-amyloid precursor protein and the genesis of amyloid β-peptide. Cell 1994; 75:1039-1042.
35. Martin BL, Schrader-Fisher G, Busciglio J et al. Intracellular accumulation of β-amyloid in cells expressing the Swedish mutant amyloid precursor protein. J Biol Chem 1995; 270:26727-26730.
36. Felsenstein KM, Hunihan LW, Roberts A. Altered cleavage and secretion of a recombinant β-APP bearing the Swedish familial Alzheimer's mutation. Nature Genet 1994; 251-258.
37. Hsiao K, Chapman P, Nilsen S et al. Correlative memory deficits, Aβ elevation and amyloid plaques in transgenic mice. Science 1996; 274:99-102.
38. Van Broeckhoven C. Presenilins. Alzheimer's disease. Nature Genet 1995; 11:230-232.
39. Levy-Lahad E, Wasco W, Poorkaj P et al. Candidate gene for the chromosome 1 familial Alzheimer's disease locus. Science 1995; 269:917-918.
40. Levitan D, Greenwald I. Facilitation of lin-12-mediated signaling by sel-12, a Caenorhabditis elegans *S182* Alzheimer's disease gene. Nature 1995; 377:351-354.
41. Artavanis-Tsanokas, Matsuno K, Fortini ME. Notch signaling. Science 1995; 268: 225-232.
42. Kovaks DM, Fausett HJ, Page KJ et al. Alzheimer-associated presenilins 1 and 2:

neuronal expression in brain and localization to intracellular membranes in mammalian cells. Nat Med 1996; 2: 224-229.
43. Scheuner D et al. Fibroblasts from carriers of familial AD linked to chromosome 14 show increased Aβ production. Nat Med 1996; 2:864-870.
44. Rogaev EI et al. Familial Alzheimer's disease in kindreds with missense mutations in a gene on chromosome 1 related to the Alzheimer's disease type 3 gene. Nature 1995; 376:775-778.
45. Corder EH, Saunders A, Strittmatter W et al. Gene dose of apolipoprotein E type 4 allele and the risk of Alzheimer's disease in late onset families. Science 1993; 261:921-923.
46. Chartier-Harlin MC, Parfitt M, Legrain S et al. Apolipoprotein E, ε4 allele as a major risk factor for sporadic early and late-onset forms of Alzheimer's disease: analysis of the 19q13.2 chromosomal region. Hum Mol Genet 1994; 3:579-574.
47. Strittmatter WJ, Weisgraber KH, Huang DY et al. Binding of human apolipoprotein E to synthetic amyloid β peptide: isoform specific effects and implications for late-onset Alzheimer's disease. Proc Natl Acad Sci USA 1993; 90:8098-8102.
48. Jarvik GP, Larson EB, Goddard K et al. Influence of apolipoprotein E genotype on the transmission of Alzheimer disease in a community-based sample. Am J Hum genet 1996; 58:191-200.
49. Poirier J, Delisle MC, Quirion R et al. Apolipoprotein E-ε4 allele as a predictor of cholinergic deficits and treatment outcome in Alzheimer disease. Proc Natl Acad Sci USA 1995; 92:12260-12264.
50. Schâchter F, Faure-Delanef L, Guenot F et al. Genetic associations with human longevity at the ApoE and ACE loci. Nature Genet 1994; 6:29-32.
51. Richey PL, Siedlak SL, Smith MA et al. Apolipoprotein E interaction with the neurofibrillary tangles and senile plaques in Alzheimer disease: implications for disease pathogenesis. Biochem Biophys Res Comm 1995; 208:657-663.
52. Ma J, Yee A, Brewer HB et al. Amyloid-associated protein α1-antichymotrypsine and apolipoprotein E promote assembly of Alzheimer β-protein into filaments. Nature 1994; 372:92-94.
53. Sanan DA, Weisgraber KH, Russell SJ et al. Apolipoprotein E associates with β amyloid peptide of Alzheimer's disease to form novel monofibrils. J Clin Invest 1994; 94:860-869.
54. Wisniewski T, Goldabek AA, Kida E et al. Conformational mimicry in Alzhelmer's disease: role of apolipoproteins in amyloidogenesis. Am J Pathol 1995; 147:238-244.
55. Strittmatter WJ, Saunders AM, Goedert M et al. Isoform specific interactions of apolipoprotein E with microtubule associated protein tau:implications for Alzheimer's disease. Proc Natl Acad Sci USA 1994; 91:11183-11186.
56. Hayek T, Oikinine J, Brook JG et al. Increased plasma and lipoprotein lipoperoxidation in apolipoprotein E deficient mice. Biochem Biophys Res Comm 1994; 201:1567-1574.
57. Miyata M, Smith JD. Apolipoprotein E allele-specific antioxidant activity and effects on cytotoxicity by oxidative insults and β-amyloid peptides. Nature Genet 1996; 14:55-61.
58. Hutchin T, Cortopassi GA. A mitochondrial DNA clone is associated with increased risk for Alzheimer disease. Proc Natl Acad Sci USA 1993; 92:6892-6895.
59. Shoffner JM, Brown MD, Torroni A et al. Mitochondrial DNA variants observed in Alzheimer disease and Parkinson disease. Genomics 1993; 17:171-184.
60. Ganrot PO. Metabolism and possible health effects of aluminium. Environ Health Perspect 1986; 65:363-441.
61. Wenstrup D, Ehmann WD, Markesbery WR. Trace element imbalances in isolated subcellular fractions of Alzheimer's disease brains. Brain Res 1990; 533: 125-131.
62. Candy JM, Oakley AE, Watt F et al. A role for aluminium, silicon and iron in the genesis of senile plaques, in: Modern Trends In Aging Research. EURAGE, John Libbey 1986; 147:443-450.
63. Wallwork JC. Zinc and the central nervous system. Prog Food Nutr Soc 1987; 11:203-247.
64. Wenk GL, Stemmer KL. Suboptimal dietary zinc intake increases aluminium ac-

cumulation into the rat brain. Brain Res 1983; 288:393-395.
65. Markesbery WR, Ehmann WD. Brain trace metals in Alzheimer disease, in Alzheimer Disease In: Terry RD, Katzman R, Bick KL, eds. New York: Raven Press, 1994; 353-367.
66. Zaleska MM, Floyd RA. Regional lipid peroxidation in rat brain in vitro: possible role of endogenous iron. Neurochem Res 1985; 10:397-410.
67. Dyrks T, Dyrks E, Hartman T et al. Amyloidogenicity of βA4 and βA4-bearing amyloid protein precursor fragments by metal-catalyzed oxidation. J Biol Chem 1992; 267:18210-18217.
68. Ehmann WD, Markesbery WR, Alauddin M et al. Brain trace elements in Alzheimer's disease. Neurotoxicology 1986; 7:197-206.
69. Grundke-Iqbal I, Fleming J, Tung YC et al. Ferritin is a component of the neuritic (senile) plaque in Alzheimer dementia. Acta Neuropathol 1990; 81:105-110.
70. Mantyh PW, Ghilardi JR, Rogers S et al. Aluminum, iron, and zinc ions promote aggregation of physiological concentrations of β-amyloid peptide. J Neurochem 1993; 61:1171-1174.
71. Evans PH, Klinowski J, Yano E et al. Alzheimer's disease: a pathogenic role for aluminosilicate-induced phagocytic free radicals. Free Rad Res Comms 1989; 6:317-321.
72. Evans PH, Peterhans E, Burge T et al. Aluminosilicate-induced free radical generation by murine brain glial cells in vitro: potential significance in the aetiopathogenesis of Alzheimer's dementia. Dementia 1992; 3:1-6.
73. Fraga CG, Oteiza PI, Golub MS et al. Effects of aluminum on brain lipid peroxidation. Toxicol Lett 1990; 51: 213-219.
74. Subbarao KV, Richardson JS, Ang L. Autopsy samples of Alzheimer's cortex show increased peroxidation in vitro. J Neurochem 1990; 55:342-345.
75. Palmer AM, Burns MA. Selective increase in lipid peroxidation in the inferior cerebral cortex in Alzheimer's disease. Brain Res 1994; 645:338-342.
76. Smith MA, Rudnicka-Nawrot M, Richey PL et al. Carbonyl-related posttranslational modification of neurofilament protein in the neurofibrillary pathology of Alzheimer's disease. J Neurochem 1995; 64:2660-2666.
77. Mecocci P, Mc Garvey U, Beal MF. Oxidative damage to mitochondrial DNA is increased in Alzheimer's disease. Ann Neurol 1994; 36:747-750.
78. Meccocci P, MacGarvey U, Kaufman AE et al. Oxidative damage to mitochondrial DNA shows marked age-dependent increases in human brain. Ann Neurol 1993; 34:609-616.
79. Smith MA, Taneda S, Richey PL et al. Advanced Maillard reaction end products are associated with Alzheimer disease pathology. Proc Natl Acad Sci USA 1994; 91:5710-5714.
80. Chweers O, Mandelkow EM, Biernat J et al. Oxidation of cysteine-322 in the repeat domain of microtubule-associated protein tau controls the in vitro assembly of paired helical filaments. Proc Natl Acad Sci USA 1995; 92:8463-8467.
81. Gabuzda D, Busciglio J, Chen LB. Inhibition of energy metabolism alters the processing of amyloid precursor protein and induces a potentially amyloidogenic derivative. J Biol Chem 1994; 269:13623-13628.
82. Vitek MP, Bhattacharya K, Glendening JM et al. Advanced glycation end products contribute to amyloidosis in Alzheimer's disease. Proc Natl Acad Sci USA 1994; 91:4766-4770.
83. Behl C, Davis J, Cole G, Schubert D. Vitamin E protects nerve cells from amyloid β protein toxicity. Biochem Biophys Res Commun 1992; 186:944-950.
84. Butterfield DA. A model for β-amyloid aggregation and neurotoxicity based on free radical generation by the peptide:Relevance to Alzheimer disease. Proc Natl Acad Sci USA 1994; 91:3270-3274.
85. Goodman Y, Mattson MP. Secreted forms of β-amyloid precursor protein protect hippocampal neurons against amyloid β-peptide-induced oxidative injury. Exp Neurol 1994; 128:1-12.
86. Hensley K, Hall N, Subramaniam R et al. Brain regional correspondance be-

tween Alzheimer's disease histopathology and biomarkers of protein oxidation. J Neurochem 1995; 65:2146-2156.
87. Mattson MP, Cheng B, Davis D et al. β-Amyloid peptides destabilize calcium homeostasis and render human cortical neurons vulnerable to excitotoxicity. J Neurosci 1992; 12:376-389.
88. Schreck R, Baeuerle PA. A role of oxygen radicals as second messengers. Trends Biochem 1991; 1:39-42.
89. Ceballos I, Javoy-Agid F, Hirsch EC et al. Localization of copper-zinc superoxide dismutase mRNA in human hippocampus by *in situ* hybridization. Neurosci Lett 1989; 105:41-46.
90. Delacourte A, Defossez A, Ceballos I et al. Preferential expression of copper-zinc superoxide dismutase in the vulnerable cortical neurons in Alzheimer's disease. Neurosci Lett 1988; 92:247-253.
91. Pappolla MA, Omar RA, Kim KS et al. Immunohistochemical evidence of antioxidant stress in Alzheimer's disease. Am J Pathol 1992; 140:621-628.
92. Smith MA, Kutty RK, Richey PL et al. Heme oxygenase-1 is associated with the neurofibrillary pathology of Alzheimer's disease. Am J Pathol 1994; 145:42-47.
93. Furuta A, Price DL, Pardo CA et al. Localization of superoxide dismutases in Alzheimer's disease and Down's syndrome neocortex and hippocampus. Am J Pathol 1995; 146:357-367.
94. Premkumar DRD, Smith MA, Richey PL et al. Induction of heme oxygenase-1 mRNA and protein inneocortex and cerebral vessels in Alzheimer's disease. J Neurochem 1995; 65:1399-1402.
95. Smith MA, Taneda S, Richey PL et al. Advanced Maillard reaction end products are associated with Alzheimer's disease pathology. Proc Natl Acad Sci USA 1994; 91:5710-5714.
96. Carney JM, Starke-Reed PE, Oliver CN et al. Reversal of age-related increase in brain protein oxidation, decrease in enzyme activity, and loss in temporal and spatial memory by chronic administration of the spin-trapping compound N-tert-butyl-α-phenylnitrone. Proc Natl Acad Sci USA 1991; 88:3633-363.
97. Oliver CN, Ahn BW, Moerman EJ et al. Age-related changes in oxidized proteins. J Biol Chem 1987; 262:5488-5491.
98. Smith CD, Carney JM, Tatsumo T, Stadtman ER, Floyd RA, Markesbery WR: Protein oxidation in aging brain. Ann NY Acad Sci 1992; 663:110-119.
99. Stadtman ER. Protein oxidation and aging. Science 1992; 257:1220-1224.
100. Harris ME, Hensley K, Butterfield DA et al. Direct evidence of oxidative injury by the Alzheimer's amyloid β peptide in cultured hippocampal neurons. Exp Neurol 1995; 131:193-202.
101. Hensley K, Aksenova M, Carney JM et al. Amyloid β-peptide spin trapping I: peptide enzyme toxicity is related to free radical spin trap reactivity. Neuroreport 1995; 6:489-492.
102. Mc Geer PL, Itagaki S, Boyes BE et al. Reactive microglia are positive for HLA-DR in the substantia nigra of Parkinson's and Alzheimer's disease brains. Neurology 1988; 38:1285-91.
103. Itagaki S, McGeer PL, Akiyama H, Zhu S, Selkoe D. Relationship of microglia and astrocytes to amyloid deposits of Alzheimer disease. J Neuroimmunol 1989; 24:173-182.
104. Perlmutter LS, Barron E, Chui HC. Morphologic association between microglia and senile plaque amyloid in Alzheimer's disease. Neurosci Lett 1990; 119:32-36.
105. Selkoe DJ. Alzheimer's disease: genotypes, phenotype, and treatments. Science 1997; 275:630-631.
106. Dickson DW, Lee SC, Mattiace LA et al. Microglia and cytokines in neurological disease and cytokines in neurological disease, with special reference to AIDS and Alzheimer's disease. Glia 1993; 7:75-83.
107. Wegiel J, Wisniewski HM. The complex of microglial cells and amyloid star in three-dimensional reconstruction. Acta Neuropathol 1990; 81:116-24.
108. Colton CA, Snell J, Chernyshev O et al. Induction of superoxide anion and nitric oxide production in cultured microglia. Ann NY Acad Sci 1994; 738:54-63.
109. Elkabes S, DiCicco-Bloom E, Black IB. Brain microglia/macrophages express neurotrophins that selectively regulate

microglial proliferation and function. J Neurosci 1996; 16:2508-21.
110. Shaffer LM, Dority MD, Gupta-Bansal R et al. Amyloid β protein (Aβ) removal by neuroglial cells in culture. Neurobiol Aging 1995; 16:737-45.
111. Klegeris A, Walker DG, McGeer PL. Activation of macrophages by Alzheimer beta amyloid peptide. Biochem Biophys Res Commun 1994; 199:984-991.
112. Meda L, Cassatella MA, Szendrei GI et al. Activation of microglial cells by β-amyloid protein and interferon-γ. Nature 1995; 374:647-50.
113. Marklund SL, Adolfsson R, Gottfries CG et al. Superoxide dismutase isoenzymes in normal brains from patients with dementia of Alzheimer type. J Neurol Sci 1985; 67:319-325.
114. Markesbery WR, Lovell MA, Ehmann WD. Increased lipid peroxidation and antioxidant enzyme activity in the brain in Alzheimer's disease. Neurobiol Aging 1994; 15:S139-S140.
115. Balazs L, Leon M. Evidence for an oxidative challenge in the Alzheimer's brain. Neurochem Res 1994;19:1131-1137.
116. Chen L, Richardson JS, Caldwell JE, Ang LC. Regional brain activity of free radical defense enzymes in autopsy samples from patients with Alzheimer's disease and from nondemented controls. Int J Neurosci 1994; 75:83-90.
117. Gsell W, Conrad R, Hickethier M et al. Decreased catalase activity but unchanged superoxide dismutase activity. J Neurochem 1995; 64:1216-1223.
118. Martins RN, Harper CG, Stokes GB et al. Increased glucose 6-phosphate dehydrogenase activity in Alzheimer's disease may reflect oxidative stress. J Neurochem 1986; 46:1042-1045.
119. Ceballos I, Agid F, Delacourte A. Neuronal localization of CuZnSOD protein and mRNA within the human hippocampus from control and Alzheimer's disease brains. Free Rad Res Comms 1991; 12-13:581-589.
120. Goedert M. Neuronal localization of amyloid protein precursor mRNA in normal human brain and in Alzheimer's disease. EMBO J 1987; 6:3627-3632.
121. Ceballos-Picot I, Merad-Boudia M, Nicole A et al. Peripheral antioxidant enzyme activities and selenium in elderly subjects and in dementia of the Alzheimer-type:place of the extracellular glutathione peroxidase. Free Rad Biol Med 1996; 20:579-587.
122. Sulkawa R, Nordberg UR, Erkinjuntti T et al. Erythrocyte glutathione peroxidase and superoxide dismutase in Alzheimer's disease and other dementia. Acta Neurol Scand 1986; 73:487-489.
123. Jeandel C, Nicolas MB, Dubois F et al. Lipid peroxidation and free radical scavengers in Alzheimer's disease. Gerontology 1989; 35:275-282.
124. Zubenko GS, Sauer P. SOD-1 activity and platelet membrane fluidity in Alzheimer's disease. Biol Psychiatry 1989; 25:671-678.
125. de Lustig E, Serra JA, Kohan S et al. Copper-zinc superoxide dismutase activity in red blood cells and serum in demented patients. J Neurol Sci 1989; 115:18-25.
126. Serra JA, Famulari AL, Kohan S et al. Copper-zinc superoxide dismutase activity in red blood cells in probable Alzheimer's patient and their first-degree relatives. J Neurol Sci 1994; 122, 179-188.
127. Percy ME, Dalton AJ, Markovic VD. Red cell superoxide dismutase, glutathione peroxidase and catalase in Down syndrome patients with and without manifestations of Alzheimer disease. Am J Med Genet 1990; 35:459-467.
128. Perrin R, Briancon S, Jeandel C. Blood activity of CuZn superoxide dismutase, glutathione peroxidase and catalase in Alzheimer's disease. A case-control study. Gerontology 1990; 36:306-313.
129. Fernandes MAS, Santana I, Januario C et al. Decreased superoxide dismutase activity in erythrocytes from patients with Alzheimer's disease. Med Sci Res 1993; 21:679-682.
130. Anneren G, Gardner A, Lundin T. Increased glutathione peroxidase activity in erythrocytes in patients with Alzheimer's disease/senile dementia of the Alzheimer type. Acta Neurol Scand 1986; 73: 586-589.

131. Basun H, Forssell LG, Wetterberg L et al. Metals and trace elements in plasma and cerebrospinal fluid in normal ageing and in Alzheiemer's disease. J Neural Trans 1991; 4:231-258.
132. Zubenko GS, Cohen BM, Boller F et al. Platelet membrane abnormality in Alzheimer's disease. Ann Neurol 1987; 22:237-244.
133. Hajjimohammadreza I, Brammer M. Brain membrane fluidity and lipid peroxidation in Alzheimer's disease. Neurosci lett 1990; 112:333-337
134. Avissar N, Ornt DB, Yagil Y et al. Human kidney proximal tubules are the main source of plasma glutathione peroxidase. Am J Physiol 1994; 266: C367-375.
135. Amstad P, Moret R, Cerutti P. Glutathione peroxidase compensates for the hypersensitivity of CuZn-superoxide dismutase overproducers to oxidant stress. J Biol Chem 1994; 269:1606-1609.
136. Ceballos I, Delabar JM, Nicole A et al. Expression of transfected human CuZnSOD in mouse L cells and NS20Y neuroblastoma cells induces enhancement of glutathione peroxidase activity. Biochim Biophys Acta 1988; 949, 58-64.
137. Toussaint O, Houbion A, Remacle J. Relationship between the critical level of oxidative stresses and the glutathione peroxidase activity. Toxicology 1993; 81:89-101.
138. Berr C, Nicole A, Godin J, Ceballos-Picot I et al. Selenium and oxygen metabolizing enzymes in elderly community residents: a pilot epidemiological study. J Am Geriatr Soc 1993; 41:143-148.
139. Bunker VW, Lawson MS, Stansfield MF et al. Selenium balance studies in apparently healthy and housebound elderly people eating self-selected diets. Br J Nutr 1988; 59:171-180.
140. Lloyd B, Lloyd RS, Clayton BE. Effect of smoking, alcohol, and other factors on the selenium status of a healthy population. J Epidemiol Coll Health 1983; 37:213-217.
141. Neve J. Physiological and nutritionnal importance of selenium. Experientia 1991; 47:187-193.
142. Selkoe DJ. Cell biology of the amyloid protein precursor and the mechanism of Alzheimer's disease. Annu Rev Cell Biol 1994; 10:373-403.
143. Zheng H, Jiang M, Trumbauer ME et al. β-amyloid precursor protein-deficient mice show reactive gliosis and decreased locomotor activity. Cell 1995; 81: 525-531.
144. Van Nostrand WE, Wagner SL, Suzuki M et al. Protease nexin-II, a potent antichymotrypsin, shows identity to amyloid β-protein precursor. Nature 1989; 341:546.
145. Mattson MP, Cheng B, Smith-Swintosky VL. Mechanisms of neurotrophic factors protection against calcium- and free radical-mediated injury: implications for treating neurodegenerative disorders. Exp Neurol 1993; 124:89-95.
146. Mucke L, Abraham CR, Masliah E. Neurotrophic and neuroprotective effects of hAPP in transgenic mice. Ann NY Acad Sci 1996; 777:82-88.
147. Citron M, Olterstorf T, Haas C et al. Mutation of the β-amyloid precursor protein in familial Alzheimer's disease increases β-protein production. Nature 1992; 360:67-674.
148. Kounnas MZ, Moir RD, Rebeck GW et al. LDL receptor-related protein, a multifunctional ApoE receptor, binds secreted beta-amyloid precursor protein and mediates its degradation. Cell 1995; 82:331-340.
149. Bush AI. Multhaup G, Moir RD et al. A novel zinc(II) binding site modulates the function of the beta A4 amyloid protein precursor of Alzheimer's disease. J Biol Chem 1993; 268:16109-16112.
150. Hesse L, Beher D, Masters CL et al. The beta A4 amyloid precursor protein binding to copper. FEBS Lett 1994; 349: 109-116.
151. Yamazaki T, Selkoe DJ, Koo EH. Trafficking of cell surface beta-amyloid precursor protein:retrograde and transcytotic transport in cultured neurons. J Cell Biol 1995; 129:431-442.
152. Cork C, Masters K, Beyreuther K et al. Development of senile plaques. Relationships of neuronal abnormalities and amyloid deposits. Am J Pathol 1990; 137:1383-1392.

153. Multhaup G, Schlicksupp A, Hesse L et al. The amyloid precursor protein of Alzheimer's disease in the reduction of copper(II) to copper (I). Science 1996, 271:1406-1409.
154. Wong PC, Pardo, CA, Borchelt DR et al. An adverse property of a familial ALS-linked SOD-1 mutation causes motor neuron disease characterized by vacuolar degeneration of mitochondria. Neuron 1995; 14:1105-1116.
155. Fitzgerald DJ. Zinc and Alzheimer's disease. Science 1995; 268:1920.
156. Yoshikawa K, Aizawa T, Hayashi Y. Degeneration in vitro of post-mitotic neurons overexpressing the Alzheimer amyloid protein. Nature 1992; 359:64-67.
157. Rumble B, Retallack R, Hilbich C et al. Amyloid A4 protein and its precursor in Down's syndrome and Alzheimer's disease. N Engl J Med 1989; 320:1446-1452.
158. Lee RKK, Wurtman RJ, Cox AJ, Nitsch RM. Amyloid precursor protein processing is stimulated by metabotropic glutamate receptors. Proc Natl Acad Sci USA1995; 92:8083-8087.
159. Copani A, Bruno V, Vaughan PJ. Activation of metabotropic glutamate receptors protects cultured neurons against apoptosis induced by β-amyloid peptide. Mol Pharmacol 1995; 47:890-897.
160. Trejo J, Massamiri T, Deng T et al. A direct role for protein kinase C and the transcription factor Jun/AP-1 in the regulation of Alzheimer's beta amyloid precursor protein gene. J Biol Chem 1994; 269:21682-21690.
161. Dewji NN, Do C, Bayney RM. Transcriptional activation of Alzheimer's beta amyloid precursor protein gene by stress. Mol Brain Res 1995; 33:245-253.
162. Dewji NN, Do C. Heat shock factor-1 mediates the transcriptional activation of Alzheimer's β-amyloid precursor protein gene in response to stress. Mol Brain Res 1996; 35:325-328.
163. Stein-Behrens B, Mttson MP, Chang I et al. Stress-exacerbated neuron loss and cytoskeletal pathology in the hippocampus. J Neurosci 1994:5373-5380.
164. Araujo DM and Cotman CW. β-amyloid stimulates glial cells in vitro to produce growth factors that accumulate in senile plaques in Alzheimer's disease. Brain Res 1992; 569:141-45.
165. Beal MF. Aging, energy and oxidative stress in neurodegenerative diseases. Ann Neurol 1995; 38:357-366.
166. Askanas V, McFerrin J, Baqué S et al. Transfer of β-amyloid precursor protein using adenovirus vector causes mitochondrial abnormalities in cultured normal human muscle. Proc Natl Acad Sci USA 1996; 93:1314-1319
167. Mutisya EM, Bowling AC, Beal MF. Cortical cytochrome oxidase activity is reduced in Alzheimer's disease. J Neurochem 1994; 63:2179-2184.
168. Corral-Debrinski M, Horton T, Lott MT et al. Mitochondrial DNA deletions in human brain: regional variability and increase with advanced age. Nature Genet, 1992; 2:324-329.
169. Blass JP, Baker AC, Ko LW et al. Induction of Alzheimer antigens by an uncoupler of oxidative phosphorylation. Arch Neurol 1990; 47:864-869.
170. Sandhu FA, Kim Y, Lapan KA et al. Expression of the C terminus of the amyloid precursor protein alters growth factor responsiveness in stably transfected PC12 cells. Proc Natl Acad Sci USA 1996; 93:2180-2185.
171. Kawabata S, Higgind GA, Gordon JW. Amyloid plaques, neurofibrillary tangles and neuronal loss in brains of transgenic mice overexpressing a C-terminal fragment of human amyloid precursor protein. Nature 1991; 354:476-478.
172. Moran PM, higgins LS, Cordell B, Moser PC. Age-related learning deficits in transgenic mice expressing the 751-amino acid isoform of human β-amyloid precursor protein. Proc Natl Acad Sci USA 1995; 92:5341-5345.
173. Laferla FM, Tinkle BT, Bieberich CJ et al. The Alzheimer's Aβ peptide induces neurodegeneration and apoptotic cell death in transgenic mice. Nature Genet 1995; 9:215-220.
174. Smith MA, Sayre LM, Monnier VM et al. Radical AGEing in Alzheimer's disease. Trends Neurosci 1995; 18, 172-176.
175. Smith MA, Sayre LM, Vitek MP et al. Early AGEing and Alzheimer's. Nature 1995; 374:316.

176. Sell DR, Monnier VM. Structure elucidation of a senescence crosslink from human extracellular matrix. Implication of pentoses in the aging process. J Biol Chem 1989; 264:21597-21602.
177. Ledesma MD, Bonay P, Colaco C et al. Analysis of microtubule-associated protein tau glycation in paired helical filaments. J Biol Chem 1994; 269:21614-21619.
178. Yan SD, Yan SF, Chen Xi et al. Non-enzymatically glycated tau in Alzheimer's disease induces neuronal oxidant stress resulting in cytokine gene expression and release of amyloid β-peptide. Nature Med 1995; 1:693-699.
179. Gustke N, Steiner B, Mandelkow E et al.The Alzheimer-like phosphorylation of tau protein reduces microtubule binding and involves Ser-Pro and Thr-Pro motifs. FEBS Lett 1992; 307:199-205.
180. Lindwall G, Cole RD. Phosphorylation affects the ability of tau protein to promote microtubule assembly. J Biol Chem 1984; 259:5301-5305.
181. Perry G. Neuritic plaques in Alzheimer disease orginate from neurofibrillary tangles. Med Hypotheses 1993; 40: 257-258.
182. Connolly JA, Kalnins VI, Cleveland D. Immunofluorescent staining of cytoplasmic and spindle microtubules in mouse fibroblasts with antibody to tau protein. Proc Natl Acad Sci USA 1977; 74: 2437-2440.
183. Seldon SC, Pollard TD. Phosphorylation of microtubule-associated proteins regulates their interaction with actin filaments. J Biol Chem 1983; 258:7064-7071.
184. Miyata Y, Hoshi M, Nishida E et al. Binding of microtubule-associated protein 2 and tau to the intermediate filament reassembled from neurofilament 70 kDa subunit protein. J Biol Chem 1986; 261:13026-13030.
185. Smith MA, Siedlak SL, Richey PL et al. Tau protein directly interacts with the amyloid protein precursor: Implications for Alzheimer's disease. Nat Med 1995; 1:365-369.
186. Quefurth HW, Wijsman EM, St. George-Hyslop PH et al. β APP mRNA transcription is increased in cultured fibroblasts from the familial Alzheimer's disease-1 family. Mol Brain Res 1995; 28:319-37.
187. Iversen LL, Mortishire-Smith RJ, Pollack SJ et al. The toxicity in vitro of β-amyloid protein. Biochem J 1995; 311:1-16.
188. Pike CJ, Burdick D, Walencewicz AJ et al. Neurodegenration induced by β-amyloid peptides in vitro:the role of peptide assembly state. J Neurosci 1993; 13: 1676-1687.
189. Hensley K, Aksenova M, Carney JM et al. Amyloid β-peptide spin trapping II: evidence for the decomposition of the PBN spin adduct. Neuroreport 1995; 6:493-496.
190. Bush AI, Pettingell WH, Multhaup G et al. Rapid induction of Alzheimer Aβ amyloid formation by zinc. Science 1994; 265:1464-1467.
191. Paresce DM, Ghosh RN, Maxfield FR et al. Microglial cells internalize aggregates of the Alzheimer's disease amyloid β-protein via a scavenger receptor. Neuron 1996; 17:553-565.
192. El Khoury J, Hickman SE, Thomas CA et al. Scavenger receptor-mediated adhesion of microglia to β-amyloid fibrils. Nature 1996; 382:716-719.
193. Shaffer LM, Dority MD, Gupta-Bansal R et al. Amyloid β-protein removal by neuroglial cells in culture. Neurobiol Aging 1995; 16:737-745.
194. Christie RH, Freeman M, Hyman BT. Expression of the macrophage scavenger receptor, a multifunctional lipoprotein receptor, in microglia associated with senile plaques in Alzheimer's disease. Am J Pathol 1996; 148:399-403.
195. Yan SD, Chen Xi, Fu J et al. RAGE and amyloid-β peptide neurotoxicity in Alzheimer's disease. Nature 1996; 382: 685-691.
196. Zhang Z, Rydel RE, Drzeewiecki GJ et al. Amyloid β-mediated oxidative and metabolic stress in rat cortical neurons:no direct evidence for a role for H_2O_2 generation. J Neurochem 1996; 67: 1595-1606.
197. Goodman Y, Steiner MR, Steiner SM et al. Nordihydroguaiaretic acid protects hippocampal neurons against amyloid peptide toxicity, and attenuates free radical and calcium accumulation. Brain Res 1994; 654:171-176.

198. Schubert D, Behl C, Lesley R, Brack A et al. Amyloid peptides are toxic via a common oxidative mechanism. Proc Natl Acad Sci USA 1995; 92:1989-1993.
199. Bruce AJ, Malfroy B, Baudry M. β-amyloid toxicity in organotypic hippocampal cultures:protection by EUK-8, a synthetic catalytic free radical scavenger. Proc Natl Acad Sci USA 1996; 93:2312-2316.
200. Sagara Y, Dargush R, Klier FG et al. Increased antioxidant enzyme activity in amyloid β protein-resistant cells. J Neurosci 1996; 16:497-505.
201. Itagaki S, Mc Geer PL, Akiyama et al. relationship of microglia and astrocytes to amyloid deposits of Alzheimer's disease. J Neuroimmunol 1989; 24:173-182.
202. Merril JE, Ignarro LJ, Sherman MP et al. Microglial cell cytotoxicity of oligodendrocytes is mediated through nitric oxide. J Immunol 1993; 151:2132-2141.
203. Boje KM, Arora PK. Microglial-produced nitric oxide and reactive nitrogen oxides mediate neuronal death. Brain Res 1992; 587:250-256.
204. Eikelenboom P, Zhan SS, Van Gool WA et al. Inflammatory mechanisms in Alzheimer's disease. Trends Pharmacol Sci 1994; 15:447-450.
205. Le W-D, Colom LV, Xie W-J et al. Cell death induced by β-amyloid 1-40 in MES 23.5 hybrid clone: the role of nitric oxide and NMDA-gated channel activation leading to apoptosis. Brain Res 1995; 686:49-60.
206. Hu J, El-Fakahany EE. β-amyloid 25-35 activates nitric oxide synthase in neuronal clone. Neuroreport 1993; 4: 760-762.
207. Ii M, Sunamoto M, Ohnishi K, Ichimori Y. β-amyloid protein-dependent nitric oxide production from microglial cells and neurotoxicity. Brain Res 1996; 720: 93-100.
208. Araujo DM, Cotman CW. β-amyloid stimulates glial cells in vitro to produce growth factors that accumulate in senile plaques in Alzheimer's disease. Brain Res 1992; 569:141-145.
209. Walker DG, Kim SU, Mc Geer PL. Complement and cytokine gene expression in cultured microglia derived from post-mortem human brains. J Neurosci Res 1995; 40:478-493.
210. Gitter BD, Cox LM, Rydel RE et al. Amyloid β peptide potentiates cytokine secretion by interleukin-1β-activated human astrocytoma cells. Proc Natl Acad Sci USA 1995; 92:10738-10741.
211. Barger SW, Hörster D, Furukawa K et al. Tumor necrosis factors α and β protect neurons against amyloid β-peptide toxicity: Evidence for involvement of a κB-binding factor and attenuation of peroxide and Ca^{2+} accumulation. Proc Natl Acad Sci USA 1995; 92:9328-9332.
212. Butterfield DA, Hensley K, Harris M et al. β-amyloid peptide free radical fragments initiate synaptosomal peroxidation in a sequence-specific fashion; implications to Alzheimer's disease. Biochem Biophys Res Commun 1994; 200: 710-715.
213. Goodman Y, Mattson MP. Secreted forms of β-amyloid precursor protein protect hippocampal neurons against amyloid peptide-induced oxidative injury. Exp Neurol 1994; 128:1-12.
214. Goodman Y, Mattson MP. Ceramide protects hippocampal neurons against excitotoxic and oxidative insults, and amyloid β-peptide toxicity. J Neurochem 1996; 66:869-872.
215. Mark RJ, Lowell MA, Markesbery WR et al. A role for 4-hydroxynonenal, an aldehydic product of lipid peroxidation, in disruption of ion homeostasis and neuronal death induced by amyloid β peptide. J Neurochem 1997; 68:255-264.
216. Hartley DP, Ruth JA, Petersen DR. The hepatocellular metabolism of 4-hydroxynonenal by alcohol dehydrogenase, aldehyde dehydrogenase, and glutathione S-transferase. Arch Biochem Biophys 1995; 316:197-205.
217. Montine TJ, Amaranth V, Martin ME et al. 4-hydroxynonenal is cytotoxic and crosslinks cytoskeletal proteins in P19 neuroglial cultures. Am J Pathol 1996; 148:89-93.
218. Mattson MP, Tomaselli K, Rydell RE. Calcium-destabilizing and neurodegenerative effects of aggregated β-amyloid peptide are attenuated by basic FGF. Brain Res 1993; 621:35-49.
219. Arispe N, Rojas E, Pollard HB. Alzheimer disease amyloid β protein forms calcium

channels in bilayer membranes: blockade by tromethamine and aluminium. Proc Natl Acad Sci USA 1993; 90:567-571.

220. Mattson MP, Goodman Y. Different amyloidogenic peptides share a similar mechanism of neurotoxicity involving free radicals and calcium. Brain Res 1995; 676:219-224.
221. Mattson MP, Cheng B, Davis D et al. β-amyloid peptides destabilize calcium homeostais and render human cortical neurons vulnerable to excitotoxicity. J Neurosci 1992; 12:379-389.
222. Zhou Y, Gopalakrishnan V, Richardson JS. Actions of neurotoxic β-amyloid on calcium homeostasis and viability of PC12 cells are blocked by antioxidants but not by calcium channel antagonists. J Neurochem 1996; 67:1419-1425.
223. Furukawa K, Abe Y, Akaike N. Amyloid β protein-induced irreversible inward current in rat cortical neurons. Neuroreport 1994; 5:2016-2018.
224. Mark RJ, Hensley K, Butterfield DA et al. Amyloid β-peptide impairs ions-motive ATPase activities: evidence for a role in loss of neuronal Ca^{2+} homeostasis and cell death. J Neurosci 1995; 15: 6239-6249.
225. Harris ME, Carney JM, Cole PS et al. β-amyloid peptide-derived, oxygen-dependent free radicals inhibit glutamate uptake in cultured astrocytes:implications for Alzheimer's disease. Neuroreport 1995; 6:1875-1879.
226. Lafon-Cazal M, Pietri S, Culcazi M et al. NMDA-dependent superoxide production and neurotoxicity. Nature 1993; 364:535-537.
227. Mattson MP. Antigenic changes similar to those seen in neurofibrillary tangles are elicited by glutamate and calcium influx in cultured hippocampal neurons. Neuron 1990; 4:105-117.
228. Mattson MP, Barger SW, Cheng B et al. β-amyloid precursor protein metabolites and loss of neuronal Ca^{2+} homeostasis in Alzheimer's disease. Trends Neurosci 1993; 16:409-415.
229. Mattson MP. Free radicals and disruption of neuronal ion homeostasis in AD:a role for amyloid β-peptide. Neurobiol Aging 1995; 16:679-682.
230. Schearman MS, Ragan CI, Iversen LI. Inhibition of PC12 cell redox activity is a specific, early indicator of the mechanism β-amyloid-mediated cell death. Proc Natl Acad Sci USA 1994; 91: 1470-1474.
231. Loo D, Copani A, Pike CJ et al. Apoptosis is induced by beta-amyloid in cultured central nervous system neurons. Proc Natl Acad Sci USA 1993; 90: 7951-7955.
232. Schearman MS, Hawtin SR, Tailor VJ. The intracellular component of cellular 3-(4,5-dimethylthiazol-2-yl)-2,5-diphenyltetrazolium bromide (MTT) reduction is specifically inhibited by beta-amyloid peptides. J Neurochem 1995; 68:218-227.
233. Slater TF, Sawyer B, Sträubli U. Studies on succinate tetrazolium reductase systems. III. Points of coupling of four different tetrazolium salts. Biochim Biophys Acta 1963; 77:383-393.
234. Berridge MV, Tan AN. Characterization of the cellular reduction of cellular 3-(4,5-dimethylthiazol-2-yl)-2,5-diphenyltetrazolium bromide (MTT):subcellular localization, substrate dependence, and involvement of mitochondrial electron transport in MTT reduction. Arch Biochem Biophys 1993; 303:474-482.
235. Hawtin SR, Dobbins AC, Tailoer VJ et al. Beta-amyloid inhibition of MTT reduction is not mimicked by inhibitors of mitochondrial respiration. Biochem Soc Trans 1995; 23:56S.
236. Burdon RH, Gill V, Rice Evans C. Reduction of a tetrazolium salt and superoxide generation in human tumor cells (Hela). Free Rad Res Comms 1993; 18:369-380.
237. Thomas T, Thomas G, McLendon C et al. β-amyloid-mediated vasoactivity and vascular endothelial damage. Nature 1996; 380:168-171.
238. Moncada S, Palmer RMJ, Higgs EA. Nitric oxide: physiology, pathophysiology and pharmacology. Pharmacol Rev 1991; 43:109-142.
239. Stamler JS. A radical vascular connection Nature 1996, 380:108-111.
240. Stamler JS. Redox signaling: nitrosylation and related target interactions of nitric oxide. Cell 1994; 78:931-936.

241. Schreck R, Rieber P, Bauerle PA. Reactive oxygen intermediates as apparently widely used messengers in the activation of NF-κB transcription factor and HIV-1. EMBO J 1991; 10:2247-2258.
242. Grilli M, Pizzi M, Memo M et al. Neuroprotection by aspirin and sodium salycylate through blockade of NF-κB activation. Science 1996; 274:1383-1385.
243. Goodman Y, Mattson MP. Ceramide protects hippocampal neurons against excitotoxic and oxidative insults, and amyloid β-peptide toxicity. J Neurochem 1996; 66:869-872.
244. Forloni G, Chiesa R, Smiroldo S et al. Apoptosis mediated neurotoxicity induced by chronic application of beta amyloid fragment 25-35. Neuroreport 1993, 4:523-526.
245. Loo DT, Copani AG, Pike CJ et al. Apoptosis is induced by beta-amyloid in cultured central nervous system neurons. Proc Natl Acad Sci USA 1993; 90:7951-7955.
246. Su JH, Anderson AJ, Cummings BJ, Cotman CW. Immunohistochemical evidence for DNA fragmentation in neurons in the AD brain. Neuroreport 1994; 5:529-2533.
247. Cotman CW, Anderson AJ. A potential role for apoptosis in neurodegeneration and Alzheimer's disease. Mol Neurobiol 1995; 10:1-19
248. Anderson AJ, Cummings BJ, Cotman CW. Increased immunoreactivity for Jun- and Fos-related proteins in Alzheimer's disease: association with pathology. Exp. Neurol. 1994; 125:286-295.
249. Anderson AJ, Pike C, Cotman CW. Differential induction of immediate early gene proteins in cultured neurons by β-amyloid (Aβ):association of c-jun with Aβ-induced apoptosis. J Neurochem 1995; 65:1487-1498.
250. Anderson AJ, Su JH, Cotman CW. DNA damage and apoptosis in Alzheimer's disease:colocalization with c-jun immunoreactivity, relationship to brain area, and effect of postmortem delay. J Neurosci 1996; 16:1710-1719.
251. Gillardon F, Skutella T, Uhlmann E, Holsboer F, Zimmermann M, Behl C. Activation of *c-fos* contributes to amyloid β-peptide-induced neurotoxicity. Brain Res 1996; 706: 169-172.
252. Estus S, Zaks WJ, Freeman RS et al. Altered gene expression in neurons during programmed cell death: identification of *c-jun* as necessary for neuronal apoptosis. J Cell Biol 1994; 127:1717- 1727.
253. Ham J, Babij C, Whitfield J et al. A c-jun dominant negative mutant protects sympathetic neurons against programmed cell death. Neuron 1995; 14: 927-939.
254. Saitoh T, Horsburgh K, Masliah E. Hyperactivation of signal transduction systems in Alzheimer's disease. Ann NY Acad Sci 1993; 695:34-41.
255. Zhang C, Qiu HE, Krafft GA et al. Protein kinase C and F-actin are essential for stimulation of neuronal FAK tyrosine phosphorylation by G-proteins and amyloid β-peptide. FEBS Lett 1996; 386: 185-188.
256. Crawford D, Zbinden I, Amstad P, Cerutti P. Oxidant stress induces the proto-oncogenes c-fos and c-myc in mouse epidermal cells. Oncogene 1988; 3:27-32.
257. Morgan JI, Curran T. Stimulus transcription coupling in the nervous system: involvement of the inducible protooncogenes *fos* and *jun*. Annu Rev Neurosci 1991; 14:421-451.
258. Satou T, Cummings BJ, Cotman CW. Immunoreactivity for bcl-2 protein within neurons in the Alzheimer's disease brain increases with disease severity. Brain Res 1995; 697:35-43.

CHAPTER 6

Oxidative Stress in Down's Syndrome

DOWN'S SYNDROME, ALZHEIMER'S DISEASE AND THEIR RELATIONSHIPS

Down's syndrome (DS) or trisomy 21 is one of the most common human genetic disorders, occuring in 1 in 600 to 1 in 800 live births, the most common genetic cause of mental retardation, and is caused by the triplication of chromosome 21. Development of the DS brain is associated with decreased neuronal number and abnormal neuronal differentiation[1,2] and adults with DS develop Alzheimer's disease.[2] Neuropathology is encountered in potentially all DS patients over the age of 35 years, with senile plaques and neurofibrillary tangles distributed in the same brain territories in Alzheimer's disease (AD). Assessment of the AD lesions in DS patients at various age as compared to the age-related occurence of AD lesions in the general population leads to the conclusion that the neuropathological processes might start in DS brain some 50 years earlier than in normal brain.[3]

Thus, DS can be regarded as a disease model for identifying biological mechanisms leading to accelerated aging and AD-like neuropathology. The symptoms of the DS syndrome are considered to occur through damage caused by overexpression of normal gene products encoded by genes on chromosome 21. The elevated levels of these proteins, due to gene dosage effect, may disturb biochemical equilibria which ensure normal functioning. The precise mechanism by which overexpression of individual genes contributes to specific symptoms of the disorders remains largely unknown.The gene for the precursor of amyloid peptide (APP) is located on chromosome 21. On the basis of a hypothesis of overproduction of APP causing AD, one might expect that transgenic animals engineered to overexpress APP would readily model the pathology; however, this does not appear to be the case.[4] By contrast, a recent mouse model of the disease, in which a copy of APP carrying a 717 mutation is overexpressed, closely models a number of the neuropathological features that are inherent in the disease.[5] In addition a recent knockout mouse model, in which the normal expression of the APP gene was disrupted by insertion of a transcription termination sequence produced a 5- to 10-fold lower level of APP mRNA than wild type controls and severely impaired spatial learning and exploratory behavior.[6]

Although the syndrome was described more than 100 years ago and the relationship between trisomy 21 and DS has been known for more than 25 years, very little is known about the way in which the additional chromosome 21 causes the disease. In particular, the cause of the neurodegenerative process in DS is unknown but oxidative stress has been considered to be a possible contributory factor in the pathology of DS.[7-9]

The Role of Oxidative Stress in Neuronal Death, by Irène Ceballos-Picot.

Evidence of Oxidative Stress in Down's Syndrome Neurons

Cortical neurons from fetal DS and age-matched normal brain differentiate normally in culture, but DS neurons subsequently degenerate and undergo apoptosis whereas normal neurons remain viable.[10] Degeneration of DS neurons is prevented by treatment with free radical scavengers such as the free-radical spin trap N-tert-butyl-2-sulphophenylnitrone (PBN), the free radical scavengers vitamin E, butylated hydroxyanisole (BHA) and N-propylgallate, and the glutathione precursor/radical scavenger N-acetyl-L-cysteine (NAC). In addition, catalase, an enzyme that hydrolyses hydrogen peroxide, inhibited the degeneration of DS neurons. Furthermore, DS neurons exhibit a 3- to 4-fold increase in intracellular ROS and elevated levels of lipid peroxidation that precede neuronal death. These results suggest that DS neurons have a defect in the metabolism of reactive oxygen species that affect DS neuronal viability and causes neuronal apoptosis. This defect may contribute to mental retardation early in life and predispose to Alzheimer's disease in DS.

If these findings in cell culture are confirmed in vivo, then the neuroprotective effects of antioxidants may provide an important therapeutic approach to mental retardation and the prevention of Alzheimer's disease in DS individuals.

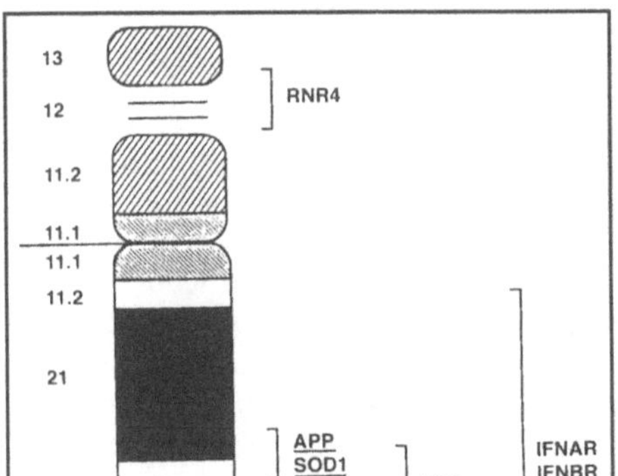

Fig. 6.1. Some important genes on chromosome 21. APP, amyloid β precursor protein; BCEI, estrogen-inducible sequence, expressed in breast cancer; CBS, cystathionine β synthetase; CRYA1, crystallin alpha 1 polypeptide; ETS2, avian erythroblastosis virus E26 oncogene homolog 2; IFNAR, receptor for interferon α; IFNBR, receptor for interferon β; PFKL, phosphofructokinase, liver type; PRGS, phosphoribosylglycinamide synthetase; RNR4, RNA ribosomal 4; SOD1 or CuZnSOD, copper-zinc superoxide dismutase.

COPPER-ZINC SUPEROXIDE DISMUTASE OVEREXPRESSION IN DOWN'S SYNDROME

The human CuZn superoxide dismutase (CuZnSOD; superoxide:superoxide oxidoreductase, EC 1.15.1.1), is encoded by a gene residing on the 21q22.1 band of chromosome 21[11] (Fig. 6.1).

This segment present in triplicate is known to be involved in the development of DS including accelerated aging and Alzheimer-like neuropathology. The 5'-flanking region of the human CuZnSOD gene contains several consensus binding sites for transcriptional factors such as AP2, SP_1 and NF-κB which may participate in the regulation of the gene expression. The human gene is composed of five exons. Two messenger RNAs are transcribed from the same gene and sequences of the 3'-untranslated region account for the difference in size between them (Fig. 6.2).

Elevated levels of the enzyme, as a result of gene dosage, were recorded in a variety of trisomy 21 cells,[7] including brain.[12] It may therefore be considered a suspect for involvement in the etiology of DS.[7–9]

CuZnSOD is a key enzyme in the metabolism of oxygen free radicals and catalyzes the conversion of superoxide $O_2^{\circ-}$ into H_2O_2 and O_2. Its overproduction may therefore disturb the existing steady-state equilibrium of active oxygen species within the cell, resulting in oxidative damage. $O_2^{\circ-}$ cannot cross the membrane's lipid bilayers except through anion channels; therefore in each cellular compartment $O_2^{\circ-}$ is converted into H_2O_2 by the colocalized SOD. The intracellular generated H_2O_2 is further metabolized to water by glutathione peroxidase and catalase. In the brain, CuZnSOD is localized in neurons[13-15] and glutathione peroxidase in glial cells (see chapter one). This cellular compartimentation suggests a loss of cooperation between these enzymes or a neuronal-glial coupling. This heterogeneous cellular distribution could contribute to a selective vulnerability of neurons which are not well protected against H_2O_2 toxicity. If the activity of CuZnSOD is increased without a concomitant increase in these enzymes, H_2O_2 accumulates and its reaction with transition metals (the so-called Fenton reaction) is facilitated. The product of Fenton reaction is hydroxyl radical (OH°), the most reactive and noxious oxygen radical species. Whenever formed in living systems it will react immediatly with almost any biological macromolecule, particularly with SH groups of proteins and polyunsaturated fatty acids of membrane lipids. The latter initiates a chain reaction, lipid peroxidation, that results in damage to membranes and to membrane-associated proteins. In addition to legitimate dismutation reactions, CuZnSOD is able to catalyze surrogate reactions such as the production of OH° using anionic scavengers and H_2O_2[16,17] and nitration of the tyrosine residues of proteins by peroxynitrite,[18,19] which may have major deleterious effects on living cells.[20]

For these above reasons, it was postulated that CuZnSOD overexpression could enhance oxidative stress and subsequent deleterious neuronal effects.[7-9] In order to test the potential neurotoxicity of CuZnSOD overexpression, cell and animal models of CuZnSOD overexpression have been designed. The use of genetic manipulations to construct cells and/or transgenic mice that specifically overexpress CuZnSOD represent the way to an understanding of the role of this gene in neurodegeneration in DS, to establish a relationship between genotype and phenotype and devising rational therapeutic approaches.

CELLULAR MODELS OVEREXPRESSING COPPER-ZINC SUPEROXIDE DISMUTASE

The SOD/GSHPx Balance and Oxidative Stress

Since CuZnSOD catalyses the dismutation of superoxide radicals $O_2^{\circ-}$ to oxygen and H_2O_2, it has been hypothesized that increased CuZnSOD in trisomy 21 might induce overproduction of H_2O_2 within cells and accelerate oxidative damages to cell components.[7] Supporting this hypothesis, the seleno-dependent glutathione peroxidase (Se-GSHPx) activity, which catalyzes the reduction of H_2O_2, has been found increased in trisomy 21 tissues like red blood cells and platelets,[22,23] whereas in fetal brain Se-GSHPx was found normal.[24] The increase in Se-GHPx activity in trisomy 21 tissues does not result from a gene dosage effect since the gene encoding Se-GSHPx is on chromosome 3. This increase is actually linked to CuZnSOD excess.[25] If CuZnSOD is upregulated out of proportion to the

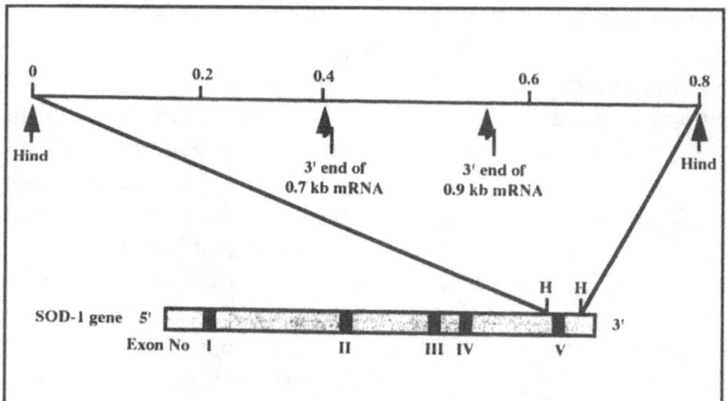

Fig. 6.2. The human *CuZnSOD* gene.

activity of glutathione peroxidase, a buildup of H_2O_2 could result. Recent reports have shown that CuZnSOD is able to catalyze hydroxyl radicals (OH°) production from H_2O_2[16,17] and that elevated levels of SOD are cytotoxic[26] and enhance the cytotoxicity of active oxygen species.[27,28] Such a mechanism may be in part responsible for certain clinical symptoms of DS patients and for the accelerated aging and AD pathology associated with the Down's phenotype.

Several reports showed that an increase in CuZnSOD activity may cause oxidative damage. Cell clones expressing elevated levels of enzymatically active human CuZnSOD had increased lipid peroxidation, which may result in alteration of the structure and function of cellular membranes[29] and higher resistance to the toxic effect of paraquat.[29–31] In stable clones producing increased amounts of CuZnSOD a marked increase in endogenous Se-GSHPx was observed[25,31] and a smaller increase in glutathione transferase activity also occurred. Manganese superoxide dismutase (MnSOD) activity was decreased in all clones, whereas catalase and NADPH reductase were not affected.[31,32] Alteration in Se-GSHPx and MnSOD correlated with increases in CuZnSOD activity. Whereas all clones were resistant to paraquat, a direct correlation between CuZnSOD activity and resistance to paraquat did not exist. Clones expressing the highest CuZnSOD activity did not display the highest resistance to paraquat. However, there was a direct correlation between the increase in Se-GSHPx activity and paraquat resistance.[31] An explanation for the decrease in MnSOD activity is that the presence of increased CuZnSOD activity leads to a reduction in the intracellular steady-state superoxide concentrations, and MnSOD is not produced when not required by the cell. Alternatively, the overproduction of CuZnSOD may be harmfull to the cell by producing increased intracellular concentrations of hydrogen peroxide, and the response of endogenous glutathione peroxidase supports this hypothesis. If this were true, the cell may try to respond by decreasing endogenous superoxide dismutase production such as MnSOD or endogenous CuZnSOD as a regulatory defense. Further studies are required to determine how these antioxidant enzymes are controlled intracellularly. The ways in which stable adaptations might occur are through heritable changes in the cell's ability to transcribe certain genes selectively, to diminish cellular turnover of specific mRNAs, or to prolong the intracellular lifespan of the enzymes themselves. The role of the balance between SOD and GSHPx for the cellular sensitivity of oxidants was recently studied.[33,34] A modest 2-3-fold increase in CuZnSOD in cellular clones transfected with the human gene sensitized them to the oxidants produced by xanthine/xanthine oxidase by increasing their hypersensitivity to DNA strand breakage. Correspondingly, a modest increase in the levels of GSHPx by introducing bovine GSHPx expression vector into CuZnSOD transfectants sufficed to compensate DNA strand breakage.[33,34] Indeed, this sensitivity to DNA strand breakage and killing by xanthine/xanthine oxidase was inversely related to the activity ratios GSHPx over SOD in SOD transfectants. Thus, the balance between several components of the antioxidant defense appears to be important for the cellular resistance to oxidative stress and a small deviation from the physiological activity ratios GSHPx/SOD has a dramatic effect on the resistance of cells to oxidant-induced damage to the genome and cell killing. Consistent with this, a high ratio of GSHPx over SOD activities was related to superior growth potential and increased resistance to killing by paraquat in NIH 3T3 cells.[31]

These results support the previous hypothesis that balanced increments of antioxidant enzymes are necessary for cell integrity and growth and that a deregulation of this complex antioxidant system may disturb the steady state equilibrium of oxyradicals within cells, resulting in oxidative damage to biologically important molecules. Such a mechanism may in part be responsible for certain clinical symptoms of DS patients such as accelerated aging, and for the AD pathology associated with the DS phenotype. Experiments support this interpretation: the transfected PC12 cells overexpressing CuZnSOD had impaired uptake of neurotransmitters such as dopamine and norepinephrine due to a diminution in the PH gradient across their chromaffin granule membrane.[35] The authors attributed the perturbation of the pH gradient to uncoupling of ATPase or increased permeability of the granule membranes to protons, likely secondary to lipid peroxidation. It was described that peroxidative damage resulting from excess CuZnSOD expression in transfected adrenocortical cells causes cell fusion, nuclear fragmentation and subsequent propagation of cell death.[26] Because the metabolism of oxygen free radicals and in particular the process of lipid peroxidation are tighly connected to the biosynthesis of prostaglandins, synthesis of prostaglandin E_2 (PGE_2) was diminished in transfected PC12 cells as well as in fibroblasts from DS patients.[36]

Disequilibrium Between SOD/GSHPx and Neuropathological Implications

This disequilibrium between CuZnSOD and GSHPx may contribute to some of the pathological features that occur as part of the Down syndrome phenotype. Consistent with this hypothesis is the finding that CuZnSOD activity increases in the brains of mice during the aging process,[21] and that this increases occurs against a background of unchanged GSHPx activity. The activity of SOD increase in all organs of aging mice, but cellular damage (assayed by lipid peroxidation and/or protein carbonyls) is only found in organs (such as the brain) that do not display a concomitant increase in GSHPx and/or catalase.[21,36] Indeed, transgenic mice overexpressing SOD without adaptative increase in GSHPx develop morphological abnormalities[37-39] similar to those in aged animals and in individuals with DS.

Thus, the balance between the activity of the first and second step antioxidant enzymes may also be an important determinant in the aging process. Furthermore this imbalance is also associated with the onset of epileptic seizures.[40] In order to further investigate the role of an altered SOD to GSHPx activity in the perturbations of cellular functions and more specifically in the induction of cellular senescence, two cell systems that have elevated SOD activity relative to GSHPx activity were recently described. In the first, SOD levels were elevated as a result of stable integration of additional copies of human SODcDNA, while the second were DS cells where SOD levels are elevated as a consequence of an altered gene dosage for SOD.[41] As we previously demonstrated, the vast majority of SOD transfectants had upregulated endogenous GSHPx[25] in response to elevated SOD levels (the adaptative response). Another population of clones had elevated SOD levels but failed to demonstrate an increase in GSHPx.[41] This population is referred to as non-adapted cell lines. Cells with altered SOD to GSHPx ratios show characteristics of cellular senescence, i.e., they proliferate more slowly, have altered morphology and size, and they have elevated Cip1. Cip1 is a cyclin-dependent kinase (Cdk)-inhibitor that forms inhibitory complexes with many different cyclins and may be a universal Cdk inhibitor, thereby inhibiting the passage of cells through specific phases of the cell cycle and also play a role in the maintenance of a senescent phenotype.[42] Furthermore, cells with elevated SOD to GSHPx activity have higher levels of H_2O_2 and, intriguingly these features of senescence can be induced by treatment of cells with H_2O_2.[41] The exact mechanism(s) by which elevated ratio of SOD to GSHPx induces Cip1 mRNA is unknown. However, H_2O_2 is known to be able to affect gene expression and increase the binding of various transcription factors such as NF-κB (see chapter 3) to their respective *cis*-binding elements in the promoters/enhancers of cellular genes. It may be that H_2O_2 acts in a similar manner to upregulate the expression of Cip1 and/or other genes whose expression is modulated during cellular senescence. Indeed, the oxidative stress hypothesis of aging extends the free radical theory of aging by proposing that the level of oxidative stress, dependent upon the rate of oxidant generation and antioxidant defenses, exerts an effect on gene expression.

The ratio of the activity of the first and second-step antioxidant enzymes, rather than the absolute levels of these enzymes per se, is the important determinant in alteration of cellular function. If H_2O_2 is an important mediator of cellular senescence and neurodegeneration, as suggested by the above studies, then therapeutic strategies that prevent the production of H_2O_2 and/or scavenge H_2O_2 may prove useful.

TRANSGENIC MICE EXPRESSING THE HUMAN CuZnSOD GENE

CuZnSOD and GSHPx in the Brain of Transgenic Mice

In order to further explore the deleterious effect of excess CuZnSOD, transgenic mice for the human CuZnSOD have been constructed.[21,49-52] The human CuZnSOD gene of sixteen kilobases with its own regulatory sequences was introduced into the germline of mice.[49]

The transgene was expressed as enzymatically active since increased activity was detected in all tissues tested and was more pronounced in the brain (a 2-fold increased activity) than in other tissues.

In brain, the neuronal-specificity of the human transgene expression is similar to that observed for the endogenous murine CuZnSOD and is particularly high in the hippocampus large pyramidal neurons and granular cells.

Transgenic mice carrying the human CuZnSOD gene are not phenotypically abnormal. If CuZnSOD excess is indeed involved in the pathogenesis of DS, we would expect its effects to be on more subtle aspects of tissue function, particularly in tissues that might be affected over long periods of time by altered metabolism of free radicals. The nervous system was the prime candidate for our investigations. Looking for biochemical modifications which could support this hypothesis, we have studied in the brain of the transgenic mice the activities of enzymes involved in oxygen radical detoxification. Despite the 2-fold increased CuZnSOD in the brains of transgenic mice, Se-GSHPx, glutathione reductase and glutathione S-transferase activities were not modified.[21] Then, it appears that, at least in young mice and in a normal oxygen environment, an increase in CuZnSOD activity

Table 6.1. CuZnSOD activity in various tissues from control and CuZnSOD transgenic mice

Tissue		Control mice (C)	Transgenic mice (T)	Comparison (t-test)	Ratio (T/C)
Brain	(n = 19)	29.55 ± 3.30	57.10 ± 6.84	P < 0.001	1.93
Heart	(n = 6)	17.21 ± 2.97	29.14 ± 3.59	P < 0.001	1.69
Lung	(n = 6)	31.52 ± 3.35	39.83 ± 5.52	P < 0.02	1.26
Thymus	(n = 6)	19.02 ± 2.68	28.34 ± 6.79	P < 0.02	1.49
Red cells	(n = 6)	10.83 ± 1.17	11.88 ± 0.94	NS	1.09
Liver	(n = 6)	57.82 ± 1.08	68.86 ± 5.36	P < 0.01	1.19
Kidney	(n = 6)	52.54 ± 4.92	61.93 ± 5.78	P < 0.02	1.18
Muscle	(n = 3)	29.92 ± 2.33	37.44 ± 4.25	NS	1.25
Spleen	(n = 5)	20.02 ± 1.94	26.98 ± 4.33	P < 0.02	1.35
Skin	(n = 8)	17.93 ± 1.93	31.06 ± 1.40	P < 0.001	1.73

Results are expressed as mean±SD; n: number of mice. Age of the mice was between 50 and 70 days. Comparison of the means was assessed by Student t-test (NS: not significant).

similar to that found in DS does not by itself produce alterations of the enzymes associated with oxygen radical detoxification. Thus, unlike what has been observed in DS erythrocytes,[22,53] in lymphoid cells and fibroblasts[54] and in transfected cells overexpressing CuZnSOD,[25,31] but similar to what is found in DS fetal brain[12] the Se-GSHPx activity is not increased in the brain of transgenic mice (Table 6.2).

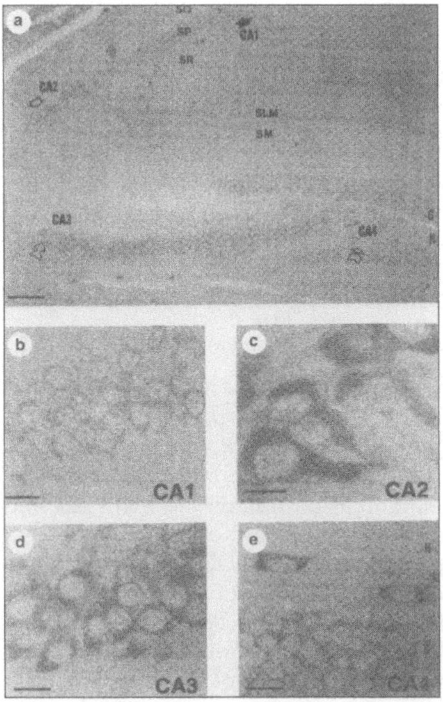

Fig. 6.3. Immunocytochemical localization by indirect immunoperoxidase of the human CuZnSOD protein in the hippocampal formation of transgenic mice. a: in sagittal brain sections, immunoreactivity was observed in the stratum pyramidale of Ammon's horn throughout the CA1-CA4 fields and in the granule cell layer of the dentate gyrus and was absent or weak in the remaining layer (bar = 100μM). b-e: photomicrographs showing at high magnification (bar = 15μM) the distribution of immunoreactive neurons in CA1 (b), CA2 (c), CA3 (d) and CA4 (e) regions of Ammon's horn. Note the confinement of immunoreactivity to the perikaryon and the proximal dendrites, the significant labeling of pyramidal cells in the CA2 (c) CA3 (d) fields and the moderate labeling of granule cells (e).

These results constrast with what has been observed in the transfected cells with the same construct.[25] In L-Cells and neuroblastoma cells an increased Se-GSHPx activity was observed and was correlated to the copper-zinc SOD overexpression.[25]

The reason for this discrepancy between the results in the brains of transgenic mice and in the transfected cells with the same human CuZnSOD gene is not clear. One possibility is that Se-GSHPx is not inducible in the brain by the increased CuZnSOD activity due to a differential cellular localization of these two enzymes in the brain of mice as previously reported in the human brain.[13-15,55-57] This leads to a loss of cooperation of these two enzymes and to increased hydrogen peroxide levels in neurons.[58] These data also suggest a heterogeneous cellular distribution of these enzymes in the brain of mice. This may form a basis for selective and/or regional expression of free radical toxicity.

Deleterious Effects of Long Term CuZnSOD Overexpression in the Brain

Immunohistochemical analysis of brain sections showed a neuronal expression of the transgene. The human SOD protein in the transgenic brain had the same expression profile as in human brain, particularly in the pyramidal cells of Ammon's horn in the CA1 to CA_4 fields of the hippocampus and in the granule cell layer of the dentate gyrus, and was absent or weak in the remaining layers. We explored whether these two types of cells (granule cells and pyramidal cells) might be particularly sensitive to an oxidative stress mediated by the excess of SOD. We focused on searching for histological changes in the hippocampus mossy fibers which are constituted by the axons of granule cells of the dentate gyrus, innervating the pyramidal neurons of the CA_3-CA_4 fields. Mossy fiber synaptic fields were significantly reduced in transgenic mice.[40] As the mossy fibers connect the granular and

Table 6.2. CuZnSOD and GSH Px enzyme activites in the brain of control and hCuZnSOD transgenic mice

		Age, Months	
	2	12	28
CuZnSOD			
Control	43.2 ± 5.6 (n = 13)	50.3 ± 5.1 (n = 12)	50.3 ± 5.4 (n = 15)
Transgenic	81.0 ± 8.3* (n = 11)	81.8 ± 9.3* (n = 13)	84.7 ± 7.0* (n = 10)
GSH Px			
Control	10.4 ± 1.9 (n = 13)	12.6 ± 1.8 (n = 12)	14.6 ± 1.4 (n = 15)
Transgenic	11.8 ± 2.4 (n = 11)	12.4 ± 1.8 (n = 13)	14.4 ± 1.5 (n = 10)

*$p < 0.001$

pyramidal cells, SOD-mediated processes are implicated in the changes observed in the mossy fibers innervation system. Further correlative experiments between mossy fibers projection and spatial memory are necessary to evaluate the functional significance of these morphometric changes.

To further delineate the relationship between peroxidative stress and increased CuZnSOD activity without an adaptative rise in Se-GSHPx activity in the brain of transgenic mice, the basal lipid peroxidation (as measured by thiobarbituric acid (TBA)-reactive material) was determined and found to be significantly higher in transgenic brains compared to controls.[21,49] This result was consistent with that reported for human fetal trisomy 21 brain[12] and for human CuZnSOD transfected cells,[29] as well as in blood platelets from human CuZnSOD transgenic mice.[59]

Other groups reported a cytotoxicity of CuZnSOD overexpression. Transgenic mice overexpressing CuZnSOD develop morphological and biochemical changes at tongue neuromuscular junctions similar to those seen in tongues of individuals with Down's syndrome.[41,42] Interestingly, these neuromuscular deteriorations are, morphologically at least, also similar to those seen in muscles of aging mice and rats.

A premature thymic involution in CuZnSOD transgenic mice was observed at the ultrastructural level, with a strong modification of the thymic microenvironment starting around 3 months and being in 7 month old mice similar to that visualized in 20 month old control mice.[61] Three other derived transgenic SOD strains had abnormal thymi showing diminution of the cortex and loss of corticomedullary demarcation, resembling thymic defects in children with DS.[62] Transgenic CuZnSOD mice were also more sensitive than control mice to in vivo injection of lipopolysaccharide (LPS), reflected by an earlier onset and enhanced apoptotic cell death in the thymus. This higher susceptibility to LPS-induced apoptosis was associated with an increased production of hydrogen peroxide and a higher degree of lipid peroxidation. These results are consistent with a possible role of CuZnSOD overexpression in premature thymic involution and immunological disorders observed in DS patients, and the defects are presumably due to increased oxidative damage resulting in enhanced cell death by apoptosis.

Moreover, platelets of transgenic CuZnSOD animals, which overexpress the transgene, contain lower levels of serotonin than nontransgenic littermate mice, due to a reduced rate of uptake of the neurotransmitter by the dense granules of the platelets.[59] In the transgenic platelets, lipoperoxidation was significantly elevated compared to nontransgenic mice and could contribute to alteration of the structure and function of cellular membranes. Furthermore, a significantly lower than normal serotonin accumulation rate was also detected in dense granules isolated from blood platelets of DS individuals. These findings suggest that CuZnSOD gene dosage is affecting the dense granule transport system and is thereby involved in the depressed level of blood serotonin found in patients with DS. These transgenic mice exhibit higher concentrations of the biogenic amines in specific brain regions, with little or no change in amino acid concentration. Furthermore, tyrosine hydroxylase activity is increased in the striatum of the transgenics whereas glutamic acid decarboxylase and choline acetyltransferase are unchanged.[63]

These findings suggest that overexpression of CuZnSOD in transgenic mice can indeed cause certain physiological abnormalities similar to those found in DS patients. CuZnSOD overexpression significantly contributes to the early aging process in Down's syndrome. Such reports are evidence that oxidant damage due to CuZnSOD excess is a factor which contributes to the aging process, and provide at least partial support for the free-radical theory of aging.[64] They also indicated that defects caused by increased CuZnSOD can have highly specific subcellular targets, and thus serve as a paradigm for how alterations in CuZnSOD expression could cause oxidative stress-mediated cell injury leading to neurodegenerative diseases.

Beneficial Effects of CuZnSOD Overexpression Towards Acute Stress

Thus, the production of these transgenic mice has opened the way for the evaluation of a number of dilemmas related to the involvement of increased CuZnSOD gene dosage in DS. In addition to their relevance to the study of DS, these transgenic mice will also be of importance in the investigation of the role of CuZnSOD in protection against the adverse effects of a variety of exogenous agents and pathophysiological situtions implicating oxygen free radicals such as hyperoxia, ischemia, radiation, chemical agents, i.e., paraquat and 1-methyl-4-phenyl-1,2,3,6-tetrahydropyridine (MPTP). These transgenic mice exhibit several differences from control mice when exposed to known or presumed forms of acute oxidative stress.

A protective effect of endogenous SOD against oxidative damage has been reported in various studies where SOD was overexpressed in transfected cells or transgenic animals.[50,65]

In vitro, primary cultures of neurons isolated from copper-zinc SOD transgenic mice were found to be more resistant to glutamate and methamphetamine-induced toxicity than controls.[66–68] Transgenic mice showed marked resistance to cold-induced brain edema and infarction.[69] In vivo, transgenic mice overexpressing CuZnSOD displayed a lower vulnerability than nontransgenic controls to focal cerebral ischemic injury[85–87] with sapring particularly the penumbra region surrounding the ischemic brain tissue, have less cerebral edema, a smaller increase in vascular permeability, and a smaller area of infarction (decreased by 50%) than do the control animals (Fig. 6.4).

Infections of striatal cells with a recombinant adenovirus expressing CuZnSOD protected these cells from glutamate-induced cell death.[70] NMDA-dependent superoxide production and neurotoxicity is abolished by SOD in vitro on cerebellar granule cells.[71] Trapping $O_2^{\circ-}$ by SOD seems to be more efficient than blocking NO production to reduce NMDA-induced neuronal death.[71] NO is necessary but not sufficient for neuronal injury and is toxic in the presence of other factors such as superoxide anions since SOD protects neurons against NO and NMDA toxicity.[72–74] Thus, neuronal susceptibility to NO and $O_2^{\circ-}$ induced damage

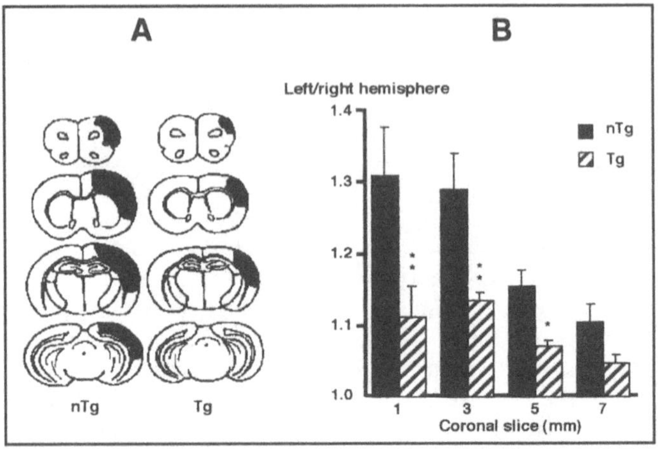

Fig. 6.4. a: Illustration of cortical infarction (black area) in brains of nontransgenic (nTg) and transgenic (Tg) mice as shown by staining with triphenyltetrazolium chloride following 24 hr of ischemia. The sizes of cortical infarction in Tg mice were significantly smaller than in nTg mice.

b: Reduced hemispheric enlargement (brain edema) in Tg mice following MCA occlusion. Data represent the ratios of left to right hemispheric areas in coronal slices 1, 3, 5, and 7 mm distal from the frontal pole. The ratios in Tg mice are significantly smaller than those in nTg mice in three of the slices. *, $P < 0.01$; **, $P < 0.05$ (versus nTg, by Student's *t* test). Reprinted with permission from Kinouchi et al. Proc Natl Acad Sci USA 1991; 88:11158. © National Academy of Science.[86]

could depend on the nature of the cells, their content in SOD, NOS and on environmental conditions.

The alterations in vascular tone and endothelial damage induced by amyloid peptides in vitro are prevented by SOD.[75] Overexpression of CuZnSOD in the brain of transgenic mice (50% increase) confers protection against the toxic effect of GSH depletion on mitochondrial complex activities. Particularly, CuZnSOD overexpression prevents inactivation of complexes II and IV (personal communication). These results strongly implicate superoxide and its metabolite peroxynitrite in neurotoxicity and in the mitochondrial impairment induced by GSH depletion and also suggest an important role for this enzyme in the protection of mitochondrial function. In this way, causing complex I inhibition by the neurotoxin MPTP[76] leads to massive damage in brains of mice with a genetic deficiency in the mitochondrial superoxide dismutase, but less in mice that overexpress the enzyme.

In humans, nonhuman primates and mice MPTP produces clinical, biochemical and neuropathological changes similar to those which occur in idiopathic Parkinson's disease.[77] Transgenic mice with higher brain CuZnSOD activity showed significant protection of the nigrostriatal dopaminergic pathway against MPTP.[78] Biochemical changes consist in a marked reduction in levels of striatal dopamine and its metabolites homovanillic acid (HVA) and dihydroxyphenylacetic acid (DOPAC). In contrast to control animals, the injection of MPTP into transgenic mice does not induce any depletion in striatal levels of dopamine and its metabolites (Fig. 6.5).

However, transgenic dopaminergic neurons do not exhibit resistance to methylphenylpiridinium (MPP+)-induced degeneration.[79] This SOD work using transgenic mice may thus help to clarify the role of superoxide in the pathogenesis of MPTP in dopaminergic cells. It appears that superoxide radicals are in fact involved in the pathogenesis of the toxic effects of MPTP on dopaminergic cells. It was recently shown

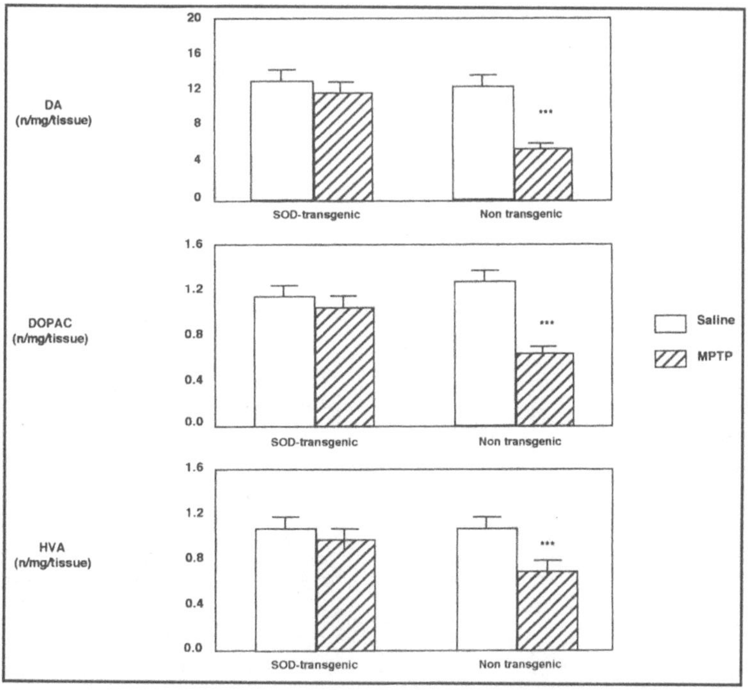

Fig. 6.5. The effects of MPTP on the content of DA, DOPAC and HVA in the striatum of SOD-transgenic mice and their non-transgenic littermates. MPTP causes significant decreases in DA, DOPAC, and HVA in nontransgenic mice. It does not affect significantly these levels in SOD-transgenic mice. *** $P < 0.001$ Reprinted with permission from Przedborski et al. J Neurosci 1992; 12:1658-1667. © Oxford University Press.[78]

that 7-nitroindazole (7-NI), an inhibitor of the neuronal isoform of NOS, can produce complete protection against MPTP-induced dopamine depletion in mice[80] and prevents MPTP-induced parkinsonism in baboons.[81] Furthermore, MPTP induced dopamine depletion was reduced in neuronal NOS knock out mice.[82] Taken together, protection by CuZnSOD overexpression and by inhibition of NOS suggest a major role for peroxynitrite in MPTP-induced neurotoxicity.

Sympathetic neurones or superior cervical ganglionic neurons microinjected with CuZnSOD expression vector or SOD protein were protected against the lethal effect of growth factor deprivation[83,84] and were made susceptible to death by decreased SOD levels. CuZnSOD delayed the time-course of death but did not completely block apoptosis.

A high survival rate of grafted dopamine neurons is crucial for reversing neurological deficits following brain tissue transplantation in Parkinson's disease. For unknown reasons the survival rate of transplanted dopamine neurons is only around 10% in experimental animals. The hypothesis that oxidative stress causes the loss of transplanted neurons was tested by grafting neurons from transgenic mice that overexpress CuZnSOD.[88] Compared with the survival of those taken from nontransgenic littermates, the dopaminergic neurons taken from CuZnSOD transgenic mouse embryonic mesencephalon survive intrastriatal transplantation by 4-fold with a concomitantly more extensive functional recovery (Fig. 6.6).

In cultured postnatal midbrain neurons of transgenic mice, the percentage of cell death was smaller in transgenic mice

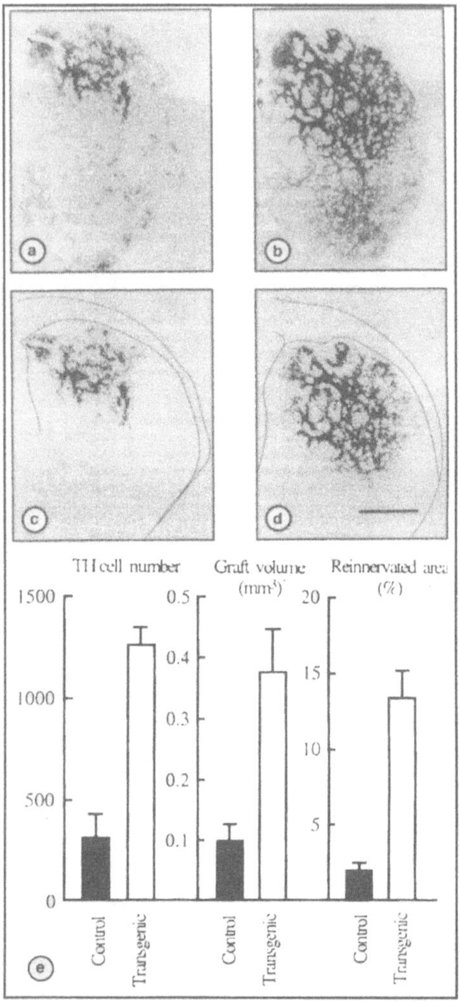

Fig. 6.6. Photomicrographs of coronal sections through the host striatum processed for tyrosine hydroxylase (TH)-immunohistochemistry (a, b) and a computerized image of the reinnervated area of the striatum in the same tissue sections (c, d). Images (a, c) represent the same section from a control rat, and panels (b, d) are taken from a representative section in the transgenic group. Scale bar, 1 mm. The bar diagrams in (e) illustrate the mean surviving TH-positive cell number, graft volume and reinnervated area for the control and transgenic groups. Reprinted with permission from Nakao et al. Nature Med 1995; 1:226-231.[88]

compared with nontransgenic.[89] Indeed, after 4 weeks, the SOD-transgenic cultures had two to three times more neurons than the nontransgenic cultures, corresponding to a slower rate of neuronal death in SOD-transgenic as compared to nontransgenic cultures. Some midbrain neurons are dying by necrosis and other by apoptosis and increased CuZnSOD levels can attenuate the both mode of cell. Furthermore, mesencephalic cell suspensions prepared from transgenic mice exhibit a significantly slower rate of decline in viability in vitro compared to preparations containing non-transgenic tissue.[79,88] This demonstration of enhanced survival of grafted dopamine neurons that overexpress CuZnSOD support other findings that the administration of antioxidant agents, 21-aminosteroids, increases the survival of grafted dopamine neurons 2.6 times and improves mesencephalic cell viability in vitro.[90] Intracellular overexpression of CuZnSOD seems more effective in protecting grafted dopamine neurons than the addition of an exogenous antioxidant compound. Furthermore, the effects can be attributed to a specific endogenous scavenging system, thereby shedding light on which free radical pathways may be crucial in mediating neuronal death. In addition to scavenging toxic superoxide anions, SOD may prevent the reaction of superoxide with nitric oxide to form peroxynitrite, which had been shown to be toxic to cultured neurons.[91] These results provide important direct support for the free radical hypothesis of dopaminergic neuron death in brain tissue grafting, suggesting that SOD may be useful when trying to achieve larger grafts in clinical transplantation trials.

Recently, a factor that prevents the inactivation of calcineurin in vitro and in vivo was identified as superoxide dismutase.[92] Calcineurin is the only protein phosphatase known to be under the control of Ca^{2+} and calmodulin. The redox state of iron provides a mechanism to regulate its activity by desensitizing the enzyme and coupling Ca^{2+}-dependent protein dephosphorylation to the redox state of the cell. The protection of calcineurin against inactivation by SOD constitutes a new physiological role for SOD in signal transduction phosphorylation cascades. Such effects could be associated with diseases caused by SOD mutations such as in ALS or SOD overexpression such as in DS.

GENETIC MANIPULATIONS OF OTHER SUPEROXIDE DISMUTASES

Interestingly, during central nervous system oxygen toxicity, protecting NO from superoxide-mediated inactivation with EC-SOD actually increased the toxicity of the hyperoxia.[93,94] This increased toxicity is apparently related to the effects of nitric oxide as an intercellular messenger (i.e., vasodilatator or neurotransmitter) in this form of oxygen toxicity. The role of extracellular superoxide in the pathogenesis of vasogenic edema was studied using transgenic mice expressing a 5-fold increase in extracellular superoxide dismutase (EC-SOD) in their brain. Increased EC-SOD expression offered significant protection against edema development after cold-induced injury. Mice treated with an inhibitor of NOS were also protected. The possible role of oxidative stress on the induction of *hsp70* and *c-fos* mRNAs following traumatic brain injury was investigated by comparing ECSOD-Tg mice and nontransgenic littermates.[95] The degree of *hsp70* induction and the degree, extent, and duration of *c-fos* induction are affected by the levels of SOD activity. These transgenic mice had an increased development of tolerance to hyperoxia associated with decreased histopathologic evidence of lung damage and mortality.[96] Alternatively, mice lacking extracellular superoxide dismutase are more sensitive to hyperoxia.[97] These results suggest that excessive superoxide, or one of its toxic products such as OH° or peroxynitrite, causes or contributes to oxygen toxicity. Transgenic overexpression of CuZnSOD[96] has previously been shown to prolong the survival of mice in high oxygen tension, as do injections of liposomes containing CuZnSOD. MnSOD, rather, CuZnSOD is the superoxide scavenger found in mitochondria. Mutant mice lacking MnSOD do not survive beyong the first 10 days of life and present severe cardiopathies.[98]

Thus, all three isoenzymes in their respective compartments have been shown to be important protectors in the complex response inflicted by oxidative stress. It may turn out that such a concerted protective action is also important in neuropathologies involving oxidant stress in the brain.

CONCLUSIONS

The use of genetic manipulations to construct cells or transgenic mice that specifically lack or overexpress a single antioxidant enzyme represent a mean to an understanding of the role of the individual antioxidant enzymes in cellular aging and degenerative diseases. These studies indicate that it is difficult to give an unequivocal answer as to whether increasing CuZnSOD activity in the brain by genetic means is, on balance, beneficial or deleterious. However, collectively, these data strongly support the idea that increased CuZnSOD provides protection towards acute oxidative stress mediated by different mechanisms. The data on neuronal survival and neuronal process development strongly suggest that increased SOD activity has a beneficial effect on neurons in culture.

However, considering possible implications of changes in CuZnSOD activity in neurological disorders, the situation is less clear. Having more SOD in the brain is not, in the long term, beneficial. The beneficial versus deleterious effects of CuZnSOD overexpression may depend on: (1) GSHPx activity and the ratio GSHPx/CuZnSOD; (2) the extent of free radical production; (3) the local concentration of transition metals; and (4) age, which could modify several of these factors.

From the above arguments, it appears that the exact mechanisms of the detrimental effect of excess CuZnSOD activity in a basal situation, where cells are not subjected to abnormal stress, remains a question of special importance for at least two reasons: (1) administration of CuZnSOD has been proposed in ALS in an attempt to slow down the progression of the disease; (2) increased CuZnSOD activity has been reported in afflicted regions of brains in DS, AD and PD post mortem,[99,100] and is located in the neuromelanin dopaminergic neurons[15] vulnerable in PD and hippocampal neurons vulnerable in Alzheimer's disease,[13,14] raising the possibility that excess CuZnSOD may play a direct role.

References

1. Wisniewski KE, Laure-Kamionowska M, Connell F et al. The neurobiology of Down's syndrome. New York: Raven Press, 1986:29-44.
2. Mann DMA. Alzheimer's disease and Down's syndrome. Histopathology 1988; 13:125-137.
3. Rumble B, Tetallack R, Hilbich C. Amyloid A4 protein and its precursor in Down's syndrome and Alzheimer's disease. New Engl J Med 1989; 22: 1446-1452.
4. Rockenstein EM, McConlogue L, Tan H et al. Levels and alternative splicing of amyloid β protein precursor (APP) transcripts in brains of APP transgenic mice and humans with Alzheimer's disease. J Biol Chem 1995; 270:47:28257-28267.
5. Games D, Adams D, Alessandrini R et al. Alzheimer-type neuropathology in transgenic mice overexpressing V717F β-amyloid precursor protein. Nature 1995; 373:523-527.
6. Zheng H et al. β-amyloid precursor protein-deficient mice show reactive gliosis and decreased locomotor activity. Cell 1995; 81:525-531.
7. Sinet PM. Metabolism of oxygen derivatives in Down's syndrome. Ann N Y Acad Sci 1982; 386:82-94.
8. Ceballos I, Nicole A, Briand P et al. Expression of the human CuZn superoxide dismutase gene in transgenic mice: model for gene dosage effect in Down's syndrome. Free Rad Res Comms 1991; 12-13:581-589.
9. Sinet PM, Ceballos-Picot I. Role of free radicals in Alzheimer's disease and Down's syndrome. In: Packer L, Prilipko L, Christen Y, eds. Free Radicals in the Brain Aging. Neurological and Mental disorders. Berlin, Heidelberg: Springer-Verlag, 1992:92-98.
10. Busciglio J, Yankner BA. Apoptosis and increased generation of reactive oxygen species in Down's syndrome neurons in vitro. Nature 1995; 378:776-779.
11. Sinet PM, Couturier J, Dutrillaux A et al. Trisomie 21 et superoxide dismutase-1. Tentative de localization sur la sous-bande 21q22.1. Exptl Cell Res 1976; 97:47-55.
12. Brooksbank BWL, Balazs R. Superoxide dismutase, glutathione peroxidase and lipoperoxidation in Down's syndrome fetal brain. Brain Res 1984; 16:37-44.
13. Ceballos I, Javoy-Agid F, Hirsch EC et al. Localization of copper-zinc superoxide dismutase mRNA in human hippocampus by *in situ* hybridization. Neurosci Lett 1989; 105:41-46.
14. Ceballos I, Javoy-Agid F, Delacourte et al. Neuronal localization of copper-zinc superoxide dismutase protein and mRNA within the human hippocampus from control and Alzheimer's disease brains. Free Rad Res Comms 1991; 12-13: 571-580.
15. Ceballos I, Lafon M, Javoy-Agid F et al. Superoxide dismutase and Parkinson's disease. Lancet 1990; 335:1035-1036.
16. Yim MB, Chock PB, Stadtman ER. Copper zinc superoxide dismutase catalyzed hydroxyl radical production from hydrogen peroxide. Proc Natl Acad Sci USA 1990; 87:5006-5010.
17. Yim MB, Chock PB, Stadtman ER. Enzyme function of copper, zinc superoxide dismutase as a free radical generator. J Biol Chem 1993; 268:4099-4105.
18. Beckman JS, Ischiropoulos H, Zhu L et al. Kinetics of superoxide dismutase and iron catalyzed nitration of phenolics by peroxynitrite. Arch Biochem Biophys 1992; 298:438-445.
19. Beckman JS, Beckman TW, Chen J et al. Apparent hydroxyl radical production by peroxynitrite: implications for endothelial injury from nitric oxide and superoxide. Proc Natl Acad Sci USA 1990; 87:1620-1624.
20. Wiedau-Pazos M, Goto JJ, Rabidazeh S et al. Altered reactivity of superoxide

dismutase in familial amyotrophic lateral sclerosis. Science 1996; 271:515-518.
21. Ceballos-Picot I, Nicole A, Clement M et al. Age-related changes in antioxidant enzymes and lipid peroxidation in brains of control and transgenic mice overexpressing copper-zinc superoxide dismutase. Mutation Res 1992; 275: 281-293.
22. Sinet PM, Lejeune J, Jerome H. Trisomy 21 (Down's syndrome), glutathione peroxidase, hexose monophosphate shunt and IQ. Life Sci. 1979; 24:29-34.
23. Crosti N, Bajer J, Gentile M et al. Catalase and glutathione peroxidase activity in cells with trisomy 21. Clin Genet 1989; 36:107-116.
24. Brooksbank BWL, Martinez M, Balazs R. Altered composition of polyunsaturated fatty acyl groups in phosphoglycerides of Down's syndrome fetal brain. J Neurochem 1985; 44:869-874.
25. Ceballos I, Delabar JM, Nicole A et al. Expression of transfected human CuZn superoxide dismutase gene in mouse L cells and NS20Y neuroblastoma cells induces enhancement of glutathione peroxidase activity. Biochim Biophys Acta 1988; 949:58-64.
26. Norris KH, Hornsby PJ. Cytotoxic effects of expression of human superoxide dismutase in bovine adrenocortical cells. Mut Res 1990; 237:95-106.
27. Scott MD, Meshnick SR, Eaton JW. Superoxide dismutase rich bacteria. Paradoxal increase in oxidant toxicity. J Biol Chem 1987; 262:3640-3645.
28. Scott MD, Meshnick SR, Eaton JW. Superoxide dismutase amplifies organismal sensitivity to ionizing radiation. J Biol Chem 1989; 264:2498-2501.
29. Elroy-Stein O, Bernstein Y, Groner Y. Overproduction of human CuZn superoxide dismutase in transfected cells: extenuation of paraquat-mediated cytotoxicity and enhancement of lipid peroxidation. EMBO J 1986; 5:615-622.
30. Krall J, Bagley AC, Mullenbach GT et al. Superoxide mediates the toxicity of Paraquat for cultured mammalian cells. J Biol Chem 1988; 263:2672-2678.
31. Kelner MJ, Bagnell R. Alteration of endogenous glutathione peroxidase, manganese superoxide dismutase, and glutathione transferase activity in cells transfected with a copper-zinc superoxide dismutase expression vector. J Biol Chem 1990; 265:10872-10875.
32. Kelner MJ, Bagnell R, Montoya M et al. Transfection with human copper-zinc superoxide dismutase induces bidirectional alterations in other antioxidant enzymes, proteins, growth factor response, and paraquat resistance. Free Rad Biol Med 1995; 18:497-506.
33. Amstad P, Moret R, Cerutti P. Glutathione peroxidase compensates for the hypersensitivity of CuZn superoxide dismutase overproducers to oxidant stress. J Biol Chem 1994; 269:1606-1609.
34. Amstad P, Peskin A, Shah G et al. The balance between CuZn superoxide dismutase and catalase affects the sensitivity of mouse epidermal cells to oxidative stress. Biochemistry 1991; 30:9305-9313.
35. Elroy-Stein O, Groner Y. Impaired neurotransmitter uptake in PC12 cells overexpressing human Cu-Zn superoxide dismutase: implications for gene dosage effect in Down's syndrome. Cell 1988; 52:259-267.
36. Minc-Colomb D, Knobler H, Groner Y. Gene dosage of CuZnSOD and Down's syndrome: diminished prostaglandin synthesis in human trisomy 21, transfected cells and transgenic mice. EMBO 1991; 10:2119-2124.
37. de Haan JB, Newman JD, Kola I. CuZn superoxide dismutase mRNA and enzyme activity, and susceptibility to lipid peroxidation, increases with aging in murine brains. Mol Brain Res 1992; 13:179-186.
38. de Haan JB, Cristiano F, Ianello RC et al. CuZn-superoxide dismutase and glutathione peroxidase during aging. Biochem Mol Biol Int 1995; 35:1281-1297.
39. Cristiano F, de Haan JB, Ianello R et al. Changes in the levels of enzymes which modulate the antioxidant balance occur during aging and correlate with cellular damage. Mech Aging Dev 1995; 80: 93-105.
40. Barkats M, Bertholet JY, Venault P, Ceballos-Picot I et al. Hippocampal mossy fiber changes in mice transgenic

for the human copper-zinc superoxide dismutase gene. Neurosci Lett 1993; 160:24-28.

41. Avraham KB, Schickler M, Sapoznokow D et al. Down's syndrome: abnormal neuromuscular junction in tongue of transgenic mice with elevated levels of human CuZn-superoxide dismutase. Cell 1988; 54:823-829.
42. Avraham KB, Sugerman H, Rotshenker S et al. Down's syndrome: morphological remodelling and increased complexity in the neuromuscular junction of transgenic CuZn-superoxide dismutase mice. J Neurocytol 1991; 20:208-215.
43. Groner Y, Elroy-Stein O, Avraham KB et al. Down syndrome clinical symptoms are manifested in transfected cells and transgenic mice overexpressing the human CuZn superoxide dismutase gene. J Physiol-Paris. 1990; 84:53-77.
44. Groner Y, Elroy-Stein O, Avraham KB et al. Cell damage by excess of CuZnSOD and Down syndrome. Biomed Pharmacother 1994; 48:231-240.
45. Singh R, Pathak DN. Lipid peroxidation and glutathione peroxidase, glutathione reductase, superoxide dismutase, and glucose-6-phosphate dehydrogenase activities in FeCl3 induced epileptogenic foci in the rat brain. Epilepsia 1990; 31:15-26.
46. de Haan JB, Tymms MJ, Cristiano F et al. Expression of copper/zinc superoxide dismutase and glutathione peroxidase in organs of developing mouse embryos, fetuses and neonates. Pediatr Res 1994; 35:188-196.
47. de Haan JB, Cristiano F, Ianello R et al. Elevation in the ratio of Cu/Zn-superoxide dismutase to glutathione peroxidase activity induces features of cellular senescence and this effect is mediated by hydrogen peroxide. Human Mol Genet 1996; 5:283-292.
48. Noda A, Ning Y, Venable SF et al. Cloning of senescent cell-derived inhibitors of DNA synthesis using an expression screen. Exp Cell Biol 1994; 211:90-98.42.
49. Ceballos-Picot I, Nicole A, Briand P et al. Neuronal-specific expression of human copper-zinc superoxide dismutase genes in transgenic mice: animal model of gene dosage effect in Down's syndrome. Brain Res 1991; 552:198-214.
50. Ceballos-Picot I, Nicole A, Sinet PM. Cellular clones and transgenic mice overexpressing copper-zinc superoxide dismutase: models for the study of free radical metabolism and aging. In: Emerit I, Chance B, eds. Free Radicals and Aging. Birhauser Verlag Basel/Switzerland 1992:89-98.
51. Ceballos-Picot I, Nicole A, Benabdesselam O et al. Modulation of brain antioxidant defenses during aging in normal and transgenic mice overexpressing copper-zinc superoxide dismutase. Bull Mol Biol Med 1993; 18:119-144.
52. Epstein CJ, Avraham KB, Lovett M et al. Transgenic mice with increased Cu/Zn-superoxide dismutase activity: Animal model of gene dosage effects in Down syndrome. Proc Natl Acad Sci USA 1987; 84:8044-8048.
53. Anneren G, Edqvist LE, Gebre-Medhin M et al. Glutathione peroxidase activity in erythrocytes in Down's syndrome. Abnormal variation in relation to age and sex through childhood and adolescence. Trisomy 21 1985; 1:9-17.
54. Anneren G, Epstein CJ. Lipid peroxidation and superoxide dismutase-1 and glutathione peroxidase activities in trisomy 16 fetal mice and human trisomy 21 fibroblasts. Pediat Res 1987; 21:88-92.
55. Ceballos I, Javoy-Agid F, Delacourte A et al. Parkinson's disease and Alzheimer's disease: neurodegenerative disorders due to brain antioxidant system deficiency? In: Emerit et al, eds. Antioxidants in Therapy and Preventive Medicine. New York: Plenum Press, 1990:493-498.
56. Pardo CA, Xu Z, Borchelt DR et al. Superoxide dismutase is an abundant component in cell bodies, dendrites, and axons of motor neurons and in a subset of other neurons. Proc Natl Acad Sci 1995; 92:954-958.
57. Delacourte A, Defossez A, Ceballos I. Preferential expression of copper-zinc superoxide dismutase in the vulnerable cortical neurons in Alzheimer's disease. Neurosci Lett 1988; 92:247-253.

58. Przedborski S, Jackson-Lewis V, Kostic V et al. Superoxide dismutase, catalase, and glutathione peroxidase activities in copper/zinc-superoxide dismutase transgenic mice. J Neurochem 1992; 58:1760-1767.
59. Schickler M, Knobler H, Avraham KB et al. Diminished serotonin uptake in platelets of transgenic mice with increased Cu/Zn-superoxide dismutase activity. EMBO 1989; 8:1385-1392.
60. Yarom R, Sapoznikov D, Havivi Y et al. Premature aging changes in neuromuscular junctions of transgenic mice with extra human CuZnSOD gene: a model for tongue pathology in Down's syndrome. J Neurol Sci 1988; 88:41- 53.
61. Nabarra B, Casanova M, Paris D et al. Transgenic mice overexpressing human CuZnSOD gene. Ultrastructural studies of premature thymic involution model of Down's syndrome. Lab Invest 1996; 74:617-626.
62. Peled-Kamar M, Lotem J, Okon E et al. Thymic abnormalities and enhanced apoptosis of thymocytes and bone marrow cells in transgenic mice overexpressing Cu/Zn-superoxide dismutase. EMBO J 1995; 14:4985-4993.
63. Schwartz PJ, Berger UV, Coyle JT. Mice transgenic for copper/zinc superoxide dismutase exhibit increased markers of biogenic amine function. J Neurochem 1995; 65:660-669.
64. Harman D. The aging process. Proc Natl Acad Sci USA 1981; 78:7124-7128.
65. Epstein CJ, Chan PH, Cadet JL et al. Resistance of SOD-transgenic mice to oxidative stress. In: FH Gage, Y Christen, eds. Gene transfer and therapy in the nervous system. Springer Verlag 1992:107-117.
66. Chan PH, Chu L, Chen S, Carlson EJ et al. Reduced neurotoxicity in transgenic mice overexpressing copper-zinc superoxide dismutase. Stroke Supp.III, 1990; 21:80-82.
67. Cadet JL et al. Attenuation of methamphetamine-induced neurotoxicity in copper/zinc superoxide dismutase transgenic mice. J Neurochem 1994; 62:380-383.
68. Beal MF, Ferrante RJ, Henshaw R et al. 3-nitropropionic acid neurotoxicity is attenuated in copper/zinc superoxide dismutase transgenic mice. J Neurochem 1995; 65:919-922.
69. Chan PH, Yang GY, Che SF et al. Cold-induced brain edema and infarction are reduced in transgenic mice overexpressing CuZn-superoxide dismutase. Ann Neurol 1991; 29:482-486.
70. Barkats M, Bemelmans AP, Geoffroy MC et al. An adenovirus encoding CuZnSOD protects cultured striatal neurones against glutamate toxicity. Neuroreport 1996; 7:497-501.
71. Lafon-Cazal M, Pietri S, Culcasi M et al. NMDA-dependent superoxide production and neurotoxicity. Nature 1993; 364:535-537.
72. Dawson TM, Dawson VL, Snyder SH. Nitric oxide mediates glutamate neurotoxicity in primary cortical cultures. Proc Natl Acad Sci USA 1991; 88:6368-6371.
73. Dawson VL, Dawson TM, Bartley DA et al. Mechanisms of nitric oxide-mediated neurotoxicity in primary brain cultures. J Neurosci 1993; 13:2651-2661.
74. Hobbs AJ, Fukuto JM, Ignarro LJ. Formation of free nitric oxide from L-arginine by nitric oxide synthase:direct enhancement of generation by superoxide dismutase. Proc Natl Acad Sci USA 1994; 91:10992-10996.
75. Thomas T, Thomas G, McLendon C et al. β-amyloid-mediated vasoactivity and vascular endothelial damage. Nature 1996; 380:168-171.
76. Cortopassi G, Wang E. Modelling the effects of age-related MtDNA mutation accumulation; complex I deficiency, superoxide and cell death. Biochim Biophys Acta 1995; 1271:171-176.
77. Chiueh CC, Miyake H, Peng M. Role of dopamine autooxidation, hydroxyl radical generation, and calcium overload in underlying mechanisms involved in MPTP-induced parkinsonism. Adv Neurol 1993; 60:251-258.
78. Przedborski S, Vladimir K, Jackson-Lewis V et al. Transgenic mice with increased CuZn-superoxide dismutase activity are resistant to MPTP-induced neurotoxicity. J Neurosci 1992; 12:1658-1667.
79. Sanchez-Ramos J, Song S, Facca A et al. Transgenic murine dopaminergic neurons expressing human superoxide

dismutase exhibit increased density in culture, but no resistance to methylphenylpyridinium-induced degeneration. J Neurochem 1997; 68:58-67.
80. Schulz JB et al. Blockade of neuronal nitric oxide synthase protects against excitotoxicity in vivo. J Neurosci 1995; 15:8419-8429.
81. Hantraye P, Brouillet E, Ferrante R, Palfi S et al. Inhibition of neuronal nitric oxide synthase prevents MPTP-induced parkinsonism in baboons. Nature Med 1996; 2:1017-1021.
82. Predborski S et al. Role of nitric oxide in MPTP-induced dopaminergic neurotoxicity. Proc Natl Acad Sci USA 1996; 93:4565-4571.
83. Greenlund LJ, Deckwerth TL, Johnson EM et al. Superoxide dismutase delays neuronal apoptosis: a role for reactive oxygen species in programmed neuronal death. Neuron 1995; 14:303-315.
84. Jordan J, Ghadge GD, Prehn JHM et al. Expression of human copper-zinc superoxide dismutase inhibits the death of rat sympathetic neurons caused by withdrawal of nerve growth factor. Mol Pharmacol 1995; 47:1095-1100.
85. Kinouchi H, Mizui T, Carlson E et al. Focal cerebral ischemia and the antioxidant system in transgenic mice overexpressing CuZn-superoxide dismutase. J Cereb Blood Flow Metab 1991; 11:S423.
86. Kinouchi H, Epstein CJ, Mizui T et al. Attenuation of focal ischemic injury in transgenic mice overexpressing CuZn superoxide dismutase. Proc Natl Acad Sci USA 1991; 88:11158-11162.
87. Yang G, Chan PH, Chen J et al. Human copper-zinc superoxide dismutase transgenic mice are highly resistant to reperfusion injury after focal cerebral ischemia. Stroke 1994; 25:165-170.
88. Nakao N, Frodl EM, Widner H et al. Overexpressing Cu/Zn superoxide dismutase enhances survival of transplanted neurons in a rat model of Parkinson's disease. Nat Med 1995; 1:226-231.
89. Przedborski S, Khan U, Kostic V et al. Increased superoxide dismutase activity improves survival of cultured postnatal midbrain neurons. J Neurochem 1996; 67:1383-1392.
90. Nakao N, Frodl EM, Duan WM et al. Lazaroids improve the survival of grafted rat embryonic dopaminergic neurons. Proc Natl Acad Sci USA 1994; 91: 12408-12412.
91. Lipton SA, Choi Y-B, Pan Z-H et al. A redox-based mechanism for the neuroprotective and neurodestructive effects of nitric oxide and related nitroso-compounds. Nature 1993; 364:626-632.
92. Wang X, Culotta V, Klee CB. Superoxide dismutase protects calcineurin from inactivation. Nature 1996; 383:434-437.
93. Oury TD, Ho Y-S, Piantadosi CA, Crapo JD. Extracellular superoxide dismutase, nitric oxide, and central nervous system O_2 toxicity. Proc Natl Acad Sci 1993; 89:9715-9719.
94. Oury TD, Piantadosi CA, Crapo JD. Cold-induced brain edema in mice: involvement of extracellular superoxide dismutase and nitric oxide. J Biol Chem 1993; 268(21):15394-15398.
95. Mikawa S, Sharp F, Kamii H et al. Expression of c-fos and HSP70 mRNA after traumatic injury in transgenic mice overexpressing CuZn-superoxide dismutase. Mol Brain Res 1995; 33:288-294.
96. White CW, Avraham KB, Shanley PF, Groner Y. Transgenic mice with expression of elevated levels of copper-zinc superoxide dismutase in the lungs are resistant to pulmonary oxygen toxicity. J Clin Invest 1991; 2162-2168.
97. Carlsson LM, Jonsson J, Edlund T, Marklund SL. Mice lacking extracellular superoxide dismutase are more sensitive to hyperoxia. Proc Natl Acad Sci USA 1995; 92:6264-6268.
98. Li Y, Huag T, Carlson EJ et al. Dilated cardiomyopathy and neonatal lethality in mutant mice lacking manganese superoxide dismutase. Nat Genet 1995; 11: 376-381.
99. Marttila RJ, Lorentz H, Rinne UK. Oxygen toxicity protecting enzymes in Parkinson's disease. Increase of superoxide dismutase-like activity in the substantia nigra and basal nucleus. J Neurol Sci 1988; 86:321-331.
100. Marklund SL, Adolfsson R, Gottfries CG et al. B Superoxide dismutase isoenzymes in normal brains and in brains from patients with dementia of Alzheimer type. J Neurol Sci 1985; 67:319-325.

CHAPTER 7

Oxidative Stress in Amyotrophic Lateral Sclerosis

INTRODUCTION AND CLINICAL PATHOLOGY

Amyotrophic lateral sclerosis (ALS) is a prototypic, age-dependent neurodegenerative disorder caused by the degeneration of motor neurons in cortex, brainstem, and spinal cord with an incidence of 1 in 10000.1 It typically afflicts individuals in middle adult life, leading to paralysis and death within 3 to 5 years. Approximately 10% of cases are familial (FALS),[2] and in a subset of cases, mutations in the *CuZnSOD* gene, which encodes CuZnSOD, have been demonstrated.[3] In both sporadic and familial forms clinical courses are highly similar.[4] The disease usually begins asymmetrically and distally in one limb, most commonly the leg, and then appears to spread within the neuraxis to involve contiguous groups of motor neurons. Pathologically, ALS is distinguished by atrophy and death of the affected neurons, with dissolution of both cytoplasm and nuclei. The motor neuron death process appears to be cell autonomous; autopsy studies reveal no evidence that other cell types participate in killing motor neurons. Many affected neurons demonstrate evidence of cytoskeletal pathology in the form of accumulations of neurofilaments,[5] both within the neuronal cell body and in axons. Aggregates of ubiquitinated, unidentified proteins are also seen, as is subtle proliferation of glial cells.[6] The motor neurons are described as the final common pathway of the central nervous system, transforming neuronal action into activation of muscle. The neuromuscular junction distinguishes lower motor neurons from all other neurons and may contribute to the selective vulnerability of motor neurons in ALS. ALS is characterized principally by degeneration of large motor neurons in the ventral horn of the spinal cord and by degeneration of upper motor neurons in the cerebral cortex. Because muscle fibers are innervated by only a single motor neuron, other motor neurons normally rewire to the muscle fiber to take over the function of a damaged motor neuron. Muscle from ALS patients contains far more neuromuscular junctions than normal muscle. Remodeling of synapses and motor endplates may cause subsequent injury of adjacent motor neurons.

Recent investigations have implicated oxidative cytotoxicity in the pathogenesis of ALS. This review considers the evidence that this disease arises from oxidative and excitotoxic injury to critical subcellular targets in motor neurons and summarizes current hypotheses concerning the mechanism of motor neuron death in ALS.

The Role of Oxidative Stress in Neuronal Death, by Irène Ceballos-Picot.

ETIOLOGY: POSSIBLE CAUSES OF MOTOR NEURON DISEASE

As with any neurodegenerative disorder, etiologic hypotheses in ALS must explain three essential aspects of the disease: the mechanisms and pathways leading to cell death, the selectivity of the process for the affected neurons, and the onset in adulthood. While many hypotheses have been advanced to explain the pathogenesis of ALS, few have adequately explained each of these aspects.

The etiology of the sporadic disease has been linked to alterations of glutamaergic neurotransmission,[7–14] environmental toxins, autoimmunity[15] and metals.[16] The motor neuron selectivity of poliovirus has prompted the theory that ALS might represent an atypical polioviral infection in the sporadic disease; to date, traditional culturing methods and in situ RNA hybridization have failed to detect poliovirus in ALS cerebrospinal fluid and brain specimens. The possibility that ALS might be a form of slow viral infection has not been supported by long-term studies of inoculation of ALS brain tissues into nonhuman primates. Because they may induce clinical paralysis and motor nerve pathology in some experimental animals, lead, aluminum, and other metallotoxins have also been implicated. Thus far, however, no convincing clinical or necropsy data support this hypothesis in conventional ALS. Both spinal and corticospinal motor neuron degeneration can be induced as autoimmune diseases. This and other observations have occasioned the hypothesis that the disease may arise on an autoimmune basis. While sera of many patients with sporadic ALS possess autoantibodies that react with constituents of the distal motor terminal, including voltage-dependent calcium channels, it has not been demonstrated that these are causal. Moreover, patients with ALS show no benefit from immunosuppressive therapy, although this does not rigorously preclude an autoimmune etiology.

Several lines of inquiry indicate that toxicity from excitatory neurotransmitters such as glutamate may be one factor in the pathobiology of ALS. Brain tissue levels of glutamate are depressed,[10] while cerebrospinal fluid glutamate and aspartate levels may be elevated both in sporadic human ALS patients[9–11] and in canine motor neuron disease.[17] Additionally, high affinity glutamate uptake and expression of the primarily glial GLT-1 glutamate transporter are diminished in synaptosomal preparations from ALS spinal cord and motor cortex.[13,14] In addition, clinical trials with riluzole and gabapentin, drugs that interfere with glutamaergic neurotransmission, showed that survival of ALS patients is extended.[18,19]

Of all cases of ALS, 5-10% are familial. FALS has been linked in some patients (fewer than 20%) to dominantly inherited mutations in the *CuZnSOD* gene[3] and in others to variants in the gene encoding the heavy subunit of the neurofilament triplet protein.[5]

As argued below, several lines of investigation prompted in part by these genetic findings suggest the hypothesis that motor neuron death in ALS reflects a complex interplay among oxidative injury, excitotoxic stimulation of the motor neuron, and dysfunction of mitochondria and critical proteins such as neurofilaments.

CuZnSOD GENE MUTATIONS IN FALS

Two forms of ALS are inherited. The predominant form, transmitted as an autosomal dominant trait, is an adult-onset disease that is clinically and pathologically indistinguishable from sporadic ALS. About 10-15% of all ALS cases fall within this category. In recent years, there has also been recognized a rare, recessively inherited childhood-onset form of ALS associated with very long survival.[20] The clinical and pathological similarity between the dominantly inherited and sporadic forms of ALS prompted a multicenter effort to identify one or more ALS genes on the premise that insights into the inherited disease might illuminate the pathogenesis of sporadic ALS.

In 1991, an ALS locus was identified on human chromosome 21q, although only a fraction of all ALS families were linked to this region.[21] In 1993, a gene within this region encoding the protein cytosolic copper-zinc SOD (*CuZnSOD*) was found to be mutated in several ALS families.[122] There is also allelic heterogeneity; at least 24 missense mutations have been found in the *CuZnSOD* gene in FALS patients.[22–25] The most common point mutation is an Ala4→Val substitution at codon 4 (A4V); this mutation account for 38% of all familial ALS.

The association with *CuZnSOD* makes it possible, for the first time, to generate credible theories of the pathogenesis of FALS. CuZnSOD is encoded by five exons and forms a globular protein that functions as a homodimer. Within each monomer there is a cave-like active site containing one ion each of copper and zinc. The major function of this enzyme is believed to be the detoxification or dismutation of the superoxide anion to form hydrogen peroxide and oxygen (see chapter 2). This, in turn, is detoxified by either glutathione peroxidase or catalase to form water. As shown in Figure 7.1, under normal circumstances superoxide anion can also react with nitric oxide to form peroxynitrite.

One major implication of this biochemical profile of CuZnSOD reactivity is that FALS may be a consequence of disturbed free radical homeostasis and cytotoxicity by ROS. CuZnSOD induction,[26] increased protein carbamylation,[27] and upregulated glutathione receptors[28] suggest that oxidative stress may also be implicated in the sporadic form of the disease. If SOD were inactive, the superoxide radical would accumulate and might itself be cytotoxic because it can act as either an oxidant or a reductant. Superoxide can also react with nitric oxide to form peroxynitrite which nitrates tyrosine residues to form a nitronium-like intermediate and might, thereby, interfere with the functions of SOD or other proteins.[29,30]

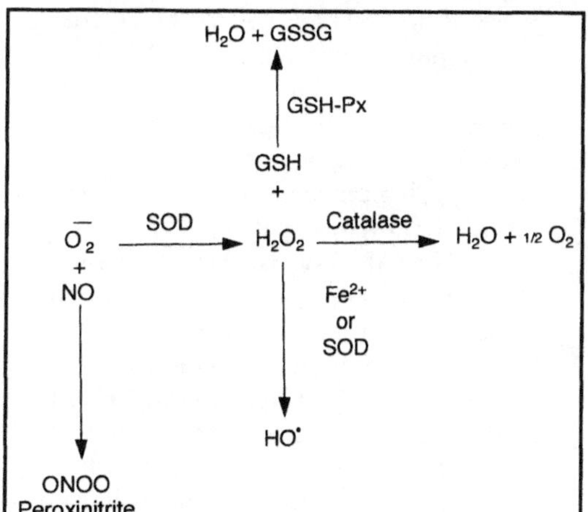

Fig. 7.1. Metabolism of reactive oxygen species.

THE PEROXYNITRITE NITRATION HYPOTHESIS OF ALS

It was proposed that the CuZnSOD mutants have a greater intrinsic ability to catalyze nitration by peroxynitrite of tyrosine residues on a key target in motor neurons. Because afflicted ALS patients have one normal SOD allele, superoxide scavenging should decrease by about 50%, which would at most double the steady state concentration of superoxide. Superoxide reacts with nitric oxide three times faster than with native CuZnSOD, to form the powerful oxidant, peroxynitrite ($ONOO^-$).[31] A doubling of the steady-state superoxide concentration will also double the rate of peroxynitrite formation. Peroxynitrite in turn reacts with SOD to form a nitronium-like intermediate, which nitrates tyrosine residues.[32] The mutations identified in ALS patients do not directly affect conserved amino acids forming the active site, but could slightly disrupt the active site pocket to allow greater access of peroxynitrite to the copper. Chemical modification of several amino acids in the active site allows SOD to react with peroxynitrite but greatly diminishes superoxide scavenging.[32] Thus, the SOD mutations may increase both the rate of peroxynitrite formation due to reduced scavenging of superoxide and nitration by peroxynitrite of critical cellular targets. In addition to nitric oxide produced by neurons, nitric oxide as well as superoxide may be produced by reactive astrocytes and microglia that remove the cellular debris from degenerating motor neurons. Peripheral nerve injury and chronic inflammation have recently been shown to induce nitric oxide synthase in the spinal cord.[33-35] Furthermore, human skeletal muscle contains high concentrations of the neuronal form of nitric oxide synthase, which is probably localized in the neuromuscular junction.[136] Nitration of proteins will slowly injure motor neurons as well as other cells. However, motor neurons, among the largest in the central nervous system, cannot regenerate. The neuromuscular junction may also be another target that makes the motor neuron particularly susceptible. Phosphorylation of tyrosine residues is a major signal transduction mechanism for most growth factors and is inhibited by nitration.[37-39] One mechanism leading to ALS would involve disruption of signal transduction. Skeletal muscle has been shown to produce two nerve growth factors, brain-derived neurotrophic factor (BDNF) and neurotrophin-3 (NT-3), that support the survival of motor neurons after injury. BDNF and NT-3 bind to the trkB receptor,[40] which is a tyrosine kinase in the neuromuscular junction of motor neurons. Like most growth factor receptors, trkB phosphorylates itself, which is necessary for activation. The trkB receptor also is known to activate phospholipase Cγ-1 by phosphorylation.[41] The trkB receptor complex with BNDF is taken up by the motor neuron and appears to be transported to the cell body by retrograde transport.[42] The trkB receptor is strongly upregulated after spinal cord injury.[43] There is a clear proliferation of motor end units in ALS, with some end units containing synaptic vesicles but not making contact with muscle. This suggests that the motor neuron in ALS might be suffering a deficiency of muscle-derived growth factors. It is also reminiscent of the proliferation of neuromuscular junctions observed in transgenic SOD mice.[41] Other potential tyrosine targets may affect kinases responsible for phosphorylation of neurofilaments, leading to impaired neurofilament assembly, which have also been implicated in the pathogenesis of ALS. In this regard, it is important to note that defective neurofilament assembly in transgenic mice leads to aggregation of neurofilaments and impaired axonal transport.[44,45]

Thus, nitration of key tyrosine residues may lead to the gradual loss of motor neurons by several potential but as yet unknown mechanisms that need to be investigated. Understanding whether such a nitration cycle occurs in ALS can lead to better therapeutic intervention by suggesting methods to break the cycle, such as selective inhibition of the nitric oxide synthase found in or around motor neurons.

A GAIN-OF-FUNCTION OF CuZnSOD IN FALS?

Over the last year, several studies have begun to define the effects of the FALS mutations on SOD1 function, although many critical questions remain unanswered. It is now clear that the mutations reduce total cellular activity of the enzyme by 25-50%, as analyzed in red blood cells, lymphoblastoid cell lines, and brain.[20,27,46,47] Analysis of cytoplasmic CuZnSOD activity in both lymphoid and COS cells has established that the loss of activity does not arise from a dominant negative interaction of mutant and wild type protein in SOD heterodimers. Rather, the loss of function is due to diminished stability of the mutant protein per se; depending on the specific mutations, the half-lives of the mutant proteins are reduced by 30-75%.[48]

It has been hypothesized that the mutations in *CuZnSOD* lead to FALS by decreasing the enzymatic activity of CuZnSOD. However, several findings have raised questions about this interpretation.

(1) The determination of low CuZnSOD activity were carried out in vitro, in some cases following isolations of the protein in nonphysiological conditions, without correlation of in vivo activity.

(2) The FALS-associated *SOD* mutations affect amino acid residues involved in enzyme dimerization or β-barrel turns, rather than those corresponding to the active site. The single exception, His46→Arg (H46R) in which the active site is directly involved, is reported to lead to a very mild form of FALS; in contrast, severe forms result from mutations far removed from the active site.

(3) All mutations identified to date have been missense, rather than nonsense, mutations.

(4) It is unusual for a null or hypomorphic mutation to lead to a dominantly inherited disease when the mutated gene encodes an enzyme.

(5) CuZnSOD activity measurements from lymphoblasts of patients with the G37R mutation have demonstrated active heterodimeric CuZnSOD, arguing that, at least for the G37R mutation, the total SOD activity is not compromised by heterodimerization of the wild-type monomer with the mutant monomer (Fig. 7.2).

(6) Transgenic mice expressing the G93A mutation demonstrate motor neuron degeneration despite an increased level of SOD activity. Most diseases arising from inactivity of a protein from so-called loss of function are recessive, as most proteins exist in sufficient abundance that their levels must be significantly reduced before a cellular phenotype results. Half-normal enzyme activity is compatible with normal life and is encountered in the heterozygous carriers of autosomal recessive genes, and these carriers are typically asymptomatic.

On the other hand, dominantly inherited diseases, like FALS, are usually thought to arise because a single mutant allele produces a mutant protein with a novel property that may be injurious to the cell. In the context of FALS, the fundamental question is whether the motor neuron degeneration arises purely from loss of CuZnSOD activity, which clearly occurs for some of the mutations, or from gain of a novel adverse function. This has an important therapeutic ramification: if the disease is a consequence of diminished function of the CuZnSOD protein (i.e., decreased dismutation of the superoxide anion or diminished binding of copper and zinc), therapies that augment levels of CuZnSOD activity may be beneficial in FALS. If the disease results from a toxic gain-of-function effect, it may be necessary to reconsider plans already in motion to administer SOD as treatment. To analyze the problem, cellular models with downregulation of CuZnSOD and transgenic mice expressing the human mutant CuZnSOD enzymes were generated.

CELLULAR AND ANIMAL MODELS WITH DOWNREGULATION OR MUTATION OF COPPER-ZINC SUPEROXIDE DISMUTASE

Downregulation of CuZnSOD in Neuronal Cells Induces Apoptosis

It has recently been shown in chronic, organotypic spinal cord cultures that partial reduction of activity of CuZnSOD by chronic application of *CuZnSOD* antisense oligonucleotides triggers neuronal death, including fulminant motor neuron death. Strikingly, the death process, which is reversed by agents that enhance intracellular antioxidant defenses (e.g., N-acetylcysteine), is apoptotic.[49] Similar results have been observed in cultures of PC12 cells.[150,151] In this model NOS inhibitors abolish cell death,[51] suggesting that formation of peroxynitrite is necessary for this process. Because they are chronically incubated in

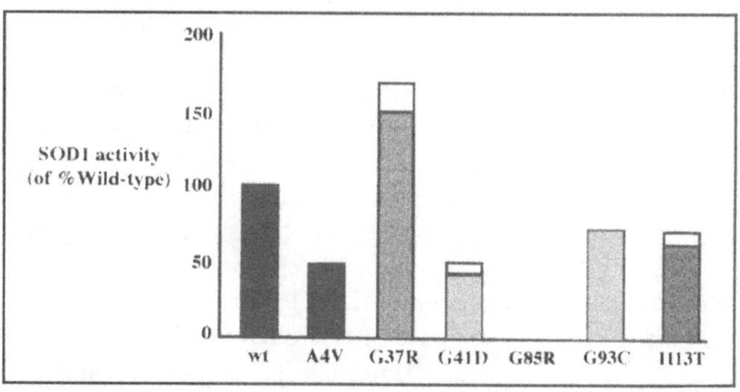

Fig. 7.2. Specific enzymatic activities of CuZnSOD polypeptides encoded by FALS-linked mutations. Reprinted with permission from Borchelt et al. Proc Natl Acad Sci USA 1994; 91:4155-4159.[48]

supraphysiologic levels of oxygen, it is probable that these cultured neurons and neuronal cell lines are under chronic oxidative stress and therefore not fully representative of neurons in vivo.

Knock Out of CuZnSOD in Transgenic Mice

Reaume et al[152] describe the creation and characterization of mice completely deficient for this enzyme. These animals develop normally and show no overt motor deficits by 6 months in age. Histological examination of the spinal cord reveals no signs of pathology in animals 4 months in age. However CuZnSOD-deficient mice exhibit marked vulnerability to motor neuron loss after axonal injury. The extent of motor neuron survival after axatomy is correlated with the levels of CuZnSOD activity. These results indicate that CuZnSOD is not necessary for normal motor neuron development and function but is required under physiologically stressful conditions following injury. To assess whether the CuZnSOD null animals are subjected to increased oxidative stress, brain tissue levels of lipid peroxidation and protein carbonyl content, two common end-products of oxidative damage, were measured in 120 to 130 day old animals. No significant differences from $SOD^{+/-}$ or $SOD^{-/-}$ animals in either lipid peroxidation or protein carbonyl content were detected. To determine whether the lack of detectable oxidative damage was due to compensatory anti-oxidant mechanisms, levels of MnSOD and the ratio of reduced to oxidized glutathione were measured. No significant differences among the three genotypes ($SOD^{+/+}$, $SOD^{-/+}$ and $SOD^{-/-}$) were found in animals 110 to 130 days in age.

Motor Neuron Degeneration in Mice That Express Human CuZnSOD Mutations

Rather than causing a loss of function, compelling data indicate that the FALS mutations in *SOD* may act through cytotoxic, possibly novel, functions that result in neuronal degeneration.

So far, three *CuZnSOD* mutations (Gly93→Ala, Gly85→Arg and Gly37→Arg) found in FALS kindreds have caused motor neuron disease when expressed in transgenic mice.[53–56] Mice that overexpress high levels of mutant SOD protein develop a denervating, paralytic process that resembles ALS in humans both clinically and pathologically. As the paralysis develops in these animals, dendritic processes in motor neurons show microvacuolization, interpreted to result from defects in rough endoplasmic reticulum and mitochondria.[53,56,57] Some of these mice develop cytoskeletal abnormalities similar to those that occur in the human illness.[45,58] Because CuZnSOD activity in these animals are greater than normal (Table 7.1). It is difficult to argue that the disease arises from loss of function, unless there is some type of highly focal subcellular loss of function. The possibility that the high levels of CuZnSOD activity in the animals trigger the neuronal degeneration is unlikely, as mice overexpressing normal *SOD* to a similar extent do not develop motor neuron disease. Thus, a plausible interpretation of the motor neuron disease in these mice is that the mutant CuZnSOD molecule is in some way cytotoxic.

Further support for the argument that the *CuZnSOD* mutations kill motor neurons via gain of a novel function rather than loss of function is the absence, to date, of clear null mutations in *CuZnSOD* in FALS patients. If haplo-insufficiency (50% reduction in activity from loss of the product of one allele) causes the disease, one would anticipate finding some mutations that prevent protein expression from the mutant allele (e.g., by introducing premature stop codons or deletions of the gene). Such null mutants are conspicuously absent. One candidate for a null *SOD* mutation, a 2 bp deletion, is predicted to introduce a premature stop codon that would truncate 23 amino acids from the 3' end of the molecule.[24] However, it is not yet clear whether this is a true null mutation. These considerations suggest it is unlikely that FALS is caused solely, if at all, by a decrease in CuZnSOD activity. The central question then is: what are the mechanisms of cytotoxicity of the mutant SOD1 molecule? Is it possible, as suggested,[59] that different mutations may act via different mechanisms? While experimental data concerning this point are sparse, several possibilities are noted. First, the mutant enzyme may be so unstable that it precipitates to form toxic cytoplasmic aggregates. Second, the mutant SOD molecule may have an altered substrate affinity that leads to high levels of toxic reaction products. In support of the second hypothesis, a recent in vitro study examined two different mutations that generated increased hydroxyl radicals, as measured using spin-trapping techniques.[20] It has been proposed that the mutant SOD molecule may have an enhanced affinity for peroxynitrite and, in consequence, produce

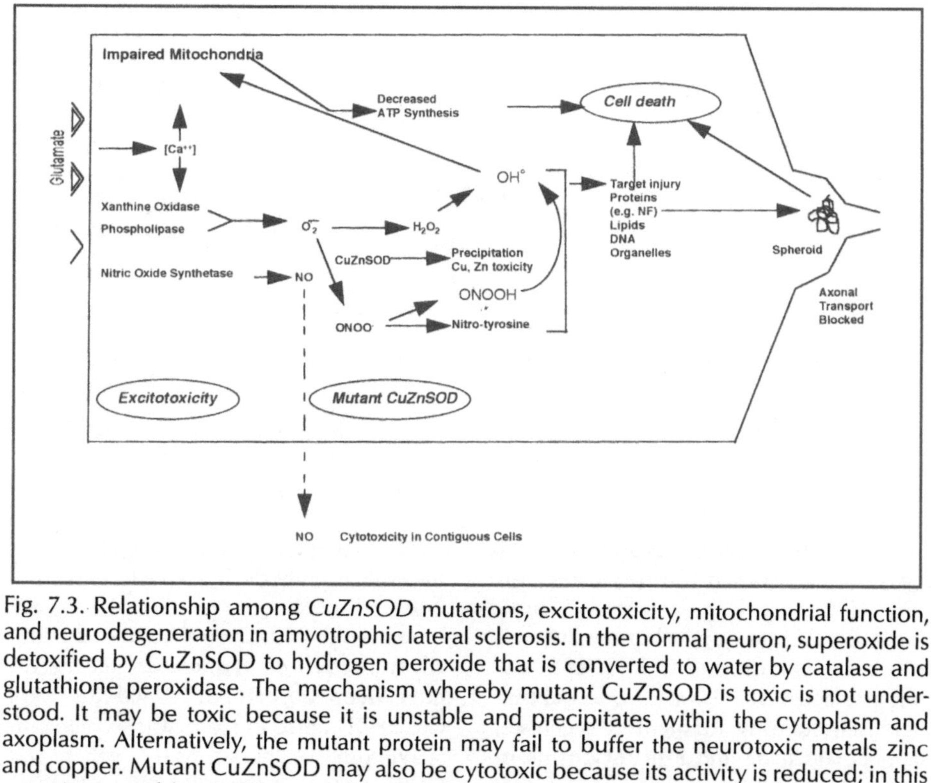

Fig. 7.3. Relationship among *CuZnSOD* mutations, excitotoxicity, mitochondrial function, and neurodegeneration in amyotrophic lateral sclerosis. In the normal neuron, superoxide is detoxified by CuZnSOD to hydrogen peroxide that is converted to water by catalase and glutathione peroxidase. The mechanism whereby mutant CuZnSOD is toxic is not understood. It may be toxic because it is unstable and precipitates within the cytoplasm and axoplasm. Alternatively, the mutant protein may fail to buffer the neurotoxic metals zinc and copper. Mutant CuZnSOD may also be cytotoxic because its activity is reduced; in this case, increased levels of superoxide anion enhance formation of peroxynitrite that may induce oxidative injury by itself or through either the formation of hydroxyl radicals or nitration of tyrosine residues on critical motor neuron proteins. One such set of protein targets are the neurofilaments (NF) whose injury (by mutation, oxidative degradation, or extrinsic insults) results in formation of fibrillary spheroids that may disrupt axonal transport. In this model, excitatory neurotransmitters such as glutamate augment any of these forms of toxicity by increasing levels of free cytosolic calcium and thereby enhancing formation of superoxide and nitric oxide and impairing mitochondrial function. This process of neuronal injury may be self-sustaining. Free radicals may compromise the structural integrity of mitochondria and the neuronal membrane. This leads, respectively, to enhanced mitochondrial release of hydroxyl radicals and other reactive oxygen species and to increased fluxes of calcium into the neuron from the extracellular space. Moreover, because it is highly lipophilic, nitric oxide can diffuse from one cell to another, thereby triggering cell-to-cell spread of the pathologic process. Reprinted with permission from Brown. Cell 1995; 80:687-692.

Table 7.1. Expression of human mutated CuZnSOD in brain of transgenic mice

Line	Mutation	SOD (U)/total protein (μg)
G1	Glu93→Ala	42.6 ± 2.1
G5		27.0 ± 2.9
G12		19.5 ± 0.8
G20		16.9 ± 0.4
A1073	Ala4→Val	14.6 ± 0.4
A1074		9.1 ± 0.4
N1029	Wild-type	37.3 ± 1.9
N1026		18.6 ± 0.9
N1030		11.8 ± 0.3
Non transgenic		10.4 ±0.5

From Gurney et al., 1994.

elevated levels of nitronium ions that, in turn, catalyze nitration of tyrosines on proteins such as neurofilaments and tyrosine kinase receptors.[60] The possibility of increased protein nitration is supported by significant increases in 3-nitrotyrosine concentrations in the spinal cords of both transgenic ALS mice and human subjects with both sporadic and *CuZnSOD*-mutation-associated ALS. It was also found that increased staining for heme oxygenase, malonedialdehyde and 3-nitrotyrosine in the anterior horn motor neurons of ALS patients and transgenic ALS mice.[61] Third, if the stability of the mutant enzyme is reduced, there may be subnormal copper or zinc buffering by the enzyme and resulting neurotoxicity from one or both of these metals. It is well established that copper is a neurotoxin;[62] indeed, as a transition metal, it can participate in several types of oxidative biochemical reactions. Similarly, zinc also has the potential to be neurotoxic.[63]

Mutations Associated with ALS Convert Superoxide Dismutase from an Antiapoptotic Gene to a Proapoptotic Gene

Two complementary systems that allow the evaluation of the in vivo effects of wild-type and mutant CuZnSOD activity: yeast mutants null for *CuZnSOD*, and conditionally immortalized mammalian neural cells have been described.[64-65] Gly85→Arg CuZnSOD failed to rescue the oxygen sensitive phenotype of *SOD*$^-$ yeast, but Gly93→Ala and Lys100→Gly CuZnSOD were apparently fully functional in vivo. The Gly85→Arg mutant have metal binding properties and CuZnSOD activity significantly altered relative to wild type. The Gly93→Ala, in contrast, demonstrated metal-binding comparable to the wild-type and activity 80% of the wild-type. These results showed that SOD activity of human FALS mutants may vary considerably in vivo, with some of them retaining a considerable amount of activity.

In conditionally immortalized mammalian neural cells, the overexpression of wild-type *CuZnSOD* inhibited apoptosis in association with an increase in CuZnSOD activity. However, expression of the FALS *SOD* mutants A4V and G37R enhanced, rather than inhibited, apoptosis in a dominant fashion, despite increasing SOD activity, and nearly as much as the wild type.[65] This finding dissociates the mutants' SOD enzymatic activity from their effect on neural cell death, showing that FALS associated with *SOD* mutations may not be simply the result of a decrease in CuZnSOD enzymatic function.

Role for Mitochondria in ALS

A role for mitochondria is supported by observations in transgenic ALS mice as well as in ALS patients. An early abnormality in the transgenic ALS mice is mitochondrial swelling, leading to splitting of the inner and outer mitochondria membrane and vacuolization.[56] Presynaptic terminals synapsing on ALS anterior horn cells show aggregated dark mitochondria.[66] Ultrastructural studies of muscle biopsies of ALS patients have shown that motor nerve terminals have significant increases in $[Ca^{2+}]i$, increased mitochondrial volume and increased numbers of synaptic vesicles relative to controls.[67]

NEUROFILAMENTS AND ALS

Another central question is why mutations in an enzyme that is ubiquitously expressed cause an illness so exquisitely specific to motor neurons, at least as gauged on clinical grounds. While no clear answer is available, it is likely that some property of the motor neuron confers a particular susceptibility to the effects of the SOD defects. Like all neurons, the motor neuron is a postmitotic, highly differentiated cell with a distinctive, polarized cellular anatomy. Unlike most neurons, motor neurons tend to be very large, with extremely high ratios of axonal length to cell width (up to 10,000:1 in the case of a lumbosacral motor neuron in an adult). This suggests that particular constituents of the motor neuron required to maintain this distinctive morphology, such as cytoskeletal proteins, may be essential for long-term motor neuron viability and may be critical targets for injury, perhaps by reactive oxygen species in FALS. Indeed, some authors report an early finding in sporadic ALS and FALS of neurofilamentous accumulations in cell bodies and proximal axons.[67] In this context, it is striking that two different mutations in the heavy chain of neurofilaments have recently been associated with ALS in three patients.[68] It remains to be established whether these neurofilament mutations will be present in large series of either sporadic ALS or FALS patients. However, it is pertinent that transgenic mice overexpressing either normal or mutant neurofilaments develop a motor neuron disease phenotype.[69,70] Moreover, mice that express a point mutation in the light neurofilament subunit reproduce even more fully aspects of human ALS pathology.[68] Further, animals intoxicated with the agent β-β-imino-dipropionitrile develop severe neurofibrillary pathology, particularly in the proximal axon, which disrupts axonal transport and thereby impairs motor neuron function.[71] Together, these findings in humans and experimental animals underscore the importance of cytoskeletal protein function for motor neuron viability.

OXIDATIVE STRESS, EXCITOTOXICITY AND MOTOR NEURON SELECTIVITY IN ALS

Both in the mouse models and in human ALS, it is likely that an additional factor beyond the complex cellular morphology and cytoskeletal function is a critical determinant of susceptibility to ALS. As stated above, data have implicated toxicity from normal excitatory inputs in the pathogenesis of ALS.[8-14] A systemic defect in glutamate metabolism has been documented in patients with ALS in whom glutamate plasma concentration was 100% higher than matched controls; these patients also demonstrated abnormal glutamate clearance following oral glutamate challenge. Elevated glutamate levels have also been detected in the cerebrospinal fluid of ALS patients.[9,11] Furthermore, post-mortem examination of neuronal tissue has revealed selective depletion of glutamate in the central nervous system.[110] At the cellular level, glutamate is rapidly eliminated from the synapse by an energy-dependent high-affinity transport system present in both astrocytes and nerve terminals. In ALS patients, glutamate transport velocity was found to be significantly impaired in the spinal cord, motor cortex, and cortex. This transport defect is due to a decrease in the relative number of transport proteins[12–14] and new evidence indicates that this decrease was due to a loss of astroglial specific transporters.[12] Whatever the underlying pathway, it is likely that glutamate contributes to ALS development and progression by accumulating to toxic concentrations at synapses and causing neuronal death.

There are several ways in which such glutamatergic excitatory synaptic input enhances the sensitivity of the motor neuron to free radical injury. The central element in neuronal excitotoxicity is an elevation of cytosolic free calcium levels that, in turn, activates calcium-dependent enzymes, including proteases and enzymes such as xanthine oxidase and phospholipase A_2 (which may elevate levels of superoxide anion) or nitric oxide synthase (which forms nitric oxide).[72] Indeed, the motor neuron may be particularly sensitive to increases in cytosolic calcium levels because, by comparison with other types of neurons, it is relatively poor in some calcium-binding proteins such as calbindin D-28k and parvalbumin.[73] At least in cultured neurons, concomitant elevation of nitric oxide and superoxide anion is a sine qua non of excitotoxicity, presumably through their combination to form peroxynitrite.[74]

Figure 7.3 from Brown[23] attempts to integrate within a single model the relationship among mutations in *CuZnSOD*, excitotoxic stimulation, mitochondrial function, and the integrity of the cytoskeleton. It offers an overview of possible mechanisms of pathogenesis (oxidative injury to or mutations in critical proteins such as neurofilaments, failure of axonal transport) and a possible basis for motor neuron selectivity (complex anatomy of the motor neuron and excitotoxicity). The model does not inherently provide an explanation for the timing of onset of the disease, although one may speculate that age-dependent factors modify the scheme.[75] For example, the normal accumulation of age-dependent oxidative injury may determine the time at which a superimposed oxidative challenge suffices to trigger irreversible pathology in the motor neuron. It is also possible that age-related decrements in mitochondrial function progressively reduce ATP production until motor neuron viability is marginal.

Three features of this model are noted. First, the processes it depicts are potentially self-reinforcing. For example, if superoxide anion or other reactive radicals impair mitochondrial function, these organelles may release reactive species that, in turn, injure other cellular structures as well as other mitochondria. Indeed, elevation of cytosolic calcium can, by itself, trigger hydroxyl radical release from mitochondria.[76–79] Further, evidence suggests that as mitochondrial function deteriorates, the potency of excitotoxic stimuli increases.[76] The possibility that mitochondrial pathology may be an element early in the pathogenesis of ALS is predicated on the observation (discussed above) that vacuolar distortion of mitochondria occurs early in the course of the motor neuron degeneration in the transgenic mice. While there is little microscopic evidence for specific pathology of mitochondria in ALS in humans, activity of mitochondrial complex I activity is increased in brains of patients with FALS.[27] The processes depicted in Figure 7.3 are also self-sustaining simply because the progression of cell death is likely to entail loss of membrane integrity and increased cytosolic free calcium levels that may, in turn, signal further cytotoxic activity. A second feature of this model is that it allows for cell-to-cell spread of the pathologic process, an aspect strongly suggested by the clinical pattern of disease progression. Thus, the model attributes a central role to nitric oxide in the development of neuronal toxicity. This highly lipophilic substance is documented in vitro to pass through membranes and cells as readily as oxygen. Once generated in injured motor neurons, it may diffuse into neighboring cells wherein it may interact with superoxide to trigger an analogous series of self-sustaining cytotoxic events. A third point inherent within this model is a role for extrinsic cellular phenomena to modulate the process. Extracellular factors that stimulate oxidative stress, such as exogenous toxins or perhaps trauma to the motor neuron or spinal cord, may trigger or accelerate this sequence of pathological reactions.

In sum, modulation of the glutamaergic system, through receptor antagonism, glutamate release inhibition, or other mechanisms may have future therapeutic benefit in the treatment of ALS.[80] The medical community is actively researching therapeutic solutions based on this approach, an approach that holds promise to all those afflicted with this devastating disease.

CONCLUDING QUESTIONS

These investigations of sporadic ALS and of FALS highlight the importance of several possible determinants of motor neuron viability, including free radical homeostasis, critical motor neuron organelles and proteins, and excitatory neural inputs on the motor neuron. Exactly how important each of these factors is should become apparent from studies addressing several critical questions. One is whether, in any form of motor neuron disease, there is unequivocal oxidative injury within the nervous system, beyond that which accumulates during normal aging. Quite simply, the question remains: is ALS a free radical disease? If FALS is caused by a novel function of the mutant CuZnSOD protein, what other genes are defective in FALS cases not arising from SOD mutations? Will these also implicate oxidative pathology? Will the pathogenetic mechanisms delineated in FALS prove relevant to understanding the more common sporadic forms of the disease? Is the motor neuron death process in ALS apoptotic or necrotic? If it is apoptotic, can it be slowed with antiapoptotic agents such as Bcl-2 or cysteine protease inhibitors? One hopes that the insights gleaned through these questions will be pertinent not only to ALS but also to other neurodegenerative diseases. Clinical and laboratory investigations have revealed several potential mechanisms of ALS initiation and progression including free radical damage, autoimmunity, and abnormal glutamate metabolism. Each of these may contribute to the overall syndrome of ALS, and pharmacologic intervention could help reduce the disease burden on ALS patients.

REFERENCES

1. Mulder DW. Clinical limits of amyotrophic lateral sclerosis. In: Rowland LP, ed. Human Motor Neuron Diseases. New York: Raven Press 1982:15-22.
2. Rowland LP. Ten central themes in a decade of ALS research. Adv Neurol 1991; 56:3-23.
3. Rosen DR, Siddique T, Patterson D et al. Mutations in Cu/Zn superoxide dismutase are associated with familial amyotrophic lateral sclerosis. Nature 1993; 362:59-62.
4. Rowland LP. Amyotrophic lateral sclerosis: human challenge for neuroscience. Proc Natl Acad Sci 1995; 92:1251-1253.
5. Figlewicz DA, Krizus A, Martinoli MG et al. Variants of the heavy neurofilament subunit are associated with the development of amyotrophic lateral sclerosis. Hum Mol Genet 1994; 3:1757-61.
6. Leigh PN, Swash M. Cytoskeletal pathology in motor neuron diseases. Adv Neurol 1991; 56:115-24.
7. Choi DW. Glutamate neurotoxicity and diseases of the nervous system. Neuron 1988; 1:623-634.
8. Perry TL, Hansen S, Jones K. Brain glutamate deficiency in amyotrophic lateral sclerosis. Neurology 1987; 37: 1845-48.
9. Perry TL, Krieger C, Hansen S et al. Amyotrophic lateral sclerosis: amino acid levels in plasma and cerebrospinal fluid. Ann Neurol 1990; 28:12-17.
10. Plaitakis A, Constantakakis E. Altered metabolism of excitatory amino acids, N-

acetyl-aspartate and N-acetyl-aspartylglutamate in ALS. Brain Res Bull 1993; 30:381-386.
11. Rothstein JD, Tsai GT, Kuncl RW et al. Abnormal excitatory amino acid metabolism in amyotrophic lateral sclerosis. Ann Neurol 1990; 28:18-25.
12. Rothstein J, Martin IJ, Kuncl RW. Decreased glutamate transport by the brain and spinal cord in ALS. N Engl J Med 1992; 326:1464-1468.
13. Rothstein JD, Jin L, Dykes-Hoberg M et al. Chronic inhibition of glutamate uptake produces a model of slow neurotoxicity. Proc Natl Acad Sci USA 1993; 90:6591-6595.
14. Rothstein JD, Van Kammen M, Levey AI et al. Selective loss of glial glutamate transporter GLT-1 in amyotrophic lateral sclerosis. Ann Neurol 1995; 38:73-84.
15. Appel SH, Smith RG, Engelhardt JI et al. Evidence for autoimmunity in amyotrophic lateral sclerosis. J Neurol Sci 1993; 118:169-174.
16. Yasui M, Yase Y, Ota K et al. Aluminium deposition in the central nervous system of patients with amyotrophic lateral sclerosis from the kii peninsula of Japan. Neurotoxicology 1991; 12:615-620.
17. Tsai G, Cork LC, Slusher BS et al. Abnormal acidic amino acids and N-acetylaspartylglutamate in hereditary canine motorneuron disease. Brain Res 1993; 629:305-309.
18. Bensimon G, Lacomblez L, Meininger V, and the ALS/riluzole study group. A controlled trial of riluzole in amyotrophic lateral sclerosis. N Engl J Med 1994; 330:585-591.
19. Gurney ME, Cutting FB, Zhai P et al. Benefit of vitamin E, riluzole, and gabapentin in a transgenic model of familial amyotrophic lateral sclerosis. Ann Neurol 1996; 39:147-157.
20. Ben Hamida M, Hentati F, Ben Hamida C. Hereditary motor system diseases (chronic juvenile amyotrophic lateral sclerosis): conditions combining a bilateral syndrome with limb and bulbar atrophy. Brain 1990; 113:347-63.
21. Siddique T, Figlewicz DA, Pericak-Vance MA et al. Linkage of a gene causing familial amyotrophic lateral sclerosis to chromosome 21 and evidence of genetic locus heterogeneity. N Engl J Med 1991; 324:1381-84.
22. Mitchell J, Gatt J, Phillips T, Houghton E et al. CuZn Superoxide dismutase and motorneurone disease. Lancet 1993; 342:1051-1052.
23. Brown RH Jr. Amyotrophic lateral sclerosis: recent insights from genetics and transgenic mice. Cell 1995; 80:687-692.
24. Pramatarova A, Goto J, Nanba E et al. A two base pair deletion in the SOD1 gene causes familial amyotrophic lateral sclerosis. Hum Mol Genet 1994; 3:2061-62.
25. Pramatarova A, Figlewics D, Krizus A et al. Identification of new mutations in the CuZnSOD gene of patients with familial amyotrophic sclerosis. Am J Human Genet 1995; 56:592-596.
26. Bergeron C, Muntasser S, Somerville MJ et al. Copper zinc superoxide dismutase mRNA levels are increased in sporadic amyotrophic lateral sclerosis motorneurons. Brain Res 1994; 659:272-276.
27. Bowling AC, Schulz JB, Brown RH et al. Superoxide dismutase activity, oxidative damage and mitochondrial energy metabolism in familial and sporadic amyotrophic lateral sclerosis. J Neurochem 1993; 61:2322-25.
28. Lanius R, Krieger C, Wagey R et al. Increased ^{35}S glutathione binding sites in spinal cords from patients with sporadic amyotrophic lateral sclerosis. Neurosci. Lett 1993; 163:89-92.
29. Ischiropoulos H, Chen J, Tsai HM et al. Peroxynitrite (ONOO-) reacts with superoxide dismutase to give the reactive nitronium ion. Free Rad Biol Med 1990; 9(Suppl.1):131.
30. Ischiropoulos H, Zhu L, Chen J et al. Peroxynitrite-mediated tyrosine nitration catalyzed by superoxide dismutase. Arch Biochem Biophys 1992; 298:431-437.
31. Huie RE, Padmaja S. The reaction rate of nitric oxide with superoxide. Free Rad Res Commun 1993; 18:195-199.
32. Beckman JS, Carson M, Smith CD et al. ALS, SOD and peroxynitrite. Nature 1993; 364:584.
33. Solodkin A, Traub RJ, Gebhart GF. Unilateral hindpaw inflammation produces a bilateral increase in NADPH-diapho-

rase histochemical staining in the rat lumbar spinal cord. Neuroscience 1992; 51:495-499.
34. Verge V, Xu Z, Xu XJ et al. Marked increase in nitric oxide synthase mRNA in rat dorsal root ganglia after peripheral axotomy: in situ hybridization and functional studies. Proc Natl Acad Sci USA 1992; 89:11617-11621.
35. Boje KM, Arora PK. Microglial-produced nitric oxide and reactive nitrogen oxides mediate neuronal cell death. Brain Res 1992; 587:250-256.
36. Nakane M, Schmidt H, Pollock J et al. Cloned human brain nitric oxide synthase is highly expressed in skeletal muscle. FEBS Lett 1993; 316:175-180.
37. Martin BL, Wu D, Jakes S et al. Chemical influences on the specificity of tyrosine phosphorylation. J Biol Chem 1990; 265:7108-7111.
38. Kong S-K, Yim MB, Stadtman ER et al. Peroxynitrite disables the tyrosine phosphorylation regulatory mechanism: lymphocyte specific tyrosine kinase fails to phosphorylate nitrated cdc2(6-20)NH_2 peptide. Proc Natl Acad Sci USA 1996; 93:3377-3382.
39. Berlett BS, Friguet B, Yim MB et al. Peroxynitrite-mediated nitration of tyrosine residues in Escherichia coli glutamine synthetase mimics adenylation: relevance to signal transduction. Proc Natl Acad Sci USA 1996; 93:1776-1780.
40. Klein R, Nanduri V, Jing S et al. The trkB tyrosine protein kinase is a receptor for brain-derived neurotrophic factor and neurotrophin-3. Cell 1991; 66: 395-403.
41. Widmer H, Kaplan D, Rabin S et al. Rapid phosphorylation of phospholipase Cγ 1 by brain-derived neurotrophic factor and neurotrophin-3 in cultures of embryonic rat cortical neurons. J Neurochem 1993; 60:2111-2123.
42. DiStefano PS, Friedman B, Radziejewski C et al. The neurotrophins BDNF, NT-3 and NGF display distinct patterns of retrograde axonal transport in peripheral and central neurons. Neuron 1992; 8: 983-993.
43. Frisen J, Verge V, Cullheim S et al. Increased levels of trkB mRNA and trkB protein-like immunoreactivity in the injured rat and cat spinal cord. Proc Natl Acad Sci USA 1992; 89:11282-11286.
44. Collard JF, Cote F, Julien J-P. Defective axonal transport in a transgenic mouse model of amyotrophic lateral sclerosis. Nature 1995; 375:61-64.
45. Tu P-H, Raju P, Robinson KA et al. Transgenic mice carrying a human mutant superoxide dismutase transgene develop neuronal cytoskeletal pathology resembling human amyotrophic lateral sclerosis lesions. Proc Natl Acad Sci USA 1996; 93:3155-3160.
46. Deng HX, Hentati A, Tainer JA et al. Amyotrophic lateral sclerosis and structural defects in Cu,Zn superoxide dismutase. Science 1993; 261:1047-51.
47. Robberecht W, Sapp P, Viaene M et al. Cu/Zn superoxide dismutase activity in sporadic and familial amyotrophic lateral sclerosis. J Neurochem 1994; 62:384-87.
48. Borchelt DR, Lee MK, Slunt HS et al. Superoxide dismutase 1 with mutations linked to familial amyotrophic lateral sclerosis possesses significant activity. Proc Natl Acad Sci USA 1994; 91: 8292-96.
49. Rothstein JD, Bristol LA, Hosler B et al. Chronic inhibition of superoxide dismutase produces apoptotic death of spinal neurons. Proc Nat Acad Sci USA 1994; 91:4155-59.
50. Troy CM, Shelanki M. Down-regulation of copper/zinc superoxide dismutase causes apoptotic death in PC12 neuronal cells. Proc Natl Acad Sci USA 1994; 91:6384-87.
51. Troy CM, Derossi D, Prochiantz A et al. Downregulation of SOD1 leads to cell death by the NO-peroxynitrite pathways. J Neurosci 1996; 16:253-261.
52. Reaume AG, Elliott JL, Hoffman EK et al. Motor neurons in Cu/Zn superoxide dismutase-deficient mice develop normally but exhibit enhanced cell death after axonal injury. Nat Genet 1996; 13:43-47.
53. Gurney ME, Pu H, Chiu AY et al. Motor neuron degeneration in mice that express a human Cu, Zn superoxide dismutase mutation. Science 1994; 264: 1772-75.

54. Price DL, Cleveland DW, Koliatsos VE. Motor neuron disease and animal models. Neurobiol Dis 1994; 1:3-11.
55. Ripps ME, Huntley GW, Hof PR et al. Transgenic mice expressing an altered murine superoxide dismutase gene provide an animal model of amyotrophic lateral sclerosis. Proc Natl Acad Sci USA 1995; 92:689-693.
56. Wong PC, Pardo CA, Borchelt DR et al. An adverse property of a familial ALS-linked SOD1 mutation causes motor neuron disease characterized by vacuolar degeneration of mitochondria. Neuron 1995; 14:1105-1116.
57. Dal Canto MC, Gurney ME. The development of central nervous system pathology in a murine transgenic model of human amyotrophic lateral sclerosis. Am J Pathol 1994; 145:1271-79.
58. Dal Canto MC, Gurney ME. Neuropathological changes in two lines of mice carrying a transgene for mutant human CuZnSOD, and in mice overexpressing wild type human SOD: a model of familial amyotrophic sclerosis. Brain Res 1995; 676:25-40.
59. Tsuda T, Munthasser S, Fraser PE et al. Analysis of functional effects of a mutation in SOD1 associated with familial amyotrophic lateral sclerosis. Neuron 1994; 13:727-736.
60. Beckman JS. The double edged role of nitric oxide in brain function and superoxide-mediated pathology. J Dev Physiol 1991; 15:53-59.
61. Beal MF. Mitochondria, free radicals, and neurodegeneration. Curr Opinion in Neurobiol 1996; 6:661-666.
62. Scheinberg H. The neuroxicity of copper. In: SC Bondy, KN Prasad, eds. Metal neurotoxicity. Boca Raton, Florida: CRC Press. 55-60.
63. Ebadi M, Murrin LC, Pfeiffer RF. Hippocampal zinc thionein and pyridoxal phosphate modulate synaptic functions. Ann NY Acad Sci 1990; 585:189-201.
64. Nishida CR, Gralla EB, Valentine JS. Characterization of three yeast copper-zinc superoxide dismutase mutants analogous to those coded for familial amyotrophic lateral sclerosis. Proc Natl Acad Sci USA 1994; 91:9906-9910.
65. Rabizadeh S, Gralla EB, Borchelt DR et al. Mutations associated with amyotrophic lateral sclerosis convert superoxide dismutase from an antiapoptotic gene to a proapoptotic gene: studies in yeast and neural cells. Proc Natl Acad Sci USA 1995; 92:3024-3028.
66. Sasaki S, Iwata M. Ultrastructural study of synapses in the anterior horn neurons of patients with amyotrophic lateral sclerosis. Neurosci lett 1996; 204:53-56.
67. Hirano A. Cytopathology in amyotrophic lateral sclerosis. Adv Neurol 1991; 56: 91-101.
68. Lee MK, Cleveland DW. A mutant neurofilament subunit causes massive, selective motor neuron death:implications for the pathogenesis of human motor neuron disease. Neuron 1994; 13:975-8.
69. Côté F, Collard JF, Julien JP. Progressive neuronopathy in transgenic mice expressing the human neurofilament heavy gene: a mouse model of amyotrophic lateral sclerosis. Cell 1993; 73:35-46.
70. Xu Z, Cork LC, Griffin JW et al. Increased expression of neurofilament subunit NF-L produces morphological alterations that resemble the pathology of human motor neuron disease. Cell 1993; 73:23-33.
71. Griffin J, Cork LC, Troncoso JC et al. Experimental neurotoxic disorders of motor neurons: neurofibrillary pathology. In: LP Rowland, ed. Human motor neuron diseases. New York: Raven Press 419-433.
72. Coyle JT, Puttfarcken P. Oxidative stress, glutamate and neurodegenerative disorders. Science 1993; 262:689-95.
73. Ince P, Stout N, Shaw P et al. Parvalbumin and calbinden D-28k in human motor system and in motor neuron disease. Neuropath Appl Neurobiol 1993; 19:291-299.
74. Dawson VL, Dawson TM, Bartley DA et al. Mechanisms of nitric oxide: mediated neurotoxicity in primary brain cultures. J Neurosci 1993; 13:2651-61.
75. Chiu AY, Zhai P, dal Canto MC et al. Age-dependent penetrance of disease in a transgenic model of familial amyotrophic lateral sclerosis. Mol Cell Neurosci 1995; 6:349-362.

76. Dykens JA. Isolated cerebral and cerebellar mitochondria produce free radicals when exposed to elevated calcium and sodium: implications for neurodegeneration. J Neurochem 1993; 63:584-91.
77. Siklos L, Engelhardt J, Harati Y et al. Ultrastructural evidence for altered calcium in motor nerve terminals in amyotrophic lateral sclerosis. Ann Neurol 1996; 39:203-219.
78. McCord JM. Mutant mice, CuZn superoxide dismutase, and motor neuron degeneration. Science 1995; 266:1586-1587.
79. Beal MF, Hyman B, Koroshetz W. Do defects in mitochondrial energy metabolism underlie the pathology of neurodegenerative diseases? Trends Neurosci 1993; 16:125-31.
80. Rothstein JD, Kuncl RW. Neuroprotective strategies in a model of chronic glutamate-mediated motor neuron toxicity. J Neurochem 1995; 65:643-651.

CHAPTER 8

Oxidative Stress in Parkinson's Disease

INTRODUCTION AND NEUROPATHOLOGY

Parkinson's disease (PD) is one of the major neurodegenerative diseases with a prevalence of approximately 150/100,000.The causes of PD remain unknown.

PD is characterized by a gradual deterioration of neurons that synthetize the neurotransmitter dopamine (DA) within the substantia nigra pars compacta of the brain. The resultant decrease of DA-containing fibers within the striatum is associated with several sensory and motor impairments, particularly tremor, rigidity and akinesia. Initially, these symptoms can be effectively controlled by oral administration of levodopa (L-dopa), a molecular precursor to dopamine that can penetrate the blood-brain barrier. As the L-dopa diffuses throughout the brain, residual nigral terminals within the striatum are capable of capturing the L-dopa and converting it to usable dopamine. As the disease progresses, however, and more nigrostriatal terminals are lost, patients typically show diminished, unpredictable and fluctuating responsiveness to L-dopa. Many patients also develop disabling dyskinesia, loss of voluntary motor control. This may reflect overstimulation of striatal cells or the inappropriate delivery of L-dopa to other brain regions.

Although other neuronal systems may be involved in PD, the main clinical and pharmacological effects are due to the death of dopaminergic neurons in the substantia nigra. Based on striatal dopamine loss at the time of presentation,[1] it is generally estimated that the patient's nigro-striatal dopaminergic neuronal population has been depleted by at least 80%. Careful analysis of a well-documented patient with PD suggests that the symptoms, and, therefore presumably the initiating factor(s), are present 5-10 years before medical advice is sought.[2] The distribution of neuronal cell loss in PD appears distinct from that of aging, with the ventral tier of substantia nigra (SN) zona compacta being most affected in PD, whilst the dorsal tier bears the brunt of pigmented cell death in aging.[3,4] Examination of the ventral tier post-mortem also reveals the presence of reactive microglia phagocytosing dopaminergic neurons in PD patients.[5] These observations taken together would suggest that a toxic process, distinct from that of aging, begins several years before presentation and is still active at the time of death.

Oxidative stress and mitochondrial dysfunction are two proposed mechanisms of cell death in PD that have actually received the most attention.[6–13]

ETIOLOGY OF PD

Oxidative damage and complex I deficiency may be important if not pre-eminent among the terminal events which are the ultimate cause of dopaminergic cell death in PD. An important question is: what initiates the cascade of reactions that results in these abnormalities?

The Role of Oxidative Stress in Neuronal Death, by Irène Ceballos-Picot.

In this, the oxidative defect or complex I deficiency may be primary or secondary and due to genetic or toxic factors.

The case for a clear genetic cause for PD is not strong. Twin studies have not shown increased concordance for PD among identical twins, although results remain inconclusive.[14] Occasional families with autosomal dominant Parkinsonism have been described.[15] There is no evidence for maternal inheritance in PD[16] and this is supported by failure to identify any MtDNA mutations of significance in PD substantia nigra.[17–19] An increase in the incidence of the 2D6 allele of cytochrome P_{450} in PD patients would suggest that an abnormality of xenobiotic metabolism might lead to the production of an endogenous toxin.[20,21] Indeed previous studies have demonstrated changes in the handling of sulphur compounds[22] and an increase in the production of *N*-methylated compounds[23] in PD patients. Superimposed upon these possibilities is the occasional report of parkinsonism induced by environmental toxins, some of which are identified (MPTP, carbon monoxide, manganese) or unidentified, e.g., those associated with well-water drinking and rural living.[24]

THE "FREE RADICAL HYPOTHESIS"

According to this so called "free radical hypothesis,"[6–11] the degeneration of the nigro-striatal system in PD is related to the relatively high exposure of these neurons to ROS, in particular H_2O_2 produced during both the enzymatic (monoamine oxidase-catalized) and nonenzymatic (auto-oxidative) breakdown of DA. Thus, DAergic cell death in PD may be caused by an overproduction of ROS and/or diminished protection against them.

There is growing evidence that oxidative damage and mitochondrial respiratory failure with attendant decrease in energy outpout are implicated in nigral neuronal death in PD. One of these indices, preceding the loss of DAergic neurons, is a decrease in the level of GSH in the Parkinsonian SN[25-27] and a mitochondrial complex I deficiency.[12,28–33]

To prevent breakdown of DA before its release from presynaptic endings, the transmitter is taken up and stored in presynaptic granules. This process is effected through the action of a transporter protein. Interference with granular storage, for instance via blockade of the ligand-binding site on the transporter, leads to an enhanced turnover of DA, inducing considerable oxidative stress as a consequence of H_2O_2 production.[33] Apart from generating oxidative stress, H_2O_2 is able to inhibit storage of DA in the nigro-striatal system through an as yet undefined interaction with the granular uptake mechanism.[34] Considering the observation that H_2O_2 preferentially interferes with presynaptic storage of DA and mitochondrial function and the fact that, in the mammalian brain, H_2O_2 is detoxified mainly by GSH whose levels are decreased in PD, recent experiments showed a direct link between granular uptake of DA and GSH deficiency in PC12 cells.[35]

Thus, the degeneration of the nigro-striatal system in PD may be the outcome of a self-sustaining cycle that is initiated by lowering of the capacity to scavenge the H_2O_2 released during normal DA metabolism, leading to a decrease in the granular storage and enhanced breakdown of DA, inducing more oxidative stress and cell death.

EVIDENCE OF OXIDANT STRESS IN THE SUBSTANTIA NIGRA

The theory that free radicals may cause or promote neuronal loss in PD was advanced over 10 years ago. The theory is based on a combination of several factors, all of which co-exist in the PD nigra and may promote free radical generation.

METABOLISM OF DOPAMINE AND FORMATION OF NEUROMELANIN

Dopamine is oxidized both enzymatically and nonenzymatically. Monoamine oxidases (MAO)A and B are both involved in the deamination of DA to 3,4-dihydroxyphenylacetaldehyde in the substantia nigra. H_2O_2 is a by-product of this reaction and this has suggested a mechanism for the production of oxidative stress in PD.[36–38] Dopamine also undergoes direct auto-oxidation with the formation of quinones, H_2O_2 and oxy-radicals.[39,40]

Neuromelanin is formed by the autooxidation of DA[39] and is found in high concentrations in the dopaminergic neurons of the substantia nigra. There is at present some debate as to whether neuromelanin-containing nigral cells are more likely to degenerate in PD.[4,41] Recently, it was shown that the iron enzyme tyrosine hydroxylase (TH), confined to dopaminergic and noradrenergic neurons/neuroendocrine cells, also catalyzes the generation of ROS.[42] The active site of TH contains redox-active nonheme iron and probably forms highly reactive dioxygen derivatives as intermediates in the hydroxylation reaction. A partial uncoupling of tetrahydropterin oxidation from hydroxylation of L-tyrosine is demonstrated for the recombinant human enzyme, promoting a Fenton type reaction.

Role of Iron in the Genesis of Oxidant Stress

Several studies have now demonstrated an increase in brain iron in PD[25,43–46] and its potential role in the etiology of PD has been reviewed.[47] An increase in iron, however, is seen in several neurodegenerative disorders and may be a nonspecific secondary phenomenon of neuronal cell loss. However, the potential for iron available in a free and reactive form to participate in oxidative reactions will depend upon its quenching by ferritin. One study has shown a decrease in nigral ferritin in PD,[49] whilst another found an increase.[25] Patient selection and antibody specificity may have accounted for some of the differences in these reports. The most recent analysis of iron and ferritin in PD substantia nigra using several anti-ferritin antibodies confirmed the earlier reports of increased iron but found no change in ferritin concentrations.[27] If free iron is available in the PD nigra, it could enhance free radical production through Fenton and Haber-Weiss type reactions.The hydroxyl radical produced by these reactions is short-lived but highly damaging. The finding that iron may bind to neuromelanin[46] enhances the potential for these factors to act together to produce free radicals.

Thus all these factors, dopamine, neuromelanin and iron, are capable of producing oxidative stress in the nigra in a complementary and amplifying manner.

Alterations in Glutathione Levels in the Substantia Nigra

Perry et al were some of the first to provide evidence for oxidative stress in postmortem PD substantia nigra by demonstrating a decrease in GSH concentration.[26,27] GSH, like ascorbate and α-tocopherol, is a reducing agent and a decrease in its concentration would suggest enhanced free radical production. This observation has subsequently been confirmed by others[25] and appears to be specific for the PD substantia nigra.

One study has reported a decrease in levels of GSHPx in PD nigra, putamen and globus pallidum.[47] A decrease in GSHPx would suggest a limitation to the metabolism of H_2O_2. In contrast, an increase in the activity of mitochondrial MnSOD has been identified in PD substantia nigra.[48] This would suggest an increase in the requirement or capacity to metabolize superoxide ions.

There is very little evidence available thus far for direct damage of biomolecules in the PD brain. One report has demonstrated an increase in lipid hydro-peroxides in PD substantia nigra[50] and this is associated with an increase in malondialdehyde levels.Thus the evidence for oxidative damage in PD is mainly indirect. Even the premise that dopamine oxidation may contribute to oxidative stress is not universally accepted. For instance, it might be expected that the use of L-dopa for symptomatic relief in PD patients might exacerbate oxidative stress, promote dopaminergic cell death and accelerate the course of the disease. However, there is no hard evidence to support this supposition, and indeed more evidence against it. Although L-dopa and dopamine may be toxic to dopaminergic neurons in culture[51,52] and perhaps decrease their survival in the ventral tegmental area of rats previously treated with 6-hydroxydopamine,[53] there is no evidence that it accelerates disease progression in patients. Furthermore, chronic administration of L-dopa to rats or mice does not lead to loss of dopaminergic neurons.[26,54] However, it can be argued that L-dopa will only participate in a cycle of free radical production and cell damage that has already been established by other factors. The lack of disease acceleration due to L-dopa, and the reversibility of even its side-effects, argue against even this hypothesis.

Differential Cellular Localization of CuZnSOD and GSHPx in the Human Substantia Nigra

CuZnSOD gene expression was investigated at the cellular level by in situ hybridization in the human mesencephalon.[6] The labelled cells were densely distributed in substantia nigra, pars compacta. No labelled cells were seen in the cerebral peduncle and superior cerebral peduncle. The grains were densely packed over the cytoplasm of the melanized neurons (Fig. 8.1).

Topography of each cell type (the CuZnSOD containing cells and the neuromelanin + CuZnSOD containing cells) was obtained with a computer assisted microscope system in which each cell type was plotted only once (Fig. 8.2a) Among positively hybridized cells, 95% were the large neuromelanin containing neurons (Fig. 8.2b).

These data indicate that the CuZnSOD gene is preferentially and highly expressed in these neurons, a subset of cells vulnerable to degenerative processes in PD. The presence of neuromelanin, due to autooxidation of dopamine, is in favor of a high local production of ROS in these cells. These results suggest that within these cells biochemical pathways leading to generation of superoxide anions are especially active, thus needing a high CuZnSOD content to facilitate removal of these radicals. Alternatively, a high cellular CuZnSOD activity, by promoting H_2O_2 production, might also contribute to the vulnerability of these neurons.

In the parkinsonian mesencephalon, the density of GSHPx-positive cells increased in relation to the severity of the neuronal injury.[55] This may be an attempt to protect the still-surviving dopaminergic neurons against an excess of toxic species. Such a buffer function of GSHPx-positive cells would be consistent with the decreased content found on homogenates of parkinsonian substantia nigra. These data contrast with the absence of increase in biochemical activity of the enzyme on tissue homogenate.[48] This discrepancy may be explained by the fact that estimation of GSHPx activity on tissue homogenates from substantia nigra

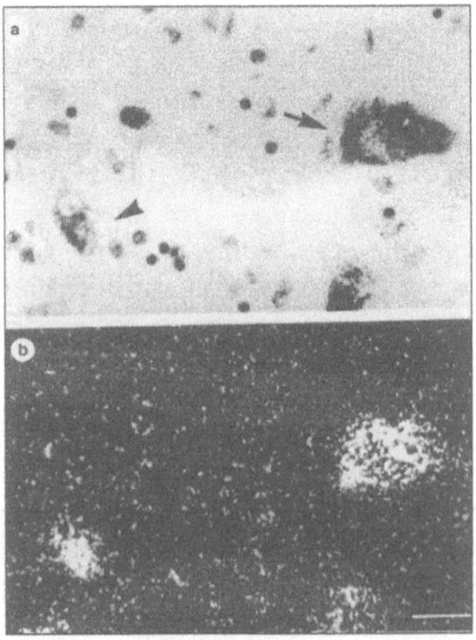

Fig. 8.1. Photomicrograph of neurons of the substantia nigra, pars compacta after hybridization with ^{35}S-labeled CuZnSOD human cDNA probe viewed using bright-field (a) and dark-field (b) illumination of grains over the neuromelanized neurons. The dark deposit is constituted only by neuromelanin.

provides average measurements of variations on heterogenous populations of cells, but cannot inform on changes at cellular resolution. Thus, a possible increase in enzyme activity resulting from the GSHPx-positive gliosis may be important at a cellular level, but slight at a regional level and therefore, difficult to detect with a global assay of enzymatic activity.

Increase in the Biomarkers of Oxidative Stress in the Substantia Nigra

Dexter et al[50] demonstrated an increase in basal levels of malondialehyde, a product of lipid peroxidation in the substantia nigra and suggested that lipid peroxidation could be a cause of nigral cell death. A histochemical approach was utilized to examine the site of oxidative damage associated with PD. Specifically, antiserum raised against 4-hydroxy-2-nonenal (HNE)-modified proteins, which recognizes HNE-protein epitopes, was used to detect HNE-protein adducts in midbrain sections of normal individuals and of patients with PD.[56] Fifty-eight percent of nigral neurons were positively stained for HNE-modified proteins in PD; in contrast only 9% of nigral neurons were positive in the control subjects. HNE is an α,β-unsaturated aldehyde that is produced during oxidation of membrane lipid polyunsaturated fatty acids. It is one of the major products of membrane peroxidation, and is considered to be largely responsible for cytotoxic effects observed under oxidative stress. Thus, HNE is considered to be the most reliable index of lipid peroxidation and the significantly higher proportion of nigral melanized neurons positively stained for HNE-modified proteins in PD indicates a significant increase of lipid peroxidation in this disorder. Glial cells were not positively immunostained, indicating that lipid peroxidation is taking place mostly within nigral neurons in PD.

These results indicate that lipid peroxidation is increased within nigral neurons in PD. This could seriously disrupt many important membrane functions such as oxidative phosphorylation, signal transduction, and regulation of electron, iron and metabolite transport and thus contribute to nigral neurodegeneration in PD. Future research

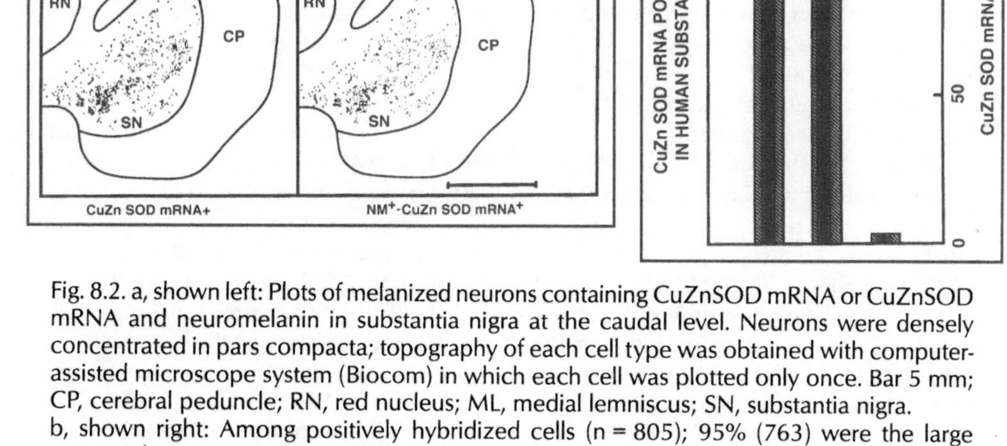

Fig. 8.2. a, shown left: Plots of melanized neurons containing CuZnSOD mRNA or CuZnSOD mRNA and neuromelanin in substantia nigra at the caudal level. Neurons were densely concentrated in pars compacta; topography of each cell type was obtained with computer-assisted microscope system (Biocom) in which each cell was plotted only once. Bar 5 mm; CP, cerebral peduncle; RN, red nucleus; ML, medial lemniscus; SN, substantia nigra.
b, shown right: Among positively hybridized cells (n = 805); 95% (763) were the large neuromelanin-containing neurons.

to develop drugs or a better nutrition based on polyunsaturated fatty acid-rich oils or meals would be pertinent to overcome lipid peroxidation within nigral neurons and such drugs or nutrition strategies may turn out to be a neuropreventive method for PD.

MITOCHONDRIAL DYSFUNCTION IN PARKINSON'S DISEASE

The idea that mitochondrial dysfunction may play some role in the pathogenesis of PD came from study of the action of the neurotoxin MPTP.

Action of MPTP

MPTP induces parkinsonism in man and other primates[57,58] through a selective uptake and conversion pathway which targets the dopaminergic neurons of the substantia nigra. In primates, it produces hypokinesia, bradykinesia, increased muscle tone, and resting tremor, the cardinal clinical manifestations of PD. These are accompanied by depletions of striatal dopamine as well as a loss of tyrosine hydroxylase-positive neurons in the substantia nigra. MPTP is first converted to 1-methyl-4-phenylpyridinium (MPP^+) by MAO_B.[59] MPP^+ is selectively accumulated into dopaminergic neurons by the dopamine re-uptake pathway[60] and is then accumulated into mitochondria by an energy dependent system for concentrating lipophilic cations.[61]

The pathogenesis of the lesions induced by MPP^+ involves inhibition of complex I of the electron transport chain as well as free radical production.[62] The consequent fall in ATP levels correlates with the action of MPP^+ in killing cell cultures.[63] The precise mechanism by which MPP^+ inhibits complex I is unclear but there is evidence that it binds to the protein at or near the rotenone site.[64] There is increasing evidence that MPP^+ may also be involved in free radical production. The binding of MPP^+ to complex I leads to a progressive decline in enzyme activity which is irreversible but can be prevented by free radical scavengers.[65] Also, the toxic effects of MPP^+ can be prevented in transgenic mice with increased superoxide dismutase activity[66] and MPTP toxicity is enhanced in brains of mice with a genetic deficiency in the MnSOD,[67] suggesting a major role of superoxide or its related products such as peroxynitrite in MPTP toxicity.

A consequence of energy defect as a result of MPTP toxicity may be activation of excitatory amino acid receptors, followed by an influx of intracellular calcium. This can lead to both increased free radical production by mitochondria as well as activation of NOS. NO has been implicated in cell death following excitotoxic insults both in vitro and in vivo. The critical mediator appears to be peroxynitrite, which produces oxidative damage to lipids, protein and DNA and also to mitochondria.

Complex I Deficiency

The demonstration of a 30-38% deficiency of complex I activity in PD substantia nigra provided a direct biochemical link between the MPTP neurotoxin model and the idiopathic disease.[29,30] Analysis of other brain areas in PD failed to show any consistent abnormality of mitochondrial function, implying that the complex I deficiency in PD is localized to the substantia nigra.[31]

An obvious concern, however, is whether the complex I defect is due to L-dopa therapy. Also, levels of L-dopa toxic to a human neuroblastoma cell line did not affect complex I activity, although activities of complexes II/III were increased.[68] However, the chronic administration of L-dopa to control rats produced a 20-25% decrease in substantia nigra and striatal complex I activity.[69] This decline in striatal complex I activity is not seen in PD caudate or putamen.[31] As has been discussed above, chronic L-dopa does not cause dopaminergic cell death in rats, suggesting that the decline in complex I nigral activity observed above is insufficient to compromise the metabolism of these neurons and induce cell death. In the context of PD, it could be argued that the L-dopa exacerbates an underlying complex I deficiency.

There has been much controversy concerning the tissue distribution of the mitochondrial defect in PD.[28] Recently further studies have found no defect of mitochondrial function in muscle,[70] whilst another found multiple complex I-IV defects.[71] Likewise the presence of a complex I defect in PD platelets remains controversial. The simplest explanation for these often conflicting results is that while there may be uniformity in the presence of a complex I defect in nigra, the variable expression of the defect in other tissues is a reflection of the potential heterogeneity of the disease in terms of etiology. Thus, some patients (a minority) may have a systemic complex I defect resulting from a genetic defect or systemically distributed and active toxin. The predominance of the deficiency in the nigra, however, reflects a summation of all the other biochemical abnormalities present in that area. It is this latter concept which has particular appeal, for it seems increasingly likely that all the factors outlined thus far may all target the nigra to induce a cycle of oxidative stress, oxidative damage and complex I deficiency.

Free Radicals and Respiratory Chain Dysfunction

The respiratory chain and oxidative phosphorylation system are, under aerobic conditions, a potent source of free radicals. In normal circumstances, the superoxide radicals generated are tightly bound to the chain and metabolized by the mitochondrial anti-oxidant system. Inhibition of complexes I or III, however, results in an increase in superoxide generation.[65,72,73] Increased free radicals can also damage the respiratory chain directly and impair function. In vitro studies with isolated mitochondria have suggested that complex I is the most sensitive to oxidative damage, although multiple deficiencies were often produced.[74–76] In vivo and cell culture studies, however, have shown that complex IV appears the most vulnerable, being the part of the respiratory chain affected first and most severely.[77–79]

These observations would suggest that oxidative damage in the PD substantia nigra would be an unlikely cause of the pure complex I deficiency observed. The complex I defect, however, could result in increased free radical production. In this situation, in vitro studies suggest that a cycle may be generated in which complex I activity is further decreased by oxidative damage, resulting in more free radical production. This cycle would be enhanced by the presence of free nigral iron.

NEURONAL CONSEQUENCES OF A SELECTIVE DEPLETION OF GLUTATHIONE IN EXPERIMENTAL MODELS

Relationships Between GSH, Oxidative Stress and Dopamine Deficiency

It is well established that GSH plays a fundamental role in protecting cells from damage by H_2O_2 or organic peroxides.[33,80-82] In addition, GSH is an antioxidant that can reduce reactive free radicals, including semiquinones and the hydroxyl radicals. Numerous experiments have been described in which depletion of GSH leads to oxidant damage[83–85] and apoptosis (see chapter 3). On the other hand an elevation of GSH can exert a protective effect that counteract an oxidative stress.[80,85–88]

In experimental studies, treatment with the neurotoxin MPTP, which results in damage to nigrostriatal dopaminergic neurons, was found to cause a decrease in GSH levels in C_{57}Bl mice.[82] This decrease was preventable by treatment with antioxidants, which also prevents the DA loss seen in the striatum and substantia nigra following MPTP administration. Decreases in GSH levels in the brain of young mice had profound morphological effects, including decreased catecholamine fluorescence per cell, increased levels of lipid peroxidation and lipofuscin accumulation, and increased numbers of dystrophic axons in dopaminergic neurons of the nigrostriatal pathway.[89] These morphological changes in the murine nigrostriatal system were similar to those previously reported to occur during normal rodent aging[90] and showed that reducing GSH content in the brain of young mice produces morphological and biochemical effects that are strikingly similar to those seen during normal aging.

Studies on fetal rat mesencephalic cultures showed that addition of L-dopa to cell cultures induced an increase in GSH levels.[86,87] The same results were observed for mouse neuroblastoma (neuro-2A), human neuroblastoma (SK-N-MC), and glia from newborn rat brain.This ability to upregulate GSH may serve a protective role in vivo. When mixed cultures were pretreated with both L-DOPA and ascorbate, which prevents the rise in GSH level, protection was lost, in accord with the failure to upregulate GSH. These results show that the upregulation of cellular GSH evoked by autooxidable agents is associated with significant protection of cells. Therefore, an oxidative mechanism leading to the rise of GSH content is implicated. Consistent with this possibility is the rise in levels of glutathione disulfide (GSSG), that precedes the rise in levels of GSH.[87] Glia play an essential role in the response of mesencephalic cell cultures.[86,91] Since the GSH is present in high concentration in astrocytes but not in neurons[92–96] and maintenance of neuronal GSH is realized by glial cells.[97]

An oxidant stress can upregulate cystine transport in cell cultures, which is the normal precursor for GSH. Recently, Sagara et al[97,98] described a transport system present in both neurons and glia that takes up cysteine. Cystine transport by way of comparison is much more prominent in glia. Following the reduction of cystine to cysteine within the glia, a portion is secreted into the medium, where it can be taken up by the neurons. In this way glia will support the biosynthesis of GSH by neurons in cell culture. The failure of pure cultured neurons to upregulate GSH may be due, in part, to the absence of glia. Up-regulation of GSH provides an antioxidant buffer for peroxides. In a paradoxical way, this antioxidant effect can be ablated by another antioxidant, ascorbate.[86,99] The reserpine-mediated emptying of the granular DA stores was accompanied by a substantial and sustained rise in the intracellular GSH content in PC12 cells.[35] Using the GSH precursor cystine to stimulate GSH formation, the extent of this reserpine-induced effect appeared to depend on the amount of GSH being synthetized in the cell and suggested the existence of a distinct pool of cytoplasmic GSH intimately linked to the DA storage compartment of PC12 cells. Alternatively lowering of PC12 GSH content, via blockade of its synthesis with BSO, led to a significantly decreased accumulation of exogenous DA without affecting uptake of the acetylcholine precursor, choline. These data suggest that GSH is involved in the granular storage of DA in PC12 cells and that, considering the molecular characterisitics of the granular transport system, it is likely that GSH is used to protect susceptible parts of this system against oxidative damage. Although the described experiments demonstrated a relationship between the granular storage of DA and cellular GSH homeostasis, additional data were needed to establish whether GSH itself plays a role in the granular transport of DA.

Taken together, these results strongly suggest that GSH is involved in the uptake of DA in the chromaffin granules and it is conceivable that oxidative damage to the proton pump due to GSH deficiency, and the consequent shift in the neuronal redox balance, forms the basis of inhibition of the vesicular storage of DA in the nigrostriatal system. Thus, an ability to upregulate GSH or other antioxidant mechanisms in the nervous system, or a failure to do so, may affect aspects of neuronal viability in vivo.

Mitochondrial Impairment Is an Early Event in the Process of Apoptosis Induced by Glutathione Depletion in Neuronal Cells: Relevance to Parkinson's Disease

An approach to determining the role of GSH in mitochondrial function and in neurodegeneration was to create a selective depletion of GSH in a neuronal cell line in culture by inhibiting its biosynthesis with BSO, a specific inhibitor of γ-glutamylcysteine synthetase. This treatment led to a nearly complete GSH depletion and induced cellular death via an apoptotic pathway after 5 days of BSO treatment (see chapter 3). By using the ROS-sensitive probe 2',7'-

dichlorofluorescin (DCF), we observed that the rapid GSH depletion is accompanied, early in the process, by a strong intracellular increase in ROS evidenced after one hour with BSO, culminating after 3 hours when GSH levels are down to about 30% of normal (Fig. 8.3).

GSH depletion induced a loss of mitochondrial function after 48 hours of BSO treatment. In particular, the activities of complexes II and IV of the respiratory chain were decreased by 70 percent as compared to controls (Fig. 8.4).

These results showed the crucial role of GSH for maintaining the integrity of neuronal mitochondrial function. Oxidative stress and mitochondrial impairment were upstream of DNA fragmentation in the apoptotic cascade and are, thereby, early events in the neuronal apoptotic process. Our data are consistent with the hypothesis that GSH depletion contributes to neuronal apoptosis in PD through mitochondrial dysfunction. Regulation of neuronal redox status may, therefore, prove to be a useful stategy to prevent mitochondrial impairment and excessive apoptosis in Parkinson's disease.

Recently, another study showed that the glutathione redox system is important for the viability of rat and mice embryonic dopamine neurons in vitro.[85,89] GSH depletion was achievied by exposing dissociated mesencephalic cell cultures to BSO. Consequently, dopamine neuron survival was significantly reduced by 65-94% in a concentration-dependent manner by 10-1000 μM BSO. The neurotoxic effects of BSO were completely prevented by supplementing the culture medium with lazaroid, which does not modify the GSH depletion.[85] Also in this insult model of severe GSH depletion, the lazaroid can afford efficient neuroprotection that does not seem to be mediated by a direct interaction with BSO or GSH.[100,101] Despite increasing evidence suggesting that the

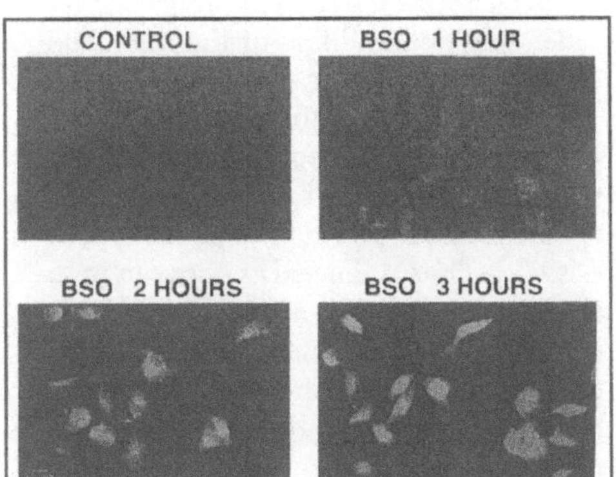

Fig. 8.3. Time-course of reactive oxygen species generation as a consequence of glutathione depletion. Note that BSO induces a time-dependent increase in DCF fluorescence (Dichlorofluorescin, a ROS-sensitive probe). A maximum DCF fluorescence is detected after 3 hours of BSO treatment when GSH levels are down to 30%.

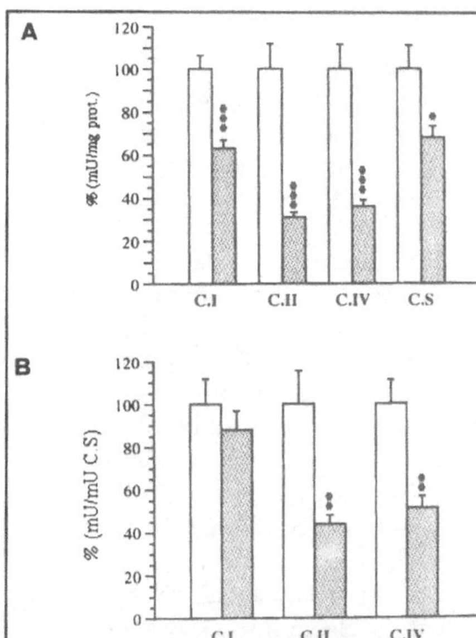

Fig. 8.4. Activities of respiratory chain complexes after two days of GSH depletion in living NS20 Y cells. A: data expressed as mU per milligram of proteins. B: data expressed as mU per mU of citrate synthase. Data are means± SD (n = 6). The experiment was duplicated. * $P < 0.05$; ** $P < 0.01$; *** $P < 0.001$. CI (complex I; NADH: ubiquinone oxidoreductase); CII (complex II; succinate: ubiquinone oxidoreductase); CIV (complex IV; cytochrome oxidase); CS (citrate synthase).

glutathione redox system can be involved in dopamine neuron death in human disease, relatively little experimental work has addressed this issue. It has recently been shown that GSHPx and NAC are both able to increase the survival of cultured mesencephalic dopamine neurons.[88]

Insight into the mechanism by which the GSH depletion leads to ROS production may also be gleaned from the observation that GSH depletion is accompanied by increased neuronal nitric oxide synthase activity and nitric oxide (NO) production.[102] DCF can also detect NO and related compounds[103] it is thus possible in this model that ROS detected by DCF-DA were NO and related compounds such as peroxynitrite. GSH depletion could intervene in the complex redox chemistry of NO because a simple change in the ambient redox milieu could allow NO interconversion from a neuroprotective to a neurotoxic species.[104] This finding has implications for the understanding of the pathogenesis of Parkinson's disease since it is now demonstrated that GSH depletion[81] and NO/peroxynitrite induce neurotoxicity through an apoptotic mechanism.[104–107]

Mitochondrial damage was an important consequence of glutathione deficiency in many tissues, particularly in the brain and neuronal cells.[84,108–109] The intracellular ROS accumulation could be responsible for the mitochondrial abnormalities observed. All the complexes of the electrons transport chain are reported to be susceptible to damage by oxygen free radicals, in vitro.[65,75–79] Producing oxidative stress in the rat brain leads to sequential damage to the components of the electron transport chain, with complex IV appearing to be the most susceptible.[85] Previous findings reported that the mechanism for NO-mediated cytotoxicity may arise as a result of energy depletion due to inactivation of the mitochondrial respiratory chain in cultured astrocytes and neurons,[110,111] cytochrome oxydase being the most affected.[112,113] A relationship betwen oxidative stress and mitochondrial damage has been reported in other model systems; copper overloading in rats results in loss of hepatic GSH and diminished complex IV activity;[114] iron loading of PC12 cells leads to loss of cellular GSH, complex I and IV activities.[115] This cytochrome oxidase deficiency could contribute to the inhibition of complexes I and II by further promoting free radical or peroxynitrite formation, or both. Whether such mechanisms are relevant to PD obviously requires further investigation, but it is of interest to note that inhibition of complex IV is required for the irreversible loss of complexe I activity in a chemical model of PD.[65] The complex deficiencies could also arise from MtDNA mutations which associated with oxidative stress and apoptotic cell death.[116,117] A consequence of this impaired mitochondrial function may be cell death due to energy deficiency. Neuronal cells have a relative paucity of GSH and need high energy production; hence they may be particularly vulnerable to the effects of GSH depletion.

Thus, we can conclude that GSH depletion being an initial insult, a cascade of events involving increase in ROS production and then mitochondrial dysfunction could contribute to cell injury and finally to apoptosis. Thus, mitochondrial dysfunction could be a critical step in the apoptotic cascade. Specific inhibitors of the mitochondrial respiratory chain induce apoptosis in different cultured mamallian cells[118] and in neuronal cells.[119] Recent demonstration of the mitochondrial control of nuclear apoptosis also argues in favor of this hypothesis.[120,121] Thus, a link exists between mitochondrial defect and oxidative stress; GSH deficiency could represent such a link. Oxidative stress and mitochondrial dysfunction could act together in determining neurodegeneration rather than being two conflicting hypotheses on the pathogenesis of PD.

Regulation of neuronal redox status may, therefore, prove to be a useful stategy to prevent mitochondrial impairment and excessive apoptosis in PD.

NEW "RADICAL" DIRECTIONS IN PARKINSON'S DISEASE

Copper-Zinc SOD Overexpression Enhances Survival of Transplanted Neurons in a Rat Model of Parkinson's Disease

Research with animal models of PD has offered a strategy for maintaining and localizing DA delivery within the damaged striatum. Tissue dissected from the embryonic mesencephalon, the brain region containing developing substantia nigra neurons, can be transplanted directly into the DA-depleted striatum. The implanted cells survive quite well in the new environmemnt, extend fibers into the host striatum, and increase DA levels around the graft. Promising results with fetal substantia nigra transplants in rats with experimental PD encouraged the development of the grafting approach for Parkinsonian patients. Results from clinical trials have been mixed; some patients significantly improved following the transplantation procedure whereas others have shown only minor changes after surgery,[122] showing that the number of surviving nigral neurons within a graft and the extent of dopaminergic fiber penetration into the host striatum both correlate with the functional efficacy of the transplants. Nakao and colleagues[123] describe, using hSOD transgenic mice, the adverse effects of oxidative stress on grafted fetal substantia nigra neurons, and suggest an approach for significantly improving the survival of embryonic dopamine-producing cells following transplantation. They found that nigral cells from hSOD transgenic embryos maintained a much higher level of viability over a 36-hour period in vitro than those from nontransgenic littermates. For in vivo assessments, cells were transplanted into the dopamine-depleted striatum of immunosuppressed rats with experimental PD. Five weeks post-implantation, transgenic grafts contained a 4-fold greater number of surviving dopamine neurons and more robust dopamine fiber outgrowth than grafts containing non-transgenic cells. Furthermore, rats implanted with the CuZnSOD transgenic nigral neurons showed a profound reduction in abnormal motor behavior 3 weeks before the first behavioral improvements occurred in rats transplanted with non-transgenic cells. Since the only difference between grafted cells was the presence of the SOD gene, these results strongly suggest that free radicals, generated either during the preparation of mesencephalic tissue for transplantation or after nigral cells are implanted, severely compromised the survival of developing dopamine neurons.

Thus, one clear strategy for improving the viability of grafted fetal tissues is to treat cells with antioxidant agents. This approach was tried in a recent study by using 21-aminosteroids,[124] agents that block free radical disruption of membrane lipids. Administration of antioxidants increased both in vitro and in vivo viability of grafted dopamine neurons, but to a lesser extent than observed in the study using the hSOD neurons from trangenic mice. The SOD work thus help to clarify the role of superoxide and its metabolites in neuronal toxicity. Interestingly, ROS induce the expression of several growth factors in neural tissues,[125,126] suggesting that trophic proteins may also offer some protection from oxidative stress. Basic fibroblast growth factor (b-FGF) in particular has been shown to have potent effects on grafted dopamine neurons.[127] As seen with the CuZnSOD grafts, b-FGF treated grafts induced a more profound and rapid behavioral recovery than control grafts. SOD as b-FGF could modify the number of excitatory amino acid receptors synthetized by neurons, limiting their responsiveness to potential excitotoxic agents.[128]

Copper-Zinc SOD Prevents the Mitochondrial Dysfunction Induced by GSH Depletion

In the brain of mice, BSO also induced a profound decrease in complexes I, II and IV of the respiratory chain. Insight into the mechanism by which the GSH depletion leads to mitochondrial complex abnormalities may be gleaned from the observation that overexpression of copper-zinc SOD in the brain of transgenic mice (50% increase in the brain of newborn transgenics) confers complete protection against the toxic effect of GSH depletion in mitochondrial complexes (Fig. 8.5).

BSO treatment was associated with a marked reduction of brain GSH and this depletion was not altered in the transgenic brains. We conclude that in the present insult model of severe GSH depletion, the CuZnSOD overexpression can afford efficient protection that does not seem to be mediated by a direct interaction with BSO or GSH. These results strongly implicate superoxide or its metabolite peroxynitrite in the mitochondrial impairment induced by GSH depletion.

Inhibition of Nitric Oxide Synthase Prevents Parkinsonism

Inhibition of neuronal NOS with the NOS inhibitor 7-nitroindazole (7-NI) prevents MPTP-induced parkinsonism in baboons.[129] Administration of 7-NI produced complete protection against the MPTP-induced dopamine depletion. 7-NI also produced significant protection against depletion of both Nissl and tyrosine hydroxylase-positive neurons in the substantia nigra. In mice it dosage-dependently blocks MPTP-induced dopamine depletion as well as increases in 3-nitrotyrosine, a marker of peroxynitrite induced damage.[130] Furthermore, MPTP-induced dopamine depletion was reduced in neuronal NOS knock out mice.[131] A role for peroxynitrite is also supported by the observation that SOD transgenic mice are resistant to MPTP neurotoxicity,[66] since scavenging superoxide will prevent the generation of peroxynitrite.

These results showed that NO and particularly peroxynitrite play a role in MPTP neurotoxicity and probably in PD neuropathogenesis and suggest that inhibitors of NOS might be useful in slowing or halting neuronal degeneration in Parkinson's disease. This is a major clinical challenge.

PREVENTIVE OR THERAPEUTIC IMPLICATIONS

Direct implications of the studies on dopamine graft survival include deciphering the pathogenesis of PD and developing rational mechanistic interventions in the neurodegenerative process. The multicenter, controlled clinical trials with antioxidant therapy for Parkinson's disease DATATOP (Deprenyl and Tocopherol Antioxidative Therapy of Parkinsonism), conducted by the Parkinson study group has indicated that such agents can successfully slow disease progression in patients with early PD,[132] but until now no work has described the preventive role of antioxidants in the development of this disease. This could be a way for the prevention of this dopaminergic neuronal-damaging disease.

Fig. 8.5. Activities of respiratory chain complexes after five days of GSH depletion in the brains of control (dashed bar) and hSOD transgenic mice (solid bar).

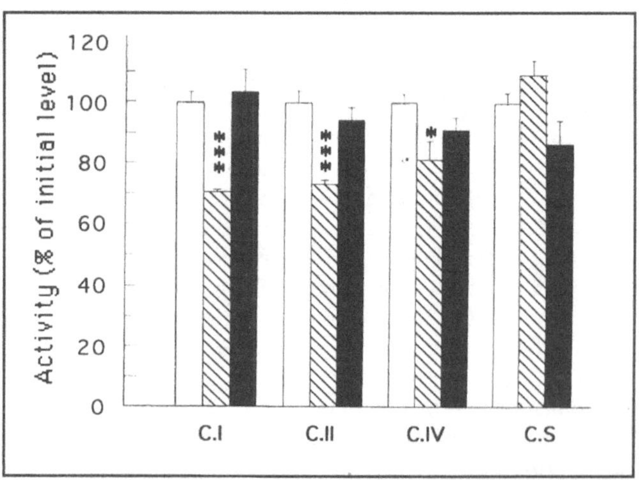

Evidence is now rather compelling that NO is a major mediator of neural damage in PD and that therapeutic strategies to block NO formation are eminently realistic. There are even suggestions that neurotoxic damage in Alzheimer's disease also involves NO. Thus, this decidedly atypical neurotransmitter may emerge as a key to treating the most serious disease confronting the neuroscience community.

References

1. Bernheimer H, Birkmayer W, Hornykiewicz OO et al. Brain dopamine and the syndromes of Parkinson and Huntington. J Neurol Sci 1973; 20:415-55.
2. Lees AJ. When did Ray Kennedy's Parkinson's disease begin? Movement Disord 1992; 7:110-16.
3. Fearnley JM, Lees AJ. Ageing and Parkinson's disease: substantia nigra regional selectivity. Brain 1991; 114:2283-301.
4. Gibb WAG, Lees AJ. Anatomy, pigmentation, ventral and dorsal subpopulations of the substantia nigra differential cell death in Parkinson's disease. J Neurol Neurosurg Psychiatry 1991; 54:388-96.
5. McGeer PL, Itagaki S, Akiyama H et al. Rate of cell death in parkinsonism indicates active neuropathological process. Ann Neurol 1988; 24:574-6.
6. Ceballos I, Lafon M, Javoy-Agid F et al. Superoxide dismutase and Parkinson's disease. Lancet 1990; 335:1035-1036.
7. Ceballos I, Agid F, Delacourte A et al. Parkinson's disease and Alzheimer's disease: neurodegenerative disorders due to brain antioxidant system deficiency? In: Emerit I, Packer L, Auclair C, eds. Antioxidants in Therapy and Preventive Medicine. Adv Exp Biol Med 1990: 493-498.
8. Jenner P, Dexter DT, Sian J et al. Oxidative stress as a cause of nigral cell death in Parkinson's disease and incidental Lewy body disease. Ann Neurol 1992; 32:S82-87.
9. Jenner P. Oxidative damage in neurodegenerative disease. Lancet 1994; 344: 796-798.
10. Olanow CW. Oxidation reactions in Parkinson's disease. Neurology 1990; 40(suppl 3):32-37.
11. Fahn S, Cohen G. The oxidant stress hypothesis in Parkinson's disease: evidence supporting it. Ann Neurol 1992; 32:804-812.
12. Di Monte DA, Chan Piu Sandy MS. Glutathione in Parkinson's disease. A link between oxidative stress and mitochondrial damage? Ann Neurol 1992; 32: S111-S113.
13. Schapira AHV. Evidence for mitochondrial dysfunction in Parkinson's disease: a critical appraisal. Movement Disord 1994; 9:125-138.
14. Johnson WG, Hodge SE, Duvoisin R. Twin studies and the genetics of Parkinson's disease—a reappraisal. Movement Disord 1990; 5:187-94.
15. Golbe LI, Di Iorio G, Bonavita V et al. A large kindred with autosomal dominant Parkinson's disease. Ann Neurol 1990; 27:276-82.
16. Maraganore DM, Harding AE, Marsden CD et al. A clinical and genetic study of familial Parkinson's disease. Movement Disord 1991; 6:205-11.
17. Lestienne P, Nelson J, Riederer P et al. Normal mitochondrial genome in brain from patients with Parkinson's disease and complex I defect. J Neurochem 1990; 55:1810-12.
18. Mann VM, Cooper JM, Schapira AHV. Quantitation of a mitochondrial DNA deletion in Parkinson's disease. FEBS Lett 1992; 299:218-22.
19. Schapira AHV, Holt IJ, Sweeney M et al. Mitochondrial DNA analysis in Parkinson's disease. Movement Disord 1990; 5:294-7.
20. Armstrong M, Daly AK, Cholerton S et al. Mutant debrisoquine hydroxylation genes in Parkinson's disease. Lancet 1992; 339:1017-8.
21. Smith CAD, Gough AC, Leigh PN et al. Debrisoquine hydroxylase gene polymorphism and susceptibility to Parkinson's disease. Lancet 1992; 339:1375-7.
22. Stevenson GB, Heafield MTE, Waring RH et al. Xenobiotic metabolism in Parkinson's disease. Neurology 1989; 39:883-7.
23. Green S, Buttrum S, Molly H et al. N-methylation of pyridines in Parkinson's disease. Lancet 1991; 338:120-1.

24. Rajput AH, Uitti RJ, Stern W et al. Early onset Parkinson's disease and childhood environment. Adv Neurol 1986; 45: 295-307.
25. Riederer P, Sofic E, Rausch WD et al. Transition metals, ferritin, glutathione and ascorbic acid in parkinsonian brains. J Neurochem 1989; 52:515-20.
26. Perry TL, Godin DV, Hansen S. Parkinson's disease: a disorder due to nigral glutathione deficiency? Neurosc Lett 1982; 33:305-310.
27. Perry TL, Yong VW. Idiopathic Parkinson's disease, progressive supranuclear palsy and glutathione metabolism in the substantia nigra of patients. Neurosci Lett 1986; 67:269-74.
28. Schapira AHV. Evidence for mitochondrial dysfunction in Parkinson's disease: a critical appraisal. Movement Disord 1994; 9:125-38.
29. Schapira AHV, Cooper JM, Dexter D et al. Mitochondrial complex I deficiency in Parkinson's disease. Lancet 1989; 1:1269.
30. Schapira AHV, Cooper JM, Dexter D et al. Mitochondrial complex I deficiency in Parkinson's disease. J Neurochem 1990; 54:823-827.
31. Schapira AHV, Mann VM, Cooper JM et al. Anatomic and disease specificity of NADH CoQ1 reductase (complex I) deficiency in Parkinson's disease. J Neurochem 1990; 55:2142-5.
32. Jenner P. Altered mitochondrial function, iron metabolism and gutathione levels in Parkinson's disease. Acta Neurol Scand 1993; 87:6-13.
33. Spina MB, Cohen G. Dopamine turnover and glutathione oxidation: implications for Parkinson's disease. Proc Natl Acad Sci USA 1989; 86:1398-1400.
34. Langeveld CH, Schepens E, Stoof JC et al. Differential sensitivity to hydrogen peroxide of dopaminergic and noradrenergic neurotransmission in rat brain slices. Free Rad Biol Med 1995; 19: 209-217.
35. Drukarch B, Jongenelen CAM, Schepens E et al. Glutathione is involved in the granular storage of dopamine in rat PC12 pheochromocytoma cells: implications for the pathogenesis of Parkinson's disease. J Neurosci 1996; 16:6038-6045.
36. Cohen G. The pathobiology of oxidative stress in Parkinson's disease: biochemical aspects of dopamine neuron senescence. J Neural Transm 1983; 9:89-103.
37. Fahn S, Cohen G. The oxidant stress hypothesis in Parkinson's disease: evidence supporting it. Ann Neurol 1992; 32:804-12.
38. Olanow CW. An introduction to the free radical hypothesis in Parkinson's disease. Ann Neurol 1992; 32:S2-S9.
39. Graham DG. Oxidative pathways for catecholamines in the genesis of neuromelanin and cytotoxic quinones. Mol Pharmacol 1978; 14:633-43.
40. Graham DG, Tiffany SM, Bell WR, Gutknecht WF. Autooxidation versus covalent binding of quinones as the mechanism of toxicity of dopamine, 6-hydroxydopamine and related compounds towards C1300 neuroblastoma cells in vitro. Mol Pharmacol 1978; 14:644-653.
41. Hirsch E, Graybiel AM, Agid Y. Melanised dopaminergic neurones are differentially susceptible to degeneration in Parkinson's disease. Nature 1988; 334:345-8.
42. Haavik J, Almas B, Flatmark T. Generation of reactive oxygen species by tyrosine hydroxylase: a possible contribution to the degeneration of dopaminergic neurons. J Neurochem 1997; 68: 328-332.
43. Dexter DT, Wells FR, Agid F et al. Increased nigral iron content in postmortem parkinsonian brain. Lancet 1987; 2:1219-20.
44. Dexter DT, Jenner P, Schapira AHV et al. Alterations in levels of iron, ferritin, and other trace metals in neurodegenerative diseases affecting the basal ganglia. Ann Neurol 1992; 32:S94-S100.
45. Sofic E, Paulus W, Jellinger K et al. Selective increase of iron in substantia nigra zona compacta in parkinsonian brains. J Neurochem 1991; 56:978-82.
46. Youdim MBH, Ben Shachar D, Riederer P. Is Parkinson's disease a progressive siderosis of substantia nigra resulting in iron and neuromelanin induced neurodegeneration? Acta Neurol Scand 1989; 80(Suppl 126):47-54.

47. Kish SJ, Morito C, Hornykiewicz O. Glutathione peroxidase activity in Parkinson's disease. Neurosci Lett 1985; 58: 343-346.
48. Saggu H, Cooksey J, Dexter DT et al. A selective increase in particulate superoxide dismutase activity in Parkinson's substantia nigra. J Neurochem 1989; 53: 692-697.
49. Dexter DT, Carayon A, Vidailhet M et al. Decreased ferritin levels in brain in Parkinson's disease. J Neurochem 1990; 55:16-20.
50. Dexter DT, Carter CJ, Wells FR et al. Basal lipid peroxidation in substantia nigra is increased in Parkinson's disease. J Neurochem 1989; 52:381-389.
51. Mena MA, Pardo B, Casarejos MJ et al. Neurotoxicity of levodopa on catecholamine-rich neurons. Movement Disord 1992; 7:23-31.
52. Michel PP, Hefti F. Toxicity of 6-hydroxydopamine and dopamine for dopaminergic neurons in culture. J Neurosci Res 1990; 26:428-35.
53. Blunt SB, Jenner P, Marsden CD. Suppressive effect of L-dopa on dopamine cells remaining in the ventral tegmental area of rats previously exposed to the neurotoxin 6-hydroxydopamine. Movement Disord 1993; 8:129-33.
54. Hefti F, Melamed E, Bhawan J et al. Long-term administration of L-dopa does not damage dopaminergic neurons in the mouse. Neurology 1981; 31: 1194-5.
55. Damier P, Hirsch EC, Zhang P et al. Glutathione peroxidase, glial cells and Parkinson's disease. Neuroscience, 1993; 52:1-6.
56. Yoritaka A, Hattori N, Uchida K et al. Immunohistochemical detection of 4-hydroxynonenal protein adducts in Parkinson's disease. Proc Natl Acad Sci USA 1996; 93:2696-2701.
57. Burns RS, Chiueh CC, Markey SP et al. A primate model of parkinsonism: selective destruction of dopaminergic neurons in the pars compacta of the substantia nigra by MPTP. Proc Natl Acad Sci USA 1983; 80:4545-50.
58. Langston JW, Ballard P, Tetrud JW et al. Chronic parkinsonism in humans due to a product of meperidine analog synthesis. Science 1983; 219:979-80.
59. Salach JI, Singer TP, Castagnoli N et al. Oxidation of the neurotoxic amine MPTP by monoamine oxidases A and B and suicide inactivation of the enzymes by MPTP. Biochem Biophys Res Comm 1984; 125:831-5.
60. Javitch JA, d'Amato RJ, Strittmatter SM et al. Parkinsonism-inducing MPTP: uptake of the metabolite MPP+ by dopamine neurones explains selective toxicity. Proc Natl Acad Sci USA 1985; 82:2173-7.
61. Ramsay RR, Dadger J, Trevor A et al. Energy driven uptake of MPP^+ by brain mitochondria mediates the neurotoxicity of MPTP. Life Sci 1986; 39:581-8.
62. Nicklas WJ, Vyas I, Heikkila RE. Inhibition of NADH-linked oxidation in brain mitochondria by MPP^+, a metabolite of the neurotoxin MPTP. Life Sci 1985; 2503-8.
63. Di Monte D, Smith TM. Free radicals, lipid peroxidation and l-methyl-4-phenyl-1,2,3,6-tetrahydropyridine (MPTP)-induced parkinsonism. Rev Neurosci 1988; 2:67-81.
64. Ramsay RR, Krueger MJ, Youngster SK et al. Interaction of MPP^+ and its analogs with the rotenone/piericidin binding site of NADH dehydrogenase. J Neurochem 1991; 51:1184-90.
65. Cleeter MJW, Cooper JM, Schapira AHV. Irreversible inhibition of mitochondrial complex I by 1-methyl-4-phenylpyridinium: evidence for free radical involvement. J Neurochem 1992; 58:786-9.
66. Przedborski S, Vladimir K, Jackson-Lewis V et al. Transgenic mice with increased CuZn-superoxide dismutase activity are resistant to MPTP-induced neurotoxicity. J Neurosci 1992; 12:1658-1667.
67. Cortopassi G, Wang E. Modelling the effects of age-related MtDNA mutation accumulation; complex I deficiency, superoxide and cell death. Biochim Biophys Acta 1995; 1271:171-176.
68. Pardo B, Mena MA, Fahn S et al. Ascorbic acid protects against L-dopa induced neurotoxicity on a catecholamine-rich human neuroblastoma cell line. Movement Disord 1993; 8:278-84.

69. Przedborski S, Jackson-Lewis V, Muthane U et al. Chronic levodopa administration alters cerebral mitochondrial respiratory chain activity. Ann Neurol 1993; 34: 715-23.
70. Di Donato S, Zeviani M, Giovannini P et al. Respiratory chain and mitochondrial DNA in muscle and brain in Parkinson's disease patients. Neurology 1993; 43:2262-8.
71. Cardellach C, Marti MJ, Fernandez-Sola J et al. Mitochondrial respiratory chain activity in skeletal muscle from patients with Parkinson's disease. Neurology 1993; 43:2258-62.
72. Hasegawa E, Takeshige K, Oishi T et al. l-methyl-4-phenylpyridinium (MPP^+) induces NADH dependent superoxide formation, and enhances NADH-dependent lipid peroxidation in bovine heart submitochondrial particles. Biochem Biophys Res Commun 1990; 170:1049-55.
73. Takeshige K, Minakami S. NADH- and NADPH-dependent formation of superoxide anions by bovine heart submitochondrial particles. Biochem 1979; 180: 129-35.
74. Hillered L, Ernster L. Respiratory activity of isolated rat brain mitochondria following in vitro exposure to oxygen radicals. J Cereb Blood Flow Metab 1983; 3:207-14.
75. Narabayashu M, Takeshige K, Minikami S. Alteration of inner-membrane components and damage to electron-transfer activities of bovine heart submitochondrial particles induced by NADPH-dependent lipid peroxidation. Biochem J 1982; 202:97-105.
76. Zhang Y, Marcillat O, Giulivi C et al. The oxidative inactivation of mitochondrial electron transport chain components and ATP. J Biol Chem 1990; 265:16330-6.
77. Benzi G, Curti D, Pastoris O et al. Sequential damage in mitochondrial complexes by peroxydase stress. Neurochem Res 1991; 16:1295-302.
78. Hartley A, Cooper JM, Schapira AHV. Iron induced oxidative stress and mitochondrial dysfunction: relevance to Parkinson's disease. Brain Res 1993; 627:349-53.
79. Thomas PK, Cooper JM, King RHM et al. Myopathy in vitamin E deficient rats: muscle fiber necrosis associated with disturbances of mitochondrial function. J Anat 1993; 183:451-61.
80. Meister A. Glutathione deficiency produced by inhibition of its synthesis and its reversal; applications in research and therapy. Pharmacol Ther 1991; 51: 155-194.
81. Merad-Boudia, M, Fideler L, Nicole A, et al. Selective modulation of brain antioxidant capacity by genetic or metabolic manipulation. In Oxidative Stress in Cell Activation and Viral Infection. Birkhaüser Verlag 1994:173-192.
82. Yong VW, Perry TL, Krisman AA. Depletion of glutathione in brainstem of mice by MPTP is prevented by antioxidant treatment. Neurosci Lett 1986; 63:56-60.
83. Ben-Yoseph O, Boxer PA, Ross BD. Assessment of the role of the glutathione and pentose phosphate pathways in the protection of primary cerebrocortical cultures from oxidative stress. J Neurochem 1996; 66:2329-2337.
84. Maker HS, Wiss C, Silides DJ, Cohen G. Coupling of dopamine oxidation (monoamine oxidase activity) to glutathione oxidation via the generation of hydrogen peroxide in rat brain homogenates. J Neurochem 1981; 36:589-593.
85. Han S-K, Mytilineou C, Cohen G. L-DOPA upregulates glutathione and protects mesencephalic cultures against oxidative stress. J Neurochem 1996; 66: 501-510.
86. Jain A, Martensson J, Stole E et al. Glutathione deficiency leads to mitochondrial damage in brain. Proc Natl Acad Sci USA; 1991; 88:1913-1917.
87. Mytilineou C, Han S-K, Cohen G. Toxic and protective effects of L-DOPA on mesencephalic cell cultures. J Neurochem 1993; 61:1470-1478.
88. Colton A, Pagan F, Snell J et al. Protection from oxidation enhances the survival of cultured mesencephalic neurons. Exp Neurol 1995; 132:54-61.
89. Andersen JK, Mo JQ, Hom DG. Effect of buthionine sulfoximine, a synthesis inhibitor of the antioxidant glutathione,

on the murine nigrostriatal neurons. J Neurochem 1996; 67:2164-2171.

90. Masuoka D, Johnson G, Finch CE. Aging and unusual catecholamine-containing structures in the mouse brain. Brain Res 1979; 169:335-341.
91. Benzi G, Moretti A. Age- and peroxidative stress-related modifications of the cerebral enzymatic activities linked to mitochondria and the glutathione system. Free Rad Biol Med 1995; 19:77-101.
92. Desagher S, Glowinski J, Premont J. Astrocytes protect neurons from hydrogen peroxide toxicity. J Neurosci 1996; 16: 2553-2562.
93. Slivka A, Mytilineou C, Cohen G. Histochemical evaluation of glutathione in brain. Brain Res 1987; 409:275-284.
94. Pileblad E, Eriksson PS, Hansson E. the presence of glutathione in primary neuronal and astroglial cultures from rat cerebral cortex and brainstem. J Neural Transm 1991; 86:43-49.
95. Philbert MA, Beiswanger CM, Waters DK et al. Cellular and regional distribution of reduced glutathione in the nervous system of the rat: histochemical localization by mercury orange and phthaldehyde-induced fluorescence. Toxicol Appl Pharmacol 1991; 107: 215-227.
96. Makar TK, Nedergaard M, Preuss A et al. Vitamine E, ascorbate, glutathione, glutathione disulfide, and enzymes of glutathione metabolism in cultures of chick astrocytes and neurons: evidence that astrocytes play an important role in antioxidative processes in the brain. J Neurochem 1994; 62:45-53.
97. Sagara JI, Miura K, Bannai S. Maintenance of neuronal glutathione by glial cells. J Neurochem1993; 61:1672-1676.
98. Sagara JI, Miura K, Bannai S. Cystine uptake and glutathione level in fetal brain cells in primary culture and suspension. J Neurochem 1993; 61:1667-1671.
99. Mena M, Pardo B, Paino CL et al. Levodopa toxicity in foetal rat midbrain neurones in culture: modulation by ascorbic acid. Neuroreport 1993; 4: 438-440.
100. Grasbon-Frodl EM, Nakao N, Brundin P. Lazaroids improve the survival of embryonic mesencephalic donor tissue stored at 4°C and subsequently used for cultures or intracerebral transplantation. Brain Res Bull 1996; 39:341-347.
101. Frodl E, Nakao N, Brundin P. Lazaroids improve the survival of cultured rat embryonic mesencephalic neurons. Neuroreport 1994; 5:2393-2396.
102. Heales SJR, Bolanos JP, Clark JB. Glutathione depletion is accompanied by increased neuronal nitric oxide synthase activity. Neurochem Res 1996; 21:35-39.
103. Gunasekar PG, Khantasamy AG, Borowitz JL et al. NMDA receptor activation produces concurrent generation of nitric oxide and reactive oxygen species: implication for cell death. J Neurochem 1995; 65:2016-2021.
104. Lipton SA, Choi Y-B, Pan Z-H et al. A redox-based mechanism for the neuroprotective and neurodestructive effects of nitric oxide and related nitroso-compounds. Nature 1993; 364:626-632.
105. Bonfoco E, Krainc D, Ankarcrona M et al. Apoptosis and necrosis: two distinct events induced, respectively, by mild and intense insults with N-methyl-D-spartate or nitric/oxide/superoxide in cortical cell cultures. Proc Natl Acad Sci USA 1995; 92:7162-7166.
106. Estevez AG, Radi R, Barbeto L et al. Peroxynitrite-induced cytotoxicity in PC12 cells: evidence for an apoptotic mechanism differentially modulated by neurotrophic factors. J Neurochem 1995; 65:1543-1550.
107. Youdim MBH, Ben-Shachar D, Eschel G et al. The neurotoxicity of iron and nitric oxide: relevance to the etiology of Parkinson's disease. In: Narabayashi H, Nagatsu T, Yanagisawa N, Mizuno, Y, eds. Advances in Neurology Vol 60. New York: Raven Press, 1993; 259-266.
108. Heales SJR, Davies SEC, Bates TE et al. Depletion of brain glutathione is accompanied by impaired mitochondrial function and decreased N-Acetyl aspartate concentration. Neurochem Res 1995; 20:31-38.
109. Bolanos JP, Peuchen SP, Heales SJR et al. Nitric-oxide-mediated inhibition of

mitochondrial respiratory chain in cultured astrocytes. J Neurochem 1994; 63:910-916.

110. Bolanos JP, Heales SJR JM et al. Effect of peroxynitrite on the mitochondrial respiratory chain: differential susceptibility of neurones and astrocytes in primary culture. J Neurochem 1995; 64:1965-1972.

111. Brown GC. Nitric oxide regulates mitochondrial respiration and cell functions by inhibiting cytochrome oxidase. FEBS Lett 1995; 369:136-139.

112. Cleeter MWJ, Cooper JM, Darley-Usmar VM et al. Reversible inhibition of cytochrome c oxidase, the terminal enzyme of the mitochondrial respiratory chain, by nitric oxide. implications for neurodegenerative diseases. FEBS Lett 1994; 345:50-54.

113. Brouillet E, Hantraye P, Ferrante RJ et al. Chronic mitochondrial energy impairment produces selective striatal degeneration and abnormal choreiform movements in primates. Proc Natl Acad Sci USA 1995; 92:7105-7109.

114. Sokol RJ, Deveraux MW, O'Brien et al. Abnormal hepatic mitochondrial respiration and cytochrome c oxidase activity in rats with long term copper overload. Gastroent 1993; 105:178-187.

115. Hartley A, Cooper JM, Schapira AHV. Iron induced oxidative stress and mitochondrial dysfunction: relevance to Parkinson's disease. Brain Res 1993; 627:349-353.

116. Cortopassi G, Wang E. Modelling the effects of age-related MtDNA mutation accumulation; complex I deficiency, superoxide and cell death. Biochim Biophys Acta 1995; 1271:171-176.

117. Yoneda M, Katsumata K, Hayakawa M et al. Oxygen stress induces an apoptotic cell death associated with fragmentation of mitochondrial genome. Biochem Biophys Res Commun 1995; 209: 723-729.

118. Wolvetang E, Johnson KL, Linnane AW. Mitochondrial respiratory chain inhibitors induce apoptosis. FEBS Lett 1994; 339:40-44.

119. Hartley A, Stone JM, Heron C et al. Complex I inhibitors induce dose-dependent apoptosis in PC12 cells: relevance to Parkinson's disease. J Neurochem 1994; 63:1987-1990.

120. Kroemer G, Petit P, Zamzani N et al. The biochemistry of programmed cell death. FASEB J 1995; 9:1277-1287.

121. Zamzani N, Susin AS, Marchetti P et al. Mitochondrial control of nuclear apoptosis. J Exp Med 1996; 183:1533-1544.

122. Brundin P, Lindvall O. Transplantation in Parkinson's disease. In: Linvall O, ed. Basic and Clinical Aspects of Neuroscience. Springer Verlag Berlin 1993; 5:5-17.

123. Nakao N, Frodl EM, Widner H et al. Overexpressing Cu/Zn superoxide dismutase enhances survival of transplanted neurons in a rat model of Parkinson's disease. Nat Med 1995; 1:226-231.

124. Nakao N, Frodl EM, Duan WM et al. Lazaroids improve the survival of grafted rat embryonic dopaminergic neurons. Proc Natl Acad Sci USA 1994; 91:12408-12412.

125. Pechan PA, Chowdhury K, Seifert W. Free radicals induce gene expression of NGF and bFGF in rat astrocyte culture. Neuroreport 1992; 3:469-472.

126. Naveilhan P, Neveu I, Jehan F et al. Reactive oxygen species influence Nerve Growth Factor synthesis in primary rat astrocytes. J Neurochem 1994; 62:2178-2186.

127. Takayama H et al. Basic fibroblast growth factor increases dopaminergic graft survival and function in a rat model of parkinson's disease. Nat Med 1995; 1:53-58.

128. Mattson MP, Keshava KN, Wang H et al. Basic FGF regulates expression of functional 71 kDa NMDA receptor protein that mediates calcium influx and neurotoxicity in hippocampal neurons. J Neurosci 1993; 13:4575-4588.

129. Hantraye P, Brouillet E, Ferrante R et al. Inhibition of neuronal nitric oxide synthase prevents MPTP-induced parkinsonism in baboons. Nat Med 1996; 2: 1017-1021.

130. Predborski S, Jackson-Lewis V, Yokoyama R et al. Role of nitric oxide in 1-methyl-4-phenyl-1,2,3,6-tetrahydropyridine (MPTP)-induced dopaminergic neurotoxicity. Proc Natl Acad Sci USA 1996; 93:4565-4571.
131. Schulz JB, Matthews RT, Muqit MMK et al. Inhibition of neuronal nitric oxide synthase by 7-nitroindazole protects against MPTP-induced neurotoxicity in mice. J Neurochem 1995; 64:936-939.
132. The Parkinson's study group. Effects of tocopherol and deprenyl on the progression of disability in early Parkinson's disease. New Engl J Med 1993; 328:176-183.

CHAPTER 9

Conclusions

As medical science has prolonged human life, it has also brought a sharp increase in the numbers of people suffering one of the most calamitous afflictions of old age. In this book, we have presented the rationale for the hypothesis that oxidative stress could be the common mediator of the entire pathologic cascade leading to neuronal death in major age-related neurodegenerative disorders. It is attractive to speculate that effective treatment, but above all prevention, of these diseases may be carried out by drugs operating at this important level of the pathogenic cascade. Lowering the ROS formation and thereby the associated ROS-damage by a large panel of antioxidants could be a challenge for the prevention of neurodegeneration.

Concomitant with this new understanding of the injurious role of ROS in neuronal death and pathology, is the increasing appreciation of the need for both fundamental and clinical research into the development of the potential preventive and therapeutic benefits that are now being foreseen for a variety of antioxidant nutritional and pharmacological interventions. The multicenter, controlled clinical trials with antioxidant therapy for Parkinson's disease DATATOP (Deprenyl and Tocopherol Antioxidative Therapy of Parkinsonism), conducted by the Parkinson study group, has indicated that such agents can successfully slow disease progression in patients with early PD,[1] but until now no work has described the preventive role of antioxidants in the development of this disease.

Evidence is now rather compelling that NO is a major mediator of neural damage in PD and that therapeutic strategies to block NO formation might be useful in slowing or halting neuronal degeneration in PD.[2] This is a major clinical challenge but also eminently realistic. There are even suggestions that neurotoxic damage in Alzheimer's disease also involves NO. Thus, this decidedly atypical neurotransmitter may emerge as a key to treating the most serious disease confronting the neuroscience community.

Studies suggest that Aβ, excitotoxicity (i.e., glutamate-induced) and iron-catalyzed, oxygen radical-induced lipid peroxidation are interactive mechanisms in regard to causing neuronal death. In any case, lipoperoxidation may be a fundamental mechanism of anterograde neuronal degeneration secondary to neuronal insult or injury, and that consequently intensive lipid antioxidant therapy may be effective in slowing neuronal degeneration in a variety of neurodegenerative disorders and that orally effective, brain-penetrating lipid antioxidants might slow the disease.[3] Among the drugs now moving into clinical trials are antioxidants such as vitamine E and selegiline, which have shown little toxicity. The furthest along is an ADCS (Alzheimer's Disease Cooperative study) trial designed to test whether vitamin E, an antioxidant, and selegeline (inhibitor of MAO) can slow Alzheimer's progression in people who already show mild to moderate disease. The data are currently being analyzed. It was reported that the metal chelator desferrioxamine reduced the rate of decline in daily living skills in patients with AD.[4] The main nationale here was to chelate excess aluminum, long considered as a possible pathogenetic factor in the disease. The drug is also a hydroxyl radical scavenger. A multicenter clinical trial is needed to establish the drug's efficacy.

The same reasoning is sparking interest in nonsteroidal anti-inflammatory drugs (NSAIDS) which can decrease the activation of microglial cells and thus the production of the deleterious ROS in Alzheimer's brains. A series of epidemiological studies over the past few years has indicated that these drugs, which include aspirin, ibuprofen and naproxen protect against Alzheimer's development.[5-7] The minimum reduction in Alzheimer's cases among people taking NSAIDs was about 50% and in some studies it was as much as 75%.[8]

By that time several additional drugs strategies may be ready to be tested. One involves the preservation of brain neurons with nerve-cell growth factors like NGF, which is needed for the survival of cholinergic brain neurons and improves memory and learning in animals. NGF may be acting through stimulation of biologic antioxidants.[9,10,11]

Other strategies are to develop drugs that target the specific proteins implicated in Alzheimer's etiology. One such stategy aims to prevent the formation of Aβ by inhibiting the protease enzymes that release it from the APP; alternatively, it might be possible to prevent Aβ from aggregating into its most toxic form. Researchers will also want to find out whether they can lower brain concentration of ApoE or reduce the phosphorylation of tau protein, both of which have been linked to Alzheimer's, and find ways to interfere with the event that actually kills the neurons. Antioxidants could be efficient by lowering crosslinking and aggregation of Aβ.

The regulation of gene expression by oxidants, antioxidants and redox state has emerged as a novel subdiscipline in molecular biology that has promising therapeutic implications in neurodegenerative disorders. Binding sites of the redox-regulated transcription factors NF-κB and AP-1 are located in the promotor region of a large variety of genes that are probably involved in the neuropathogenesis. Moreover, many of the basic events of cell regulation such as protein phosphorylation, critical steps in the signal transduction cascade and binding of transcription factors to consensus sites of the DNA, are driven by the physiological oxidant-antioxidant homeostasis, especially by the thiol disulfide balance. Thus, the efficacy of different compounds that modulate intracellular redox state to favorably influence the molecular mechanisms known to be implicated in the etiology of neurodegenerative diseases should be a critical determinant of their selection for prevention or therapy. Neuroprotection from EAA by aspirin through blockage of NF-κB activation[12] shows that strategies to pharmacologically manipulate NF-κB activation are expected to be part of future research on neurological disorders. Effective strategies to block RAGE and SR receptors should be also of preventive or therapeutic importance.

Researchers in academia and the pharmaceutical industry are now devising new drug strategies that might improve patients' symptoms or, better yet, slow down or prevent neuronal degeneration. Packer et al propose that the metabolic role and antioxidant properties of the dithiol antioxidant α-lipoate (thioctic acid) may be clinically useful in treating or preventing brain and nerve disorders.[13] α-lipoate is a low molecular weight substance that is absorbed from the diet and cross the blood-brain barrier and is reduced in cells in and tissues in dihydrolipoate. Both α-lipoate and especially dihydrolipoate have been shown to be potent antioxidants, to generate through redox cycling other antioxidants like vitamin C and vitamin E, and to raise intracellular glutathione levels. These various metabolic antioxidant properties relate to its possible therapeutic roles in a variety of brain and neuronal tissue pathologies.

A new breed of synthetic antioxidants, purported to outperform the body's endogenous free radical scavengers, are entering clinical use for stroke, Parkinson's disease and other neurodegenerative disorders.[14] The compounds under investigation are analogs of the so-called spin-trap reagents, such as a-phenyl tert-butyl nitrone (PBN). Such compounds have traditionally been used in magnetic resonance spectroscopy because they so readily react with short-lived free radical species, such as hydroxyl radicals, to form a stable detectable adduct.

Major avenues of research remain to be explored to determine the scope of a variety of dietary, nutritional and pharmacological antioxidant regimens, individually or in combination, to slow or prevent the disease. Once effective treatments are established, future prospects will include using genetic and biochemical markers to identify patients at risk for neurodegenerative ilnesses, so that one can initiate therapy at a presymptomatic stage and hopefully prevent the illness.

References

1. The Parkinson's study group. Effects of tocopherol and deprenyl on the progression of disability in early Parkinson's disease. New Engl J Med 1993; 328:176-183.
2. Hantraye P, Brouillet E, Ferrante R et al. Inhibition of neuronal nitric oxide synthase prevents MPTP-induced parkinsonism in baboons. Nat Med 1996; 2: 1017-1021.
3. Hall ED, Braughler JM. Central nervous system trauma and stroke. II. Physiological and pharmacologic evidence for involvement of oxygen radicals and lipid peroxidation. Free Rad Biol Med 1989; 6:303-313.
4. Mc Lachlan C, Dalton DR, Kruck TPA et al. Intramuscular desferrioxamine in patients with Alzheimer's disease. Lancet 1991; 337:1304-1308.
5. Breitner JS, Gau BA, Wilsh KA et al. Inverse association of antiinflammatory treatments and Alzheimer's disease: initial results of a co-twin control study. Neurology 1994; 44:227-232.
6. Rogers J, Kirby LC, Hempelman SR. Clinical trial of indomethacin in Alzheimer's disease. Neurology 1993; 43:1609-1611.
7. Mc Geer PL, Mc Geer E, Rogers J, Sibley J. Antiinflammatory drugs and Alzheimer's disease. Lancet 1990; 335:1037.
8. Marx J. Searching for drugs that combat Alzheimer's. Science 1996; 273:50-53.
9. Frim DM, Wullner U, Beal MF, Isacson O. Implanted NGF-producing fibroblasts induce catalase and modify ATP levels but do not affect glutamate receptor binding or NMDA receptor expression in the rat striatum. Exp Neurol 1994; 8: 172-180.
10. Jackson GR, Sampath D, Werrbach-Perz K, Perez-Polo JR. Effect of nerve growth factor on catalase and glutathione peroxidase in a hydrogen peroxide-resistant pheochromocytoma subclone. Brain Res 1994; 634:69-76.
11. Mattson MP, Lovell MA, Furukawa K, Markesbery WR. Neurotrophic factors attenuate glutamate-induced accumulation of peroxides, elevation of intracellular Ca^{2+} concentration, and neurotoxicity and increase antioxidant enzyme activities in hippocampal neurons. J Neurochem 1995; 65:1740-1751.
12. Grilli M, Pizzi M, Memo M et al. Neuroprotection by aspirin and sodium salicylate through blockade of NF-κB activation. Science 1996; 274:1383-1385.
13. Packer L, Trizchler HJ, Wessel K. Neuroprotection by the metabolic antioxidant α-lipoic acid. Free Rad Biol Med 1997; 22:359-378.
14. Veggeberg S. High performance antioxidants for a multitude of diseases. Mol Med Today 1996; 406-409.

INDEX

O

P

R

S

T

U

V

SPRINGER NATURE

GPSR Compliance

The European Union's (EU) General Product Safety Regulation (GPSR) is a set of rules that requires consumer products to be safe and our obligations to ensure this.

If you have any concerns about our products, you can contact us on ProductSafety@springernature.com

In case Publisher is established outside the EU, the EU authorized representative is:

Springer Nature Customer Service Center GmbH
Europaplatz 3
69115 Heidelberg, Germany

FSC
www.fsc.org
MIX
Papier | Fördert
gute Waldnutzung
FSC® C083411

Kolibri 360 GmbH
Ferdinand-Jühlke-Straße 7
99095 Erfurt, Deutschland
produktsicherheit@kolibri360.de